CW01668611

The Gardens of Their Dreams

About the Cover

The painting, "Seedlings of Africa," by Canadian Native artist Maxine Noel, is one of a collection of works done to build relations between Canadian Native and African communities. After a series of droughts and famines in Sub-Saharan Africa in the 1980s, studies of the aid donor lists showed there had been an extraordinary response from Native Canadians. Efforts were then made to put Native and African communities in direct contact. As one part of this programme, artists from Africa and Canada visited and painted each others' worlds. Maxine Noel's painting celebrates a "twinning" between Bunkpurugu, Ghana, and Thunder Bay, Ontario. It is painted using elements of Native style. It shows the people of Bunkpurugu working to renew their country under a blazing sun.

The Gardens of Their Dreams

Desertification and Culture in World History

Brian Griffith

Fernwood Publishing ▲ Zed Books

Copyright © 2001 Brian Griffith

All rights reserved. No part of this book may be reproduced or transmitted in any form by any means without permission in writing from the publisher, except by a reviewer, who may quote brief passages in a review.

Editing: Douglas Beall
Cover painting: "Seedlings of Africa," by Canadian artist Maxine Noel, who generously allowed the use of her work.
Design and production: Beverley Rach
Printed and bound in Canada by Hignell Printing Limited

Published in Canada by Fernwood Publishing Ltd.
Box 9409, Station A
Halifax, Nova Scotia, B3K 5S3

Published in the rest of the world by Zed Books Ltd.
7 Cynthia Street, London NI 9JF, UK
and Room 400, 175 Fifth Avenue,
New York, 10010, USA
Distributed in the USA exclusively by Palgrave,
a division of St. Martins Press, LLC, 175 Fifth Ave., New York, 10010, USA.

Zed Books
ISBN 1 85649 800 X Paper
ISBN 1 85649 799 2 Cloth

British CIP available from the British Library
American CIP has been applied for.

Fernwood Publishing Company Limited gratefully acknowledges the financial support of the Department of Canadian Heritage and the Canada Council for the Arts for our publishing program.

Le Conseil des Arts du Canada | The Canada Council for the Arts

For permission to reprint copyrighted material, the following acknowledgments are gratefully made: (1) Addison, Wesley, Longman, Inc, for use of material by Fatima Mernissi, The Veil and the Male Elite, © Editions Albin Michel S.A. 1987; English translation © 1991 Addison-Wesley Publishing Company Inc., pages 8, 9 and 16. (2) Little Brown and Company, London, for use of quotations from Opening the Gates: A Century of Arab Feminist Writing, edited by Margot Badran and Miriam Cooke, © 1990 Virago Press.

Canadian Cataloguing in Publication Data

Griffith, Brian

The gardens of their dreams: desertification and culture in world history

Includes bibliographical references and index.
ISBN 1-55266-044-3 (bound). ISBN 1-55266-045-1 (pbk.)

1. Desertification—Social aspects. 2. Civilization--History. 3. Human ecology—Arid regions. I. Title.

GF55.G75 2001 909'.0954 C00-901776-3

Contents

Tables, Figures and Maps

Prologue

WHILE I LIVED IN A VILLAGE IN INDIA, THE LAND GREW VISIBLY MORE BARREN every year. Each dry season there were fewer trees, and wells often gave out before the rains returned. To deepen the wells, some villagers climbed down the rock-lined shafts and worked with iron digging-spikes like miners. Others hauled out buckets of mud and broken stone. According to some villagers the water table was falling a foot per year. Across the Deccan Plateau, wells and mechanized pumps in every village were chasing the groundwater, sucking it up like a hundred thousand straws.

Later, when I lived in a Kenyan village, our source of water was a seasonal stream known as the Ikiwe River. During the November rains, roaring waterfalls on this river could be heard over a kilometer away. But by April or May the river was completely dry. To get water in that season you had to walk down to the riverbed and dig in the sand. After your hole was a foot or two deep, it would slowly fill with water and you could scoop it into a jerry can, one ladle at a time.

The first time I went to the river in April, the trees had recently lost their leaves for the hot season. The air felt like summer, but the countryside looked like winter. The whole hour of digging and dipping for water was very peaceful, like playing in the sand as a child. But my play was disturbed by a quiver of fear, like the passing shadow of a bird: What if tomorrow even this last trickle of water was gone? Growing up in North America, I had never known this kind of fear. The sensation, however, seemed strangely familiar, striking some chord of memory like the twang of a sitar.

The dirt in south-central Kenya is a cracked, hard subsoil, as ravaged as the face of an eighty-year-old farmer. It is a red, sun-baked clay twinkling with tiny mica stars. For several thousand years this land had been a vast pasture for nomadic cattle drivers like the Masai, who filtered down the Rift Valley from the north. Repeatedly their cows chewed over the grass, leaving the less palatable thornbush to grow. When the bush got too thick, the herdsmen would set it aflame, returning the soil to its bare beginnings. The following rainy season would catch the slopes denuded of plants, and our Ikiwe River would run red with runoff soil. These cycles of grazing, burning and eroding had been repeated perhaps every fifty years since cattle-herding first came to this land.

In the early 1900s the local Akamba tribe farmers had expanded from their cloudy hilltops to the savannas below. They carved out fields from the bush, planting maize and beans. Sometimes they fought with the passing Masai nomads, rekindling the ancient enmity between farmers and pastoralists. Slowly the farmers landscaped their slopes with terraces, reducing the massive soil erosion. But by that time the topsoil was basically gone. Most of the lowland is now laterite, a mineral and clay mixture used to surface roads. If this soil were rated on a scale with rich, humus-filled earth at the top and the nearly inorganic sand of the Sahara at the bottom, it would rank over halfway down.

Many people in these Indian and Kenyan villages seemed to accept the gradual decline of their world as inevitable. Some wanted to escape from the worsening situation, and escape took many forms. Some people tried to gather money and leave. Others hoped for spiritual salvation from an obviously degenerate world. Many Kenyan men were leaving the villages and moving to towns. The women commonly stayed behind, raising their children with one hand, while farming and fighting the desert with the other. Some women seemed to be slowly losing this struggle. Others were banding together and managing very well. Many women and some men possessed a powerful, old-fashioned optimism. They believed that their work with plants and animals was good for the world and were confident that the barren land could be nursed back to health.

Soon after returning to North America, I read *The Chalice & the Blade* by Riane Eisler (1987). Her account of villages overrun by warlike invaders and the resulting clashes between dominator and partnership cultures had been drawn from European history, but these themes seemed to apply to Asia and Africa as well. In both these regions, the early cultures had been concerned with nurturing and reproducing life, and had later faced incursions by warlike migrants who always seemed to come from the direction of the desert. For example, a few hundred miles northeast of our village in India lay the Great Indian Desert. Over the course of recorded history, this desert has grown to the east and south at faster than glacial speed. The path of its advance, through Pakistan and Rajasthan and into Gujarat, was roughly the same route used by most human invaders of India over the past four thousand years. These conquerors, such as the Aryans, Kushanas and Mughuls, were themselves migrants from the deserts and steppes of Inner Asia. They came bearing traditions forged in lands of scarcity, and imposed those traditions on the greener lands they conquered.

What follows is a story of those who have faced the wasteland and lived to become either its agents or its healers. For helping me tell this story, I want to thank primarily Riane Eisler, Marija Gimbutas, Min Jiayin, Fatima Mernissi, Vandana Shiva and Joseph Campbell, and secondly all the other researchers I will cite. For editorial assistance the author would like to thank Ronnie Seagren, Riane Eisler, Douglas Beall of Fernwood Publishing, and Helen Wu of the University of Toronto, East Asia Studies Department. I want to dedicate this story to its chief inspirer, a Kenyan woman known as Mama Ndolo, whom you will be introduced to later.

Chapter 1

Cultural Consequences of Desertification

OVER THE PAST SEVEN THOUSAND YEARS, A DESERT SLOWLY SPREAD THROUGH the center of the Old World. The affected regions had been wastelands in previous arid ages. But this time human civilizations witnessed and participated in the desiccation. Our ancestors watched as patches of desolation appeared in the landscape, like holes in worn-out cloth. In drought-stricken areas, the trees shed their leaves in summer, leaving the countryside barren. Where the wind once rustled through fields of foliage, now it whispered in the naked branches. Sooner or later the bubbling creeks fell silent. At dawn, choruses of songbirds no longer greeted the sun. At night the chirruping of crickets and frogs slowly faded to a hush. Herds of grazing animals migrated from the region, following the grass and rain. High in the sky, birds of prey circled without a sound.

Eventually the "true deserts," such as the Eastern Sahara, Arabia's Empty Quarter and the Chinese Taklamakan, came to resemble the sandstorm plains of Mars. This is how it looks where the web of life is stripped to the bone.

Some of the people in these collapsing environments died of hunger or thirst. Some fled to other regions and some learned to survive in the wasteland. But whether they stayed or fled, all these people suffered the virtual death of their homelands. They became exiles from their ancestors' world, inheriting the inner scars of a catastrophe. According to some desert peoples' beliefs, their ancestors were cast out from a garden of paradise, as a punishment for their sins. Since that time, they had struggled and wandered in a kind of exile, through a harsh land which served as a testing ground for souls. Their true home, if they ever found it in a promised land or another world, would resemble the gardens of their dreams.

The Wasteland's Refugees

For thousands of years, wave after wave of environmental refugees moved toward the greener lands of Europe, tropical Africa, India and China. To these lands some migrants came as beggars, and some as conquerors. According to Marija Gimbutas and Riane Eisler, the ancient farming cultures of Old Europe were invaded by nomadic warlord tribes that imposed a social order of "force-backed ranking." A similar history of "dominator" invasions can be traced in almost every country bordering the arid zone from China to eastern Europe from ancient to recent times. Feminist historians such as Eisler name these invaders as the primary sources of patriarchal and militaristic culture. Marilyn French agrees but calls for a closer understanding of the invaders themselves: "Patriarchy did not arise after three million years of peaceful hominid and human life because men suddenly became more aggressive, or suddenly decided to oppress women" (1985: 75). So who were these raiders? Why did they launch their invasions, and what forces shaped their starkly belligerent cultures? In *The Chalice & the Blade*, Eisler says that the conquerors generally came from "the less desirable fringe areas of our globe" (1987: 44). Those "fringes" were Inner Asia, Arabia and North Africa—regions then turning into deserts.

In most history books, the invaders from the harsh lands are described either as superior conquerors or scourges of the earth, depending on the writer's point of view. The ancient Chinese and Romans both looked on the steppe or desert people as subhuman barbarians. Many modern Europeans and Indians, however, still romanticize "the Aryans" from Inner Asia as the bringers of all "higher" civilization. Today, probably most people in the world are mixed descendants of the old invaders and the village gardeners they conquered. That clash of cultures and environments still lies imbedded in our minds. It seems we need to understand both kinds of ancestors better.

Social Costs of Environmental Destruction

A string of deserts now stretches almost unbroken from Mauritania to Manchuria. And across that vast expanse, each region's biological and cultural diversity has been diminished in a different way. Each group of people has adjusted to the wasteland in its own style. Yet despite regional differences, desert history holds common themes. The scarcity of food and water alone has imposed similar conditions on people across the arid belt. In coping with desert conditions, certain cultural patterns emerged in the ancient Middle East and seemed to spread with the desert itself. At the risk of overgeneralizing, here are three of those patterns:

1. *In non-productive lands, women were commonly viewed as the non-productive sex.* In the pre-desert environment, women were very often primary breadwinners. Their gardens, field gleanings, groves of trees, barnyard animals or local markets supplied most of their peoples' food. Where the land was relatively productive, women managed all these kinds of work while mothering a series of infants over most of their lives. But where the land turned to desert, household production grew much harder. As in modern India, environmental decline tended to hurt women most. Women's gardens were blighted by drought. The wild plants they gathered in nearby fields grew ever scarcer. Their domestic animals could no longer find enough grass near the yard, and care for them soon required ever longer treks through the wilderness in search of food. The more arid the grasslands grew, the more difficult it became for anyone to manage children and animal herds at the same time.

As the land grew barren, women's local production for local trade tended to die off. Since few places could support a community for long, most permanent settlements dispersed. Trade in the desert became a long-distance operation. Later, the caravan trade evolved to carry supplies through a deadly landscape, to and from distant trading centers at the desert's edge. This was hard on mothers with children, though some women, such as Muhammad's first wife, became trading company managers.

A third important activity in the desert economy was raiding. Under regularly desperate conditions, raiding was accepted as a time-honoured profession, and as crucial for survival. But of all economic activities, raiding was the least compatible with mothering. It was the men who could risk living by the sword, and therefore the men who would eventually view themselves as the most essential people.

In such a scarce and hostile landscape, women often came to be seen as mere dependents of male herdsmen, traders and warriors. The feminine dimension of humanity often became subordinated and isolated within a wider male world.

2. *Where the earth seemed hostile to humanity, culture grew hostile to the earth*. In the pre-desert nature religions of the ancient Middle East, the lord of heaven was commonly seen as a male deity whose sperm fell as rain on the female earth. The earth then brought forth life like a divine mother. But where the land appeared to die or grow barren, it was no longer seen as a living goddess. Mother Earth became the dead earth, and only the heavens above still seemed holy.

The desert landscape often seemed animated with hostility towards humanity. The frequent dust devils and storms of sand appeared as incarnations of malice. So, in the Sahara or the Taklamakan, the "forces of nature" seemed to be primarily forces of death. In the common themes of desert mythology, this earth was a fallen world and an abode of devils. It was a place of trial and exile, rather than our good home.

When desert migrants came to other lands, they brought their views of the world with them. So religions from the ancient Middle East spread north into Europe and south into Africa. Religion from the desert became "Western religion" as we know it. Out of that heritage, "Western science and economics" presumed that the planet is a dead resource, and that only the human community matters.

3. *In lands of scarcity, means of coercion were often more important than means of production*. Experience in the desert seemed to prove that there was not enough for all. Where this was accepted as a fact of life, the question became *whom* there would be enough *for*. So, the issue of pecking order emerged and often took on central importance. Of course, scarcity was a problem in every part of the world, but it was more extreme in the desert. There, production of food and supplies was often so unpredictable as to seem beyond human control. What the rulers *could* control was the distribution of whatever little there was. Traditional desert rulers therefore relied unusually little on fostering production, and

unusually much on military control of supplies. With some exceptions, this remains the political heritage of the barren lands.

Perhaps we can expect social patterns like these to appear wherever the land is denuded in the future, including in North America. We may already be seeing such trends emerging on the heels of environmental crises in nations around the world. But before pursuing this any further, there are some glaring exceptions to these patterns which should be examined.

The Different Histories of Major and Minor Deserts

Clearly the native cultures of the American, Australian and South African deserts do not fit these patterns. For some set of reasons, the people of these "minor deserts" managed to avoid the most vicious cycles of desertification. Their traditional ways of life rarely destroyed the land's fragile vegetation. For example, the Wangkangurru people of Australia's Simpson Desert lived on water from desert claypans in the wet months, then retreated to shallow wells at the desert's edges for the dry period. In their circuits through a land receiving 10–25 centimeters of rain a year (which can be ten or twenty times more than the eastern Sahara gets), they were able to gather sufficient "bush tucker" of wild onions, bush bananas, emus, goanna lizards and spiny echidnas (Vessels 1992: 70–87). Until recent centuries the people of Australia, the African Kalahari and the American Southwest did not raise herds of cows, sheep, goats or horses. Perhaps for mainly climatic reasons, these naturally arid lands rarely degenerated into true deserts. Their people seldom became raider migrants from wastelands in which they could no longer support themselves.

As the American West dried up after the most recent ice age, the Southwest natives of the "Desert Archaic" period depended ever less on hunting big animals and ever more on gathering arid-country plants, such as piñyon pine nuts. In their traditional specialization of roles, women were the primary gatherers of wild plants. As this work grew more important, women who did it grew in social status. Since their intimate knowledge of plants (and where they grew) was crucial for group survival, women seldom left their own villages to marry. Instead, husbands moved to the wives' camps. If a village had few women, the women sometimes agreed to take extra husbands, so no man would live entirely without a woman's love and support. Highly effective women sometimes took two or more husbands (Hurst 1994: 56–57).

What can we say about this glaring contrast with the desert traditions of the Middle East? Perhaps the first thing is that the Afro-Asian deserts are obviously harsher than those of North America. Water evaporates in the Sahara roughly twice as fast as it does in the Californian or Australian deserts (Langewische 1996: 121). In much of North Africa or Arabia, traditional plant gathering virtually faded for sheer lack of plants to gather. Perhaps in the future, the American, southern African and Australian deserts will degenerate further, till they match the barrenness of the Sahara. Then the problems of these minor deserts may more closely resemble those of the great Afro-Asian arid belt. Already, news reports from Haiti, the Sudan and Afghanistan show a certain convergence of emergencies, their diverse problems of famine, tyranny and civil conflict linked by environmental poverty. To some outsiders it may seem that these countries have simply remained as they were in the past. But accelerating degeneration of the land is also a modern disease, and these "undeveloped" countries could be dealing with problems of the future. Egyptian writer Mahjoub Omar certainly seemed to think so in describing his trip to a Cairo international fair in a suburb outside the city:

> First you walk a lot to get there. Then you sit and drink Saudi coffee. You have reached "yesterday." Then you walk some more until you reach "today," with all the computers and machinery in the fair. Then you carry on walking a bit until you find desert, the real desert of the Nasr district. And that is "tomorrow." (in Butt 1997: 212)

The Desert's Wisdom for the Future

When most Western urbanites talk about the environment, they tend to deal in predictions about the future. According to certain calculations, specific problems must arise if the land is degraded at a given rate. Jeremy Rifkin, for example, offers the following projections of near-gloom via global warming for the year 2035:

> Decades of drought have scorched the earth, turning once fertile agricultural lands into parched deserts.
>
> Tens of millions of people continue to trek northward to higher latitudes in the greatest mass migration of populations in recorded history. Entire nations experience massive depopulation and starvation in the wake of prolonged droughts. Canada has become overrun, its population swelling from 25 million to 80 million in less than four decades. (1989: 3)

This is a fine warning, but obviously it is speculation. Doubters can dismiss it at will. There is little need, however, for speculation about the past. Vast regions of the world already resemble the surface of the moon. People have already lived under such conditions for hundreds or thousands of years. So, with a wisdom born of ancient experience, the book of Isaiah explains,

> Without green plants, all things would perish, and the Earth would revert to its original barren state. Mankind survives on the Earth only because green plants grow here. Indeed, "all flesh is grass." (Isaiah 40:6, in McLuhan 1994: 297)

In dealing with any challenge, we are free to learn from those who have previous direct experience. This book is an attempt to paint a broad picture of our ancestors' experiences in destroying and healing the environment. Focusing on desertification seems to highlight the worst and the best in human history. The plot involves three related stories, each of which could stand alone as a book, if relating them were not so important. These substories are (1) how the expanding wastelands have shaped people's images of nature, women, politics and religion; (2) how waves of refugees from the arid lands, including the historic migrations of Kurgans, Aryans, Huns and Mongols, have influenced civilizations in the green borderlands from China to Europe; and (3) how local people have responded to threats of invasion and environmental poverty.

On the arid zone's frontiers in China, Iran and Kenya, strings of villages stand before the desert. Their people are mainly traditional farmers, whose expertise is nurturing plants and animals. Some of them have to build their own soil from virtually lifeless sand and clay, mixing in layers of wild grass and compost. They strive to protect their seedlings from the wind with shelters of earth, shrubs and trees. If their efforts succeed, they slowly generate a microclimate favorable to life. Perhaps in the future their powers of nurturance will accomplish more. A future economy could arise which enriches nature as it grows. That, as Anil Agarwal (1989) says, would be the *real* green revolution. Such a civilization would be environmentally literate. It would have a vast working knowledge of how people and environments can help or hurt each other. That sort of wisdom will grow largely by tapping our long, rich experience from the past.

Region	5000 BCE (4800, 4600, 4400, 4200)	4000 BCE (3800, 3600, 3400, 3200)	3000 BCE (2800, 2600, 2400, 2200)	2000 BCE (1800, 1600, 1400, 1200)	1000 BCE (900, 800, 700, 600)
East Asia (mainly China)	Early “Matriarchal” cultures in river & coastal areas of China	Legendary Golden Age	**Period of legendary perfect emperors**	**Hsia (Xia) rulers**; **Shang Dynasty**; * Shang conquest of North India	**Chou (Zhou) Dynasty**; * Chou (Zhou) invasion; Expansion of northern and western deserts
South Asia (mainly India)	Dravidian village cultures rise from the Hindu Kush to the Bay of Bengal		* Harappan towns rise over destroyed Dravidian villages; **Harappan civilization —Indus Valley**	* Aryan migrants overrun Harappan Civilization	Aryan invasions of South India; “Forest Books” or Upanishads signal recovery of Dravidian culture
Middle East	Village cultures flourish in areas where rain-fed farming is no longer possible	**Ancient Sumer**; Cultivated areas shrink - irrigation evolves - salinization of irrigated land - wars over watered land	**Accadian Empire**; * Saragon of Agade conquers all of Mesopotamia; Great drought	**Assyrian Empire**	**Old Kingdom of Israel**; **Persion Empire**; * Scythian invasions
Egypt		Migrations into the Nile Valley from drying Sahara	**Old Kingdom**; * Drought leads to collapse of Old Kingdom	**Middle Kingdom**; **New Kingdom**; Invasion by Hyksos	* Libyan invasion; * Assyrian invasion
N. Africa	Neolithic cultures in the old green Sahara	Migrations to the east, west, north, and south as the Sahara region dries up			**Carthage**; Trade by chariot still possible between Carthage & West Africa
Europe	Cucuteni village culture thrives in the Ukraine; * Kurgan Wave I hits old Europe; Old European “Age of the Goddess”	* Kurgan Wave II drives farming cultures from Ukraine	* Kurgan Wave III spreads “Battle-Ax” Culture across Europe	* Destruction of Old Crete; Various Indo-Aryan invasions establish military-class rulers over Europe	Greek Dark Age; **Classical Athens**; * Dorian conquest in Greece

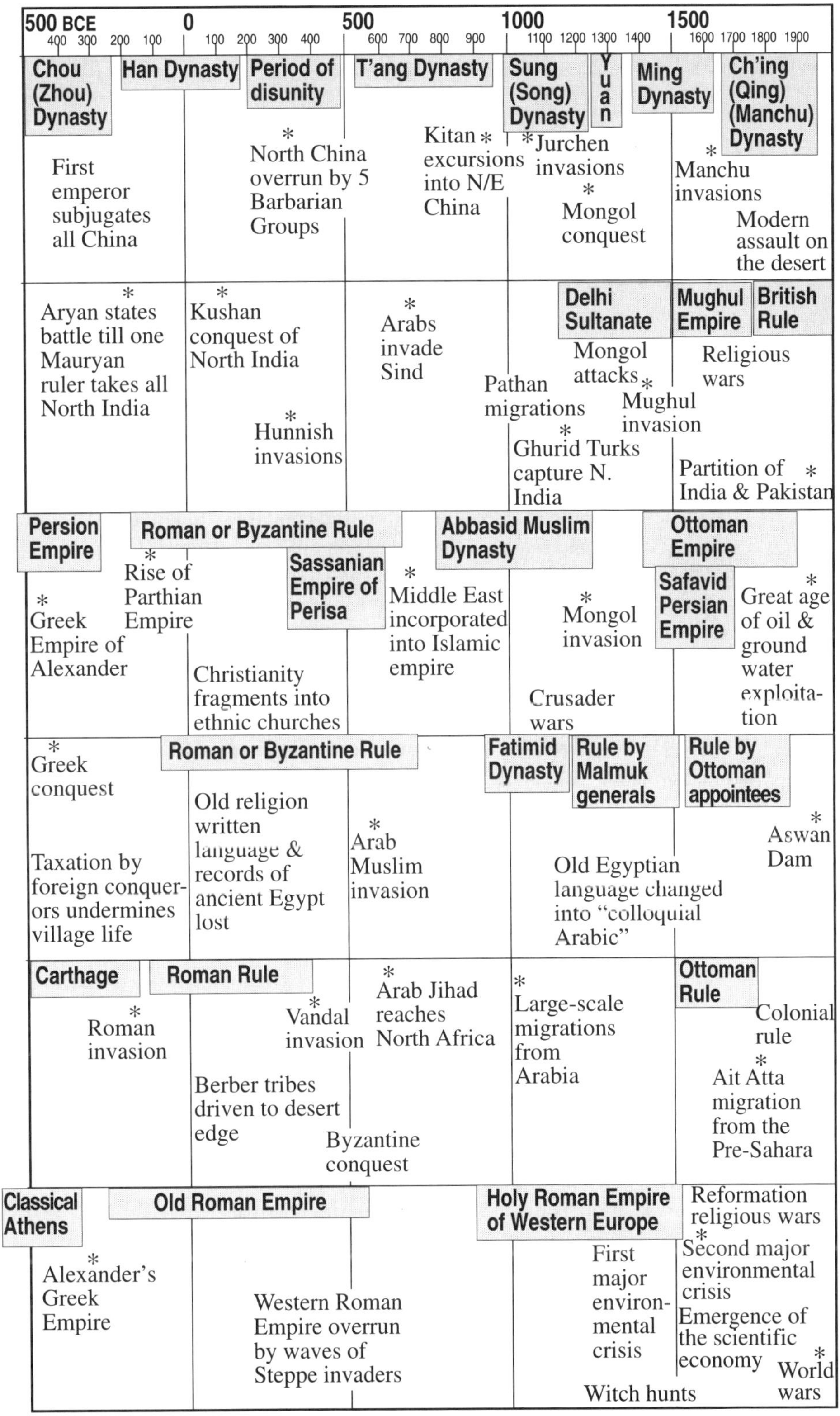

500 BCE
400 300 200 100
0
100 200 300 400
500
600 700 800 900
1000
1100 1200 1300 1400
1500
1600 1700 1800 1900
Chou (Zhou) Dynasty
Han Dynasty
Period of disunity
T'ang Dynasty
Sung (Song) Dynasty
Yuan
Ming Dynasty
Ch'ing (Qing) (Manchu) Dynasty
First emperor subjugates all China
* North China overrun by 5 Barbarian Groups
Kitan * excursions into N/E China
* Jurchen invasions
* Mongol conquest
* Manchu invasions
Modern assault on the desert
Delhi Sultanate
Mughul Empire
British Rule
* Aryan states battle till one Mauryan ruler takes all North India
* Kushan conquest of North India
* Hunnish invasions
* Arabs invade Sind
Pathan migrations
Mongol attacks
* Mughul invasion
* Ghurid Turks capture N. India
Religious wars
Partition of India & Pakistan *
Persion Empire
Roman or Byzantine Rule
Sassanian Empire of Perisa
Abbasid Muslim Dynasty
Ottoman Empire
Safavid Persian Empire
* Greek Empire of Alexander
* Rise of Parthian Empire
Christianity fragments into ethnic churches
* Middle East incorporated into Islamic empire
* Mongol invasion
Crusader wars
* Great age of oil & ground water exploitation
Roman or Byzantine Rule
Fatimid Dynasty
Rule by Malmuk generals
Rule by Ottoman appointees
* Greek conquest
Taxation by foreign conquerors undermines village life
Old religion written language & records of ancient Egypt lost
* Arab Muslim invasion
Old Egyptian language changed into "colloquial Arabic"
* Aswan Dam
Carthage
Roman Rule
Ottoman Rule
* Roman invasion
* Vandal invasion
* Arab Jihad reaches North Africa
Berber tribes driven to desert edge
Byzantine conquest
* Large-scale migrations from Arabia
Colonial rule
* Ait Atta migration from the Pre-Sahara
Classical Athens
Old Roman Empire
Holy Roman Empire of Western Europe
* Alexander's Greek Empire
Western Roman Empire overrun by waves of Steppe invaders
First major environmental crisis
Witch hunts
Reformation religious wars
* Second major environmental crisis
Emergence of the scientific economy
* World wars

Chapter 2

The End of the Green Sahara

THE SAHARA DESERT GIVES US A ROUGH IMAGE OF HOW OUR PLANET MIGHT look if we managed to destroy all life. The most common landscape in the Sahara is one of glistening rock from horizon to horizon. The rarer sand dunes provide a softening touch, with their comforting resemblance to living soil. But they are just sparkling bits of inorganic stone, ground into fragments by the wind. These vistas of rock and sand are so vast as to seem a permanent feature of the natural universe, unchanged since the world began. But the present desert formed and spread within the span of human history, and some of our ancestors helped it happen.

Those who go digging in the Sahara often find bands of sediment oscillating between light-colored sand and the darker remains of fossil forests (Walton 1969: 67; Schneider and Londer 1984: 73), showing how often the lands in these latitudes have switched from desert to forest and back. The sediments show that between 40,000 and 23,000 BCE, North Africa was a green country, populated by Old Stone Age tribes. The land was littered with ostrich eggs and the bones of elephant, waterbuck, hippopotamus, rhinoceros and giraffe. Then the desert returned to drive humans out and polish their abandoned stone tools with blowing sand (Milburn 1984). The most recent green phase of the Sahara lasted from around 12,000–4000 BCE. Soil samples from that period often hold enough organic matter to suggest 300–400 mm of rain per year, but some of the same locations today receive only 10 mm (Flon 1979: 15–28). In the wetter climate of the green Sahara, rivers flowed across North Africa from the highlands of Adrar, Aïr and Ahaggar. The waters of Lake Chad were 320 meters higher than their present level and stretched 640 kilometers north of the modern shoreline (Swift et al. 1978: 42–43).

By 9000 BCE, North Africa was the very portrait of a happy hunting ground. While much of Europe was still encased in ice, ancestors of the Europeans thrived in the green Sahara (Davidson 1991: 10). From the hilltops they surveyed panoramas of grazing antelopes and sleeping big cats. The savannas probably resembled the national parks of Kenya today. And the popularity of Kenya's game parks among Europeans today may represent a collective memory of the beauty of an ancient homeland.

By the time of Old Kingdom Egypt (after 3000 BCE), a return of arid conditions forced another exodus from North Africa. Great herds of animals moved east, west, north and south. In villages of southwest Libya, grinding stones fell from use as farming became hopeless (Barich 1987: 202). Hunting grew more difficult and dangerous, especially in the dry seasons, when travelers could go for days without finding water. After killing an animal, thirsty hunters sometimes squeezed blood and water from its stomach to drink (Hassan and Gross 1987: 91–92). As wild animals grew scarce, the locals learned to keep herds of the slowest beasts. At first they kept cows. Later, as the vegetation grew too sparse for cattle, they increasingly relied on goats and fat-tailed sheep.

Factors of Human Error

Till this point, the people of North Africa had been little more than spectators to climatic change. But pastoralism made them a force in the country, capable of either slowly healing the land or pushing it over the brink to ruin. Under favourable natural conditions, herds of grazing animals can stimulate plant growth with their cropping, saliva and manure. Wild herbivores generally avoid overgrazed areas, and roam in patterns covering hundreds of miles. So, many herders in the Sahel today still claim to simply follow their cows. Some modern ranchers manage animals in ways that improve the land. But wherever the ancient herding tribes made conflicting claims to grasslands or water sources, the animals' movements were constrained. For non-natural reasons the herds could be kept in areas missed by the rains, and the hungry animals might then destroy plant cover beyond hope of short-term recovery. During the Roman empire, for example, many Berbers in North Africa were driven from their coastal lands and pushed out into the pre-Sahara. There they had to graze their animals on the desert's edge year-round, in both damp and dry seasons, till the desert claimed their outer grasslands. Such overgrazing may not be inevitable, but it has been the prevailing trend in North African history. Perhaps assuming that pastoralism can do no better, David Attenborough (1987: 173) says, "Once goats are established, the land stands little chance of recovering its trees and regenerating its topsoil. The goats consume every seedling that sprouts and every leaf that unfurls."

The plant life of arid savannas is commonly called "fire climax" vegetation. In such powder-dry country, conflagration is a regular feature of nature, and humans tend to increase the incidence of fire. In the drier seasons, sparks from cooking fires can easily ignite the bush and burn for miles. Ancient hunters also set fires deliberately to flush out animals. Later, it was more often cattle herders who burned the range—after their cows had grazed it so repeatedly that little save inedible thornbushes remained. The pastoralists would then set the country aflame, destroying the bush and turning back the cycle of growth to point zero. Many regions of Africa were swept by such range fires, perhaps every fifty years for thousands of years. The trade winds often fanned range fires toward the forested regions. So, in 1986, uncontrolled brush fires from the Sahel burned into the forests of the Ivory Coast, destroying perhaps two-thirds of the nation's cacao plantations.

A natural rhythm of fire recycles nutrients to the soil and vitalizes plant growth. But regular natural fires plus regular manmade fires can prove too much. All fires kill some of the soil's humus and its nitrogen-fixing bacteria. And a hot fire can melt organic compounds into a sticky, waterproof coating over each particle of soil, making the earth less able to absorb moisture. The burning of shrubs and trees may also destroy the countryside's last islands of shade, in which many species of plants had survived during previous drought years. Finally, frequent fires strip the earth bare to the sky, leaving the topsoil to blow in the wind or harden in the sun (Warren and Maizels 1977: 210).

The borderline between forest and arid grassland can be sharp, even where the climate has no such boundary. So, in northern California, Mansanubu Fukuoka (1989: 35) described hills of redwood forest to his right, and pale mountainsides of dry grass to his left: "Here, under identical climatic conditions, was an expanse of green on one side and a desert on the other. Why?" Fukuoka claims (1989: 346) the increased reflection of sunlight from cleared and overgrazed ground drives up the local temperature: "If the temperature is thirty degrees C. (86° F.), the heat reflected from this ground surface raises it to forty degrees (104° F.)." The hot air over such land rises in an updraft, nudging aside most drifting rain clouds. If such conditions kill the grass, things get even worse. In the Sahara, regions of bare rock or sand reflect about 90 percent of the sun's radiation, and travelers often find themselves sunburned from below (Langewische 1996: 47).

As most of North Africa slowly turned into a nearly treeless range, the living pumps for recycling water between earth and sky disappeared. We now know that in most inland regions of the world roughly half the rainfall comes from water transpired by local plants. This can be measured, because where the forests are cut today in Brazil or Ethiopia, local rainfall generally declines by half (Newell 1971: 457–59).

In these ways and others, the most recent green phase of North Africa was probably cut short by human predation on the land. The current dry phase, which is now about six thousand years old, may also be prolonged by human intervention. At this point, the Saharan moonscape is so harsh that only the icefields of Antarctica are more hostile to life. And in recent years the desert seems to be getting worse. A resident of Adrar, Algeria, told William Langewische (1996: 114), "It's raining less. And every year it's hotter. Nomads can no longer survive in this climate."

Yet after six thousand withering years of desertification, some creatures of the old green Sahara still cling to life. Certain isolated gullies or highlands still support clusters of Laperrine's olive, Duprey cyprus, or oleander. At least until 1900 CE there were desert crocodiles; in that year the myth of crocodiles in the desert was rendered fact when a French soldier shot one of the last survivors (Swift et al. 1978: 51). Little islands of forest in the Atlas range still hold groves of cedar a hundred feet tall, maritime pine, evergreen oak, walnut, holly and ivy (Attenborough 1987: 33). Such remnants offer an occasional glimpse of how Morocco and Algeria once looked under slightly better conditions. Similar holdouts of Mediterranean or African trees are found in the Ahaggar mountains in the desert's core (Huzayyin 1972: 308). As benchmarks, these groves of trees show how far the surrounding country has fallen. As survivors, they suggest that life can endure this trial and may again prevail someday.

The Burden of Hope

As desertification intensified, the human survivors adjusted to worsening conditions as best they could. Most retreated to perennial water points or riverbanks and, as various groups converged, often fought over wells or bits of irrigated land. Often it was the losers who turned back to the parched wilderness and tried to find their living there. Naturally, they wondered what they had done to deserve their hardship. Speaking for such people in a land near the Sahara, the book of Isaiah says,

> Behold, the Lord will lay waste the earth and make it desolate, and he will twist its surface and scatter its inhabitants. And it shall be, as with the people, so with the priest; as with the slave, so with his master; … as with the buyer, so with the seller.… The earth shall be utterly laid waste and utterly despoiled; for the Lord has spoken this word.
>
> The earth mourns and withers, the world languishes and withers; the heavens languish together with the earth. The earth lies polluted under its inhabitants; for they have transgressed the laws, violated the statutes, broken the everlasting covenant. Therefore a curse devours the earth, and its inhabitants suffer for their guilt; therefore the inhabitants of the earth are scorched, and few men are left. (24:1–6)

Some farmers in the afflicted regions, such as the old Berbers or the Nabateans of the Negev, learned to make do with ever less water. They became masters of water conservation, able to cultivate rain-fed gardens in virtually barren country. At some point, however, even the Nabateans abandoned their land. Other groups of desert people mastered the arts of coercion and lived as predators on others. All told, desertification usually brought an escalating series of droughts, famines, and battles over the means of life. A succession of shocks sometimes traumatized whole societies for generations.

As Isaiah continues,

> The new wine dries up, the vines sicken, and all the revellers turn to sorrow. Silent the merry beat of tambourines, hushed the shouts of revelry, the merry harp is silent. No one shall drink wine to the sound of song; the liquor will be bitter to the man who drinks it. The city of chaos is a broken city, every house barred that no one may enter … all revelry is darkened, and mirth is banished from the land. Desolation alone is left in the city and the gate is broken into pieces. So shall it be in all the world, in every nation, as when an olive tree is beaten and stripped, as when the vintage is ended. (24:7–13)

That terrible prediction was actually meant to inspire hope, to turn people toward a path that might heal and renew the world. Our visions of that path may differ from Isaiah's. We may have different understandings of what causes environmental catastrophe, but perhaps our hopes are similar. As the deserts have continued to spread, some communities have been walking their own paths of hope, nursing their lands like parents caring for sick children. While some people have responded to catastrophe as victims, others have tried their hands as healers.

Chapter 3

Women's Place in the Desert

IN THE TENERE REGION OF NORTHERN NIGER, WHICH THE TUAREG CALL A desert within a desert, the remains of numerous villages lie thinly covered in sand. In places, flint tools and bits of pottery litter the ground for hundreds of meters in every direction. The remains include millstones, sickles, fishhooks, nets and catfish bones (Roset 1987: 211, 225). Perhaps life in these villages was hard, but likely it was also peaceful and beautiful. When the desert came, it didn't arrive as a sudden invasion of sand. Each settlement or valley changed gradually, almost softly, the way Erika Friedl describes a drought year in Iran:

> During the long and tired afternoons between milk boiling and the evening milking, Atri, aching in every bone with her fifth pregnancy, would squat in the shady doorway, looking down the dust-colored, sun-baked slopes toward the village.... She wondered how fast the grass was turning the color of sand under the blue sky and watched twisters moving hither and yon over the Deh Koh plain, twirling slender columns of dust up high into the shimmering air. (1989: 201–2)

When the land grew so desolate, what happened to the women? How did their lives change when the environment turned unproductive? What has been their place in the desert?

Women's Ways of Production Dry Up

In the now barren hills where Uganda, the Sudan and Kenya meet, the Ik people used to roam the country harvesting wild plants. The women's gathering parties and gardens produced the bulk of each band's food, while the men specialized in their glorious game of hunting, with its less predictable results. At night these people met around campfires, planning their journeys for the coming days. And since the women usually secured more food, their voices were at least equal to the men's (Turnbull 1972: 26). Robert O'Connell, in his history of warfare, *Ride of the Second Horseman*, claims that this sort of gathering and hunting life prevailed for 99 percent of human history. He describes it as "a relatively low-key existence emphasizing personal independence, general equality among group members, including women, consensus-based decision making achieved through open and protracted discussion, and freedom of movement, particularly as a means of conflict resolution" (1995: 226).

The skeletons of ancient gatherer-hunters are commonly taller and more robust than the skeletons of later farming people (O'Connell 1995: 57–58). We must suspect that farming was taken up by necessity, as the natural landscape grew less fruitful. The people of Siwa Oasis in western Egypt clung to their old wandering and gathering life as long as they could. The women harvested wild plants according to season: green shoots in early spring, fruits in June or July, nuts and grains in the cool season. In the saline pools they caught "water worms," or primitive shrimp. As the oasis water grew brackish, people used ostrich shells to collect rainwater from pools in the desert rocks (Hassan and Gross 1987: 93–98).

When the desert came, it undercut people's ways of living. Where they had mixed gardening with collecting wild plants, both activities grew less productive. But the lives of women and men changed in different ways. People were affected by desertification differently, according to who they were, the work they did and the responsibilities they bore.

Divisions of labor between men and women have always differed across cultures. But in Neolithic Africa and Asia, at least one pattern was widespread. Women usually practiced trades that allowed them to watch their children while working (Judith Brown, in Barber 1994: 29). In the old Middle East, women likely did gardening, pottery, weaving, tanning, teaching, local marketing, and care for domesticated animals—almost all the crafts of the "neolithic revolution." They did these jobs in or near their homes, usually while mothering a series of infants over most of their lives. But certain neolithic-age tasks were difficult to combine with child care. These jobs, such as hunting, long-distance trade, herding animals or fighting enemies, tended to become "men's work." Men may have joined in the kinds of work women did, including child care, but certain rather dangerous jobs were commonly characterized as tasks for mature males. Peggy Reeves Sandray (1998: 9) suggests that if tasks involving long-distance travel or danger were made exclusively male roles, it was because fathers seemed more expendable than mothers in the process of raising families.

When the desert came, what happened to this common division of labor? Generally, women's roles suffered more that those of men, because women's ways of raising and gathering food were rendered less productive. In the Tadrart Acacus Hills of Libya, the people of early settlements gathered and ground grain. Possibly they sowed some seed near their homes and collected more growing wild in the countryside. Much of the harvesting and grinding was likely women's work. But after about 5550 BCE, grinding stones and sickles fell from use (Barich 1987: 202). The land dried up, and the local women lost their main means of food production.

Those who migrated to water points and river valleys could still grow their food, but as the climate grew more severe over most of the Middle East, almost nothing would grow without irrigation. No longer could women sow seeds around their homes, let the rain fall, pull weeds and collect the fruit. Now water had to be brought, channels dug and parched land plowed. These tasks required brute strength and the use of draft animals, which could easily trample human babies (Barber 1994: 165). On top of that, irrigated land in the arid regions usually became a scarce prize and an object of war. Those who controlled watered land were increasingly those who could hold it by force or collect tribute from the farmers. So gardening continued in the green areas of the Middle East, but usually under conditions difficult for mothers.

The original tamers of sheep and goats were quite possibly women. After hunters had killed a wild mother goat, perhaps they caught the babies and brought them home, and women decided to raise the little orphans. Such keeping of domesticated animals would have been mainly women's work, as it often is today in India or China. The women carried fodder and water to the animals, as they went about feeding their own children. But if fodder and water near the settlement grew scarce, then the animals had to be taken out on the open range. We have many examples of ancient communities edged into a life of nomadic animal herding. So in the 500s BCE, the Rechabite tribe of the kingdom of Judah gave up building houses and sowing crops, and took to following its animals about the countryside while living in tents (Patai 1962: 81–82).

As pastures grew sparser and farther apart, herding animals often became men's work, because traveling beyond an oasis in arid country involved serious risks of dehydration. Around Siwa, the men generally went out hunting only in the rainy season, when scattered pools of rainwater made it safer to trek through the harsh landscape (Hassan and Gross 1987: 98). Domestic animals, however, could not wait for a better time of year, and so desert herding required journeys through dangerous country in all seasons. Over time the treks for grass and water generally grew longer. So the Kabbabish of Sudan and the Rwala of Arabia each evolved patterns of migration covering hundreds of kilometers between summer and winter pastures. Such treks involved not only the hardships of travel, but also the violent tasks of defending the herds from raiders or predatory cats. Masai herdsmen still stand guard over their cows with spears in hand, ready to fight off attacking lions. It is a job deemed unsuitable for mothers with young children.

In a desiccated landscape, trade often became long distance work as well. As the produce of local settlements declined, communities dispersed. Home production for local trade grew pointless as customers left. Barbara Barich (1987: 202) reports that the ceramic and stone tools made in pre-desert Libya were "eventually replaced by tools of much lower technical sophistication" as the craft trade among sedentary villagers faded away. After that, trade continued mainly by caravan over vast distances and under extremely dangerous conditions. This too was difficult for mothers with children and tended to become men's work.

Even where gardening grew impossible, the men could still hunt. In North Africa, late neolithic arrowheads grew smaller as larger wild creatures disappeared, but hunting went on for smaller game, including desert rodents (Barich 1987: 199–200). Later, some hunters would master falconry and supplement their food with the prey of desert hawks.

When all else failed in a drought year, people could still survive by raiding.

The old Arabian Bedouin deemed it necessary, and even virtuous, to raid the supplies of other groups. Scarcity promoted coercion as a means of meeting basic needs. This too was work best risked by men. When the men went raiding, they usually left the women and children behind. In this case, however, their women might themselves become objects of raids by other groups. So, in an economy partly based on raiding, a mother could seem to depend on male guardians for her very existence.

With all these difficulties, what work did desert women do? Some managed to hold on to their old economic roles. The Tubu women of the Tibesti hills (of northern Chad) continued as weavers, tanners, caretakers for goats and gatherers of plants. In these Saharan highlands, enough greenery remained for plant specialists to play an important role. Over the arid zone's vast borderlands from Mongolia to the African Sahel, most of the country was merely reduced to grassland. Although most people in these regions adapted a nomadic way of life, the country through which they passed was still green enough for women to operate as productive nature workers. Mongolian women usually tended sheep and goats, gathered edible plants, churned cheese and made most of the camp equipment. But they seldom had charge of the horses and camels, which were larger, more dangerous and economically important.

Within the shrinking islands of traditional farming (mainly the Atlas Mountains, the Nile Valley, Mesopotamia and the Indus Valley), many women worked the fields, weeding, grinding grain or feeding animals. Ancient skeletons show that many women ground grain till their toe, knee and shoulder bones were deformed from pushing the heavy grinding stones (Barber 1994: 96). But over vast regions, the spreading desert forced a rift between economic work and child care. Fekri Hassan and Timothy Gross (1987: 98) say that, in the desert, "none of the resources is highly productive, is found over a large area, or has a predictable occurrence in space and time." Old neolithic villages of the Sahara gave way to life on the move, in which women were generally left to carry the children and baggage from place to place. Under desert conditions, the basics of caring for children, securing water and fuel, cooking and moving camp were often all-consuming. As American archaeologist Wendell Phillips said in the 1960s,

> Whether high born or low, the Bedouin women … are all obliged to put up and prepare the tents, to strike and roll them up, and to load and unload them … also to attend to all the domestic work, such as collecting dung and wood to cook on; and lastly, to find the time as best they can to look after and bring up the children. By twenty-eight, the majority I have observed in Oman, Yemen and the former Aden Protectorates are old and worn-out grandmothers. (1966: 141)

The work of desert women was hard, and men's labors could be light by comparison. But women's roles were generally considered less economically important. Women might milk the goats, make the cheese, shear the sheep and spin the wool, but men handled transactions with the outside world across the desert. Women's products were sold through men, and men controlled the income (Minai 1981: 106). It began to appear that desert women depended on men for all the necessities of life, and a female's contribution was mainly through sexual reproduction. Thus women were increasingly treated as a sexual resource and little else.

As if the economics of scarcity were not enough, the politics of scarcity often made things worse. According to the great historian Ibn Khaldun, nomads of the wastelands often coveted the wealth and security they saw in watered lands. When afflicted by drought, they could come to the green areas as beggars or, more manfully, seize what they needed by force. More manfully yet, they could conquer and establish themselves as lords over the farmers. So islands of green land across the Middle East were repeatedly invaded and ruled by warlords from surrounding deserts. In these conquests, groups of formerly nomadic women came into the green lands as members of a dominator elite, and their warlord husbands often tried to seclude them from contact with the conquered population. Back in the desert, these women had often been regarded as a primarily sexual commodity, but at least they had been free to wander and work in the open country. Now they became a jealously guarded sexual commodity to be isolated from the surrounding community. Soon, the restricted "status" of the conqueror's women became an ideal for the general population.

The difficulties of desert women were initially environmentally imposed, but over time they became socially imposed as well. An inherited social inferiority evolved, sanctified by force of custom. Within desert regions, even women in cities, or in islands of farmland such as the Nile Valley, came under this stereotype. In the prevailing context, females were presumed to be dependents, even when obviously doing backbreaking work. Speaking in 1909, Bahithat al-Badiya objected,

> Men say to us categorically, "You women have been created for the house and we have been created to be breadwinners." Is this a God-given dictate? How are we to know since no holy book has spelled it out? ... The division of labor is a merely human creation.... Women of the villages of Upper and Lower Egypt help their men till the land and plant crops. Some women do the fertilizing, haul crops, lead animals, draw water for irrigation.... Do you have any doubt that a woman from Minufiya [a farming area in the Nile Delta] would be able to beat the strongest man from al-Ghuriya [a section of Cairo] in a wrestling match? (in Badran and Cooke 1990: 230)

In the 1920s and '30s, the English missionaries Mildred Cable and Francesca

French traveled up and down China's old Silk Road through the Gobi, Lop and Taklamakan deserts. They hoped to sow cultural seeds from Europe, partly by presenting themselves as examples of how Western women might live independently of men. But their sense of compassion involved a certain acceptance of the "official" context—that the role of local women really was one of mere dependence: "To the Tungan woman, home is the place where she is fed and clothed in exchange for bearing children, and where all that is required of her is entire submission to her husband, and that she put his pleasure before every other consideration" (Cable and French 1984: 167).

Women in a Herdsman's World

In ancient Sumer, the kings of the land were supposedly elected as the men best loved by the goddess of the earth. They were chosen, or re-chosen, each year through an enactment of sacred marriage to the goddess. The king was called to the holy bed chamber atop a ziqqurat, where the earth met the sky. There he joined in love with a priestess filled by the spirit of Inanna. He became her ever passing and returning lover, Dumuzi, and received "the godship of the land" for the coming cycle of seasons (Kramer 1990: 305).

While the soil of Mesopotamia retained its early fertility, the lover of Inanna was usually represented as a farmer-king. But, later, as primitive irrigation turned the land salty, kings were sometimes chosen from the herdsman tribes. The cuneiform literature playfully harps on the contrast between farmers and herdsmen. Dumuzi is sometimes a shepherd, vying with a farmer for the love of Inanna, and sometimes the herdsman wins. Dumuzi the shepherd becomes the beloved of the goddess, and king of the land. Then the farmer, bowing to the wishes of the goddess, allows the herdsman to water and graze his animals about the farm (Kramer 1990: 305–16).

Where animal husbandry was only supplementary to raising crops, men and women often worked together in caring for both crops and animals, as they still do in much of India. But where herdsmen became the chief providers, then by implication the most productive use the land could bear was pasturage. Thus a mainly pastoral economy grew characteristic of the whole Afro-Asian arid belt: the Mongols bred horses, the Arabs raised camels, the Masai drove cattle.

The herding life for men involved freedom, danger and courage. It was conducive to manly pride—a pride which commonly scorned the farmer's subjection to manual labor, taxes and landlords. Among nomadic men, pride in their own importance often involved contempt for the seemingly lesser role of women. Women were often associated with the weakness of the children they cared for. In a mainly pastoral economy, women might follow the herds in tents or yurts, but they seldom played leading roles in drives to remote pastures, in raids or counter-raids for animals, or in combat with predatory cats. The Tuareg and Tarqui of the Sahara, and the Mongols, may be known for relative equality between the sexes, and some societies in arid lands retained something of the ancient partnership among hunter-gatherers. But the usual role of women among pastoralists in Mauritania, the Sudan, Arabia, Baluchistan and Xinjiang was as helpmates in a man's world.

In the more fertile lands of Europe, tropical Africa, India and China, women usually engaged in a greater diversity of roles. In those richer and more complex environments, both sexes had a wider range of vocations and lifestyles. But desert women were often valued almost exclusively in sexual terms—for the pleasure they gave men, and the children (hopefully male) they bore. What other valued thing did most desert women produce? What else were they good for, or created for? Despite vast changes in the modern Middle Eastern economy, this view of women's potential remains widespread today. A female teacher in Jordan recently complained to reporter Gerald Butt (1997: 232):

> No matter how well qualified you are, unless you are married—and preferably have borne a male child—you are nothing.... In the marriage market the working girl is still judged by many as loose, immoral and, in certain cases, promiscuous, in contrast to those girls who are secluded in their homes and thereby considered paragons of virtue and chastity. Given these circumstances, it is not surprising that very few young women continue their schooling beyond age fifteen, however late they may actually marry.

At least the traditional cultures inherited from nomadic peoples had generally placed more value on women as mothers than as sex objects; at least, a mother tended to grow in status as her children matured and she aged.

Women's Rights in Arid Lands

Where men viewed women as the less productive sex, women's rights and powers were usually reduced accordingly. Traditional law in Somalia put it clearly: "Until a couple of years ago [the 1970s], taking a man's life could cost a hundred camels; a woman's was valued at fifty" (Jordan 1981: 733). And where females were considered dependents on men, they were often treated as men's legal property.

Among the largely pastoral pre-Exodus Hebrews, the tradition of bride price meant that a father decided whom his daughter married, and how much he charged for her. A daughter was an asset of her father's to be disposed of for his profit. The father's right to income from the sale of his daughters was firmly protected by law. As the book of Deuteronomy ruled:

> When a man comes upon a virgin who is not pledged in marriage and forces her to lie with him, and they are discovered, then the man who lies with her shall give the girl's father fifty pieces of silver, and she shall be his wife because he has dishonoured her. He is not free to divorce her all his life long. (22:28–29)

Riane Eisler (1987: 97) points out that the father's right to a fair price for damaged goods was insured, and this was the main moral concern. The principle of legal ownership could hardly be clearer if it was a matter of compensation to slave owners for attacks on their slaves.

A "regular union" of man and woman for the ancient Hebrews was one approved by the woman's father and clan. If a child was born outside that authority, the child was to be socially boycotted: "No descendant of an irregular union, even down to the tenth generation, shall become a member of the assembly of the Lord" (Deuteronomy 23:2). Hopefully, this would stop women from choosing their own lovers. If they did, both they and their children would be made to suffer for it all their lives.

A history of family law can be partly reconstructed for ancient Mesopotamia. In early Sumer down to 2371 BCE, laws concerning marriage and property were quite equal for men and women (Hawkes 1973: 182). Then, as a series of warlords from inland regions conquered the river valleys, and as crop yields dropped with the salinization of irrigated soil, the legal position of women slowly fell. Under the early Babylonian empire (starting c. 1750 BCE), the powers of men over their wives and children grew stronger. Jacquetta Hawkes (1973: xxv) notes an "ever-growing ferocity in Mesopotamia of the punishments to be inflicted on adulterous wives or disobedient children." Then, under the warlords of Assyria from the 1200s to 600s BCE, the laws against women grew draconian. Under Assyrian law, if a married man raped a virgin, the girl's father could give his daughter to the rapist and take the rapist's wife for himself. Women who stole from their husbands could be executed, and if they stole from another family, they could have their noses or ears cut off. If a woman gave shelter to a runaway wife, both women could lose their ears. For women who aborted themselves, the law said "they shall impale her on stakes without burying her." Concern for women's virginity was institutionalized. Married women, widows, and daughters of gentlemen were forbidden to go out unveiled. Harlots and female slaves, however, were barred from wearing veils. If either class of women violated their veiling restrictions, the punishment was flogging with staves (Hawkes 1973: 182–83).

If a proposal for such laws had been made in pre-dynastic China, Dravidian India, tropical Africa, Old Europe, Polynesia, the early Americas or even ancient Egypt, it would probably have been regarded as barbaric or laughable. Such laws were expressions of a changing social climate within a limited portion of the globe. The attitudes behind those laws rose first and took their most extreme forms near the center of the great Afro-Asian arid belt. And over that same region, customary inequality has endured stubbornly down to the present, despite powerful movements of religious protest against it.

Religious Revolts against Desert Traditions

In Arabia before Muhammad, women had been the tradable property of men. Widows had been commonly inherited as part of a dead man's estate.

Muhammad's first companions were surprised when the Prophet announced God's condemnation of the practice: "O ye who are believers, it is not permitted to you to inherit women against their will" (Quran 4:19).

The Quran also ruled that inherited property must be divided among both females and males: "Unto the men belongeth a share of that which parents and near kindred leave, and unto the women a share of that which parents and near kindred leave, whether it be little or much—a legal share" (4:7). According to the commentaries of al-Tabari, this caused the men of Medina deep concern:

> "How," say men, "can one give the right of inheritance to women and children, who do not work and do not earn a living? Are they now going to inherit just like men who have worked to earn that money?"... Then they said to themselves, "We must go ask for clarification." And they went to the Prophet and asked him some questions on the subject. (in Mernissi 1991: 125)

Ibn Hajar al-'Asqalani (in Mernissi 1991: 122), in his biographies of the companions of the Prophet, reports, "Umm Kajja ... complained to the Prophet: 'My husband is dead, and they are preventing me from inheriting.'" After comparing many similar accounts, Moroccan scholar Fatima Mernissi concludes, "The Prophet faced a constant parade of women who came to demand the application of the new law." The concept of women as productive persons who should inherit wealth was normal in the green countries of old Ireland or Siam, but in a desert region this was an explosive idea.

Islam made, or at least attempted, a range of other reforms related to women's status. For example, it modified the practice of fathers selling their daughters. According to a saying of the Prophet, "None, not even the father or the sovereign, can lawfully contract in marriage an adult woman of sound mind without her permission, whether she be a virgin or not" (Hadith, in Minai 1981: 9–10). The bride price (paid to the bride's father) and dowry (which the bride brought with her to the groom's family) were to be viewed as exchanges of gifts rather than as payments for human property. Likewise, Muhammad opposed a custom of his own Quraysh tribe in Mecca, by which a man could divorce his wife, yet retain veto power over her right to remarry. This had been a profitable tradition. To remarry, a divorcee and her suitor had to buy the consent of her former husband, for whatever price he chose (Tafsir al-Tabari, in Mernissi 1991: 123).

Islamic declarations of women's rights were powerful protests against desert tradition. Like early Christians, many early Muslims took pride in their egalitarian spirit. At first it seemed possible that freedom and spiritual worth might apply equally to male and female believers. Fatima Mernissi (1991: 10–11) describes that hope, recalling Muhammad's Medina state as "the time both far away and near at the beginning of the Hejira, when the Prophet could be a lover and a leader hostile to all hierarchies, when women had their place as unquestioned partners in a revolution that made the mosque an open place and the household a temple of debate."

But desert traditions often proved stronger than the words of departed prophets. The Islamic moral protest was forceful in itself, but little else in the desert economy or environment changed. Conditions of life in the lands of scarcity went on much as before and, with a resilience as powerful as the desert itself, the old ways often regained their hold. After Muhammad's death, the second Caliph, Umar, modified certain rules concerning Muslim women. The Quran had indicated flogging as the penalty for adultery: "The adulteress and the adulterer, you shall whip each of them one hundred lashes"(Sura 24:2). Umar increased the penalty to death by stoning. Yet earlier, Umar had urged the Prophet to order his wives veiled: "Messenger of God, you receive all kinds of people at your house, moral as well as evil. Why do you not order the *hijab* (veil) for the Mothers of the Believers?" (Tasfir al-Nisaburi, in Mernissi 1991: 184). In Arabia, the veil was a custom by which the militarily powerful "noble" tribes guarded their women from the eyes of conquered tribute-paying tribes (Patai 1962: 261). Evidently Muhammad had not required such veiling for his wives, and some of his followers felt this was shameful; if Islam's leading women went unveiled, they would be dressed like members of a non-dominant tribe.

Within several years after Muhammad's death, a new generation of religious authorities banned women from worshipping at mosques in mixed company with men (Minai 1981: 21), and the backlash over women's inheritance gathered force. The Quran said, "Give not unto the foolish your wealth, which Allah hath given you to maintain; but feed and clothe them from it, and speak kindly unto them"(Sura 4:5). Some male religious leaders now argued that the "the foolish" obviously included all women (al-Tabari, in Mernissi 1991: 126). Among many of the Muslim Bedouin, the right of inheritance for women remained unthinkable, though the Bedouin commonly claimed their traditions to be the very essence of Islam. Down to recent times, the Bedouin of Jordan or Syria generally refused inheritance to women, and modern governments starting with the Ottoman administration tried in vain to uphold Islamic law (Patai 1962: 107).

Concerning female slavery, Muhammad told his soldiers to free women captured in war. Most of his troops, however, refused to give up their female captives, even when the Prophet offered them compensation from his personal wealth. After Muhammad died, the practice he had condemned was widely reaccepted, provided that Islamic soldiers targeted only "unbelieving" women. For centuries this became a new religious sanction for the ancient Bedouin practice of raiding "foreign" villages and taking the women as slaves. Under such conditions, Muslim women often had to wear clothing deemed "Islamic" or risk being attacked as if they were plunderable enemy women. Even after the age of jihads and crusades, a shadow of this attitude remained. Many men descended from nomadic raiders, or from those conquered by desert raiders, still felt that any female who did not hide herself behind male guardians and traditional garb was virtually asking to be attacked. And this attitude of control or intimidation towards women is still widely accepted as "religious" across North Africa and West Asia. Concerning such survivals of desert tradition, Mernissi says,

> This almost phobic attitude toward women is all the more surprising since we have seen that the Prophet has encouraged his adherents to renounce it as representative of the *jahiliyya* [the "time of ignorance" before Islam].... This leads me to ask: Is it possible that Islam's message had only a limited and superficial effect on deeply superstitious seventh century Arabs?... Is it possible that the hijab, the attempt to veil women, that is today claimed to be basic to Muslim identity, is nothing but ... the persistence of the pre-Islamic mentality, the *jahiliyya* mentality that Islam was supposed to annihilate? (1991: 81)

The records of early Christianity also suggest a prominent role for women in the primitive church, and a belief in the equality of souls before God. The gospel accounts show women as prominent figures in Jesus' movement (Luke 8:1–3); Acts refers to a female disciple named Tabitha (9:36–42), and Paul praises a female apostle called Junia (Romans 16:7).

Several of the earliest Christian texts, most of which were later discarded from the official New Testament, describe the first Christians as deeply respectful toward women. The Gospel of Mary (Magdalene) quotes the apostle Peter saying to Mary, "Sister, we know that the savior loved you more than the rest of women. Tell us the words of the savior which you remember, which ... we do not [know] and have not heard" (in Pagels 1995: 67).

But if many male Christians still accepted patriarchal standards as orthodox, then to affirm women's equality seemed to negate morality as they knew it. Therefore a second generation of church leaders backtracked to defend pre-Christian Middle Eastern customs and ensure the respectability of their movement. Within a few decades of Jesus' death, church leaders were drumming women out of the movement's leadership. In a letter of doubtful authorship they had Paul of Tarsus thundering, "I do not allow any woman to teach or to exercise authority over a man"(First Timothy 2:12). And nearly two hundred years later, Tertullian (of Carthage) felt it necessary to repeat and clarify: "It is not permitted for a woman to speak in church, nor is it permitted for her to teach or baptize, nor to offer [the Eucharist], nor to claim for herself a share in any masculine function—not to mention any priestly office" (in Pagels 1989: 60).

Control of Women's One Remaining Power

In the desert, where often a woman's main productive role was literally sexual, it was widely assumed that women were preoccupied almost exclusively with sex. What else did they do that was important? What else might they ever think of? To control a woman's presumably insatiable sexual appetite, drastic preventative measures were thought to be required. Elizabeth Oram, working with midwives in rural Mauritania, described one traditional method of controlling women's lust:

> Whenever I asked Maure women why they have their daughters circumcised, their first response was almost always, "No man would marry an uncircumcised woman." They believe that if left intact, the clitoris greatly arouses a woman and causes her to chase uncontrollably after any man as soon as her husband's back is turned. Thus, they believe that an uncircumcised woman's thoughts would constantly turn to sex and that she would do anything necessary to satisfy her lust, a behaviour completely unacceptable for a woman. (in Badran and Cooke 1990: 64)

Across North Africa this custom was claimed as an Islamic tradition required by God. In 1997, Islamic "traditionalists" in Egypt tried to overturn the national law against female circumcision through court action. They were apparently shocked when the Supreme Court of Egypt upheld the existing law, basing its decision on original Islamic laws, which condemn the practice (*Globe and Mail*, January 3, 1998). It is unclear, however, if this ruling reduced the social demand for female circumcision.

Where desert women's sexuality was not suppressed in this most physical way, it was generally controlled by rigid social isolation. Possibly nowhere else on earth were the walls of sexual segregation raised so high as in the arid lands from North Africa to Central Asia. Raphael Patai (1962) feels that over much of the Middle East the concern to control women's sexuality came to pervade almost every aspect of social life. In the 1990s it was still illegal for a Saudi Arabian woman to drive a car. Where would she go? Who would she see while traveling away from her husband or father? In Afghanistan, any woman who dared to become a musician was commonly assumed to be a prostitute. Why else would women perform music other than to seduce men?

In 1952 the Mufti (Islamic religious leader) of Egypt, Shaikh Hasanain Makhluf, tried to veto the right of women to vote and hold office. He did so, citing the Quran, Sura 33:33, and interpreting it according to desert-born common sense. The passage says that the wives of the Prophet should set an example of virtue:

> And stay in your houses. Bedizen not yourselves with the bedizenment of the Time of Ignorance. Be regular in prayer, and pay the poor-due, and obey Allah and His messenger. Allah's wish is but to remove uncleanness far from you, O Folk of the Household, and cleanse you with a thorough cleansing.

In what other part of the world would such words be assumed to mean that women must be veiled from head to toe, secluded from the world and barred from the exercise of civic rights and responsibilities? Where else would honour and virtue for a man and a woman be so drastically different?

In that spring of 1952, Duriya Shafiq, a famous Egyptian writer, pointed out that the previous Mufti of Egypt, Shaikh Alam Nassar, had interpreted the Quran in an opposite way: "Islam looks at the woman as it looks on the man with respect to humanity, rights and personality." Shafiq therefore posed a question to the women of Egypt: "Whom shall we follow ... the present Mufti or the previous Mufti?" If she was to follow the current Mufti, she would then say,

> I have decided to stay at home, quit journalism, remain with my daughters in my house, and not allow them to be educated. I shall ask the state to expel women students from the universities and fire women employees as well as nurses, doctors, lawyers, and engineers because these professions are suitable only for men. All this, in response to the appeal of his eminence, the Mufti of Egypt, Shaikh Hasanain Makhluf. (Shafiq 1952, in Badran and Cooke 1990: 355)

The women in Muhammad's life were business managers (Khadija), religious leaders (Fatima) and generals (A'isha), and virtue for them involved far greater things than avoiding contact with men. It was not the visionaries of Islam, Christianity or Judaism who reduced women to a primarily sexual identity, but the overriding conditions of life in the desolate lands.

Where men valued women for little else besides their sexuality, a woman's sexual misconduct usually merited her death. Islam tried to control and equalize this draconian penalty. In theory at least, Caliph Umar reinstated death by stoning for both the woman and the man, but it was almost impossible to convict. The accuser had to produce four sworn eyewitnesses to the crime, and if this highly unlikely proof was lacking, the accuser would be lashed: "Those who defame honourable women and cannot produce four witnesses shall be given eighty lashes. No testimony of theirs will be admissible [in the future], for they are great transgressors" (Quran, Sura 24:4). And if, in spite of this warning, an accuser dared to bring the charge and won a conviction, then the death sentence still might not be carried out—unless the accuser was willing to personally throw the first stone.

In common practice, rules against sexual "theft" were less equitable, and the hard but even hand of Islam was modified by older ways. Across North Africa and West Asia, a pre-Islamic Bedouin code lived on. Women could be killed by any of their male relatives without trial, on the mere suspicion of sexual misconduct. During British colonial rule of Iraq in the 1920s the CID officer in Baghdad reported an average of around one hundred murders a week. The majority of the victims were women killed by male relatives for alleged sexual misconduct. Reportedly, accused male paramours were often neglected in this vengeance (Phillips 1966: 143–44). By such threats of death, women were to be held in line from even looking at another man. In a similar vein, a Marionite Christian in Syria told reporter Gerald Butt (1997: 216), "I would murder my daughter rather than see her marry a Muslim."

In Syria, Bouthaina Shaaban's girlfriend in high school was murdered by her brother in 1968. The deed was considered a "crime of honour." Having destroyed a "loose woman" who threatened the family's reputation, the brother handed himself in to the police, spent several months in jail and emerged with "a mysteriously heroic air." Shaaban (1988: 5, 9) writes that before that, as a little girl with no knowledge of sex, "I desperately wanted to have tangible proof that I was a virgin, to the extent that I wished virginity were a physical deformity that could be seen in full view on a woman's body."

In 1990, Saddam Hussein proposed re-legalizing "crimes of honour" in Iraq. He equated local tradition with Islam and then sought to win "Islamic" credentials for his dictatorship by urging that judicial authority over Iraqi women should be given to Iraqi men. Fortunately, the national legislature did not enact this proposal.

Such traditional "moral discipline" posed a considerable danger to women because suspicion of female motives was so overpowering. If a woman was seen in the company of a man, the reason was often held to be obvious. If rumor tainted a woman's reputation, perhaps only her blood could clear the family name. In Baluchistan, Pathan women accused of unauthorized social relations were expected to hang themselves in their tents. Otherwise, their families would simply refuse them food. In case of rape, a woman was often presumed guilty of adultery unless the rapist said otherwise. Against such a traditional reign of terror, women could not justifiably object. Their reason for protest would be assumed to be desire for illicit sex, and for this desire there could be no excuse. Safety among their custodians, therefore, lay in strict conformity to a role of service for one man and his offspring and relatives. Yet this was the very role for which women were judged less valuable than men.

Of course, the enforcement of male control over women was far less absolute in many regions of the Middle East, and in most rural areas women could never be strictly confined. The Bedouin women had to work in open country and commonly went without veils; their domestic roles required wandering in search of fuel and bringing water. In the Middle East's pockets of farmland, most poorer women worked outside in the fields. Independent women in rural areas were threatened mainly by "talk." As an Iranian village widow told Erika Friedl,

> A young widow without brothers has a hard life. If I went to the town … or to go to the doctor, because I have no brother to accompany me, people talked. If I was out haggling with the shopkeepers … people talked. If I rented my back rooms, people talked.… And all because I

> am on my own. I am not afraid of anybody—I can answer, I can argue, be it with the governor himself. (1989: 196)

Perhaps the ideals for guarding women's "purity" could never be seriously enforced, save among wealthier and more powerful families living in urban areas (Patai 1983: 132–33). The ruling elites of desert countries were often descendants of nomadic warlords who had established themselves in the towns. Women of these families were a tiny minority of all Middle Eastern women, but it was these families who tended to set the official tone and the stereotypical image for desert society. While millions of rural women toiled in rural workplaces, a relative few lived behind veils and latticed windows. Naila Minai (1981: 42) says, "It is these few who were remembered and emulated ... as the epitome of aristocratic femininity and ... morality."

The Perils of Educating Desert Women

"What if," an Omani man asked, "one of my wives learned to write—she might write to someone who was not her husband." In that case, the purpose of *purdah* would be defeated. The woman would have a new means of willful contact with anyone in the world, right through the walls. According to an Arab saying, "A woman who is taught to write is like a serpent which is given poison to drink" (Phillips 1966: 140).

Partly due to such concern over the moral effects of female literacy, as late as 1962 United Nations statistics showed that three nations of the world—Saudi Arabia, North Yemen and Somalia—had 100 percent female illiteracy (Patai 1983: 331). Reportedly, these countries had not even one literate native woman.

When Middle Eastern women did begin publishing for each other, starting with Warda al-Yaziji's *The Rose Garden* in 1867, the fears of many men were seen to be justified. Some men worried that literate women might tell their sexual fantasies to the world or tell stories to shame their husbands in public. But women's thoughts on society often proved more disturbing than any breach of privacy. In the public pages of books and magazines, desert women began speaking to each other, comparing their lives, problems and answers. In 1909, for example, Bahithat al-Badiya disturbed the protective men of Egypt by calling for free economic competition between men and women:

> Men say when we become educated we shall push them out of work and abandon the role for which God has created us. But isn't it rather men who have pushed women out of work?... Before, women winnowed the wheat and ground flour on grinding stones for the bread they used to make with their own hands.... Then men established bakeries employing men.... [Several other examples are given.] By what I have just said ... I simply wanted to show that men are the ones who started

> to push us out of work and that if we were to edge them out today we would only be doing what they have already done to us. (in Badran and Cooke 1990: 228–29)

By 1914, the Middle East sported fourteen magazines on women's affairs, all founded and edited by women (Butt 1997: 227). In 1928, Nazira Zain al-Din of Lebanon opened her book *Unveiling and Veiling* by comparing her Middle Eastern realm with a rather Victorian outside universe:

> Ladies and Gentlemen, in the beginning I compared opposites, the numbers of the veiled and the unveiled. I found that the veiled are no more than a few million Muslims living in towns. Those in the villages of the Islamic world and more that one thousand seven hundred million in other nations are not veiled.... The unveiled nations are the ones that have discovered through research and study the secrets of nature and have brought the physical elements under their control. (in Badran and Cooke 1990: 272)

As a result of her views, Al-Din was accused being a paid agent for Christian colonial interests.

Marie-Aimee Helie-Lucas of Algeria urged critical comparisons among North African and West Asian cultures:

> We are not even aware of the differences between one Muslim country and another.... Let Muslim women step out of their national ghettoes. Let them see that the clitoridectomy practiced in Africa is unthinkable in Asia, that the veil worn in Arab countries is absent in Sub-Saharan Africa, that none of these practices are based on religious precepts, but that religion everywhere backs such practices whenever they allow for greater control of women. (in Badran and Cooke 1990: 113)

Ghada Samman of Syria wrote in 1961:

> Islam forbade us to be dolls decorating tables and playthings for the god of petrol, and butterflies around the colored lamps of vanities.... From today they will not plant us as worms in cocoons of fog. (in Badran and Cooke 1990: 143)

During the twentieth century, desert nations found it profitable to educate girls, but many fathers and husbands still felt honour-bound to control how women used their new skills. As Bouthaina Shaaban writes,

> When I was a little girl, I used to scribble a few words on paper.... I used

> to hide what I had written in a safe place because I had a vague idea that my father and brothers would not approve of it.... At high school I began to understand that what I was writing was poetry. (1988: 10–11)

Once, however, Shaaban's father and brothers caught her reading her thoughts on men's morality to some friends. They dragged her from the room and ripped her writing to shreds. Her brother screamed in her face, "You think you are something. I shall prove to you that you are nothing" (Shaaban 1988: 11).

Costs of Devaluing Women

In the arid lands from North Africa to Inner Asia, men have traditionally outlived and outnumbered women (Patai 1983: 329). It might be simplistic and hurtful to generalize about the reasons why, but such facts beg questions. Possible factors include overwork for poor women, and unequal access to resources. Raphael Patai (1983: 329–30) claims patterns of female mortality had three main causes. The first was neglect of girl infants and the relative pampering of boys. Second was mortality from female circumcision, which was still widely practiced in Africa in the late twentieth century, despite its illegality. Third was the high death rate of women in childbirth.

Girls in old Arabia were commonly cut off from breastfeeding after one year to eighteen months, while boys often continued to ages three or four (Phillips 1966: 69). One reason for this was the belief that lactating mothers would not get pregnant, and the mother of a girl was often under social pressure to get pregnant again, in hopes of having a more valuable, male child. Relatively early weaning of girls reduced contact with their mothers and created a greater danger of exposure to thirst and hunger (Patai 1983: 30). In conditions of scarcity this often meant a major difference in early nutrition between the sexes.

According to Marvin Harris (in Sandray 1988: 173), male supremacy evolved mainly in situations of food scarcity. Where the environment did not provide enough food, people often felt their survival depended on strong providers to compete for what little there was. When there was not enough for all, the stronger men prevailed in taking what they wanted and left the rest for the "less productive" non-warriors. This may relate to what some Omani men told Wendell Phillips (1966: 37): "What we cannot eat ourselves we give to the women—women can eat anything."

Where environmental constraints were tight, "neglect" of girls often included female infanticide. In times of hardship, "non-productive" female mouths were sometimes deemed unaffordable. Unwanted baby girls in ancient Arabia were buried alive, and sometimes girls as old as six were disposed of in this way. The Quran attacked the practice in Sura 16:58–59, and Sura 81 warns that on the Day of Judgment the buried girl child shall be asked for what sin she was put to death. According to another saying of the Prophet, "Whoever hath a daughter and doth not bury her alive, or scold her, or prefer his male children to her, may God bring him into paradise" (Hadith, in Minai 1981: 15).

Where a girl was seen as only indirectly productive and of help only to her future husband's family, she was usually married off at puberty or before. To keep her longer would run up the father's expenses. So, for generation after generation, for thousands of years, most desert girls of ten, eleven or twelve were handed over, often to older men, for what most people would now term childhood sexual abuse.

Child brides were likely to begin bearing children from the earliest possible age and, over their lifetimes, women in the traditional Middle East averaged a large number of pregnancies. As the rate of maternal mortality per birth was also high by world standards, a West Asian woman's risk of death in childbirth was many times greater than that faced by a European woman (Patai 1983: 329). Of course the situation improved considerably in the late twentieth century. In Tunisia, for example, both the birth rate and the rate of death in childbirth have fallen dramatically since the 1970s. To some extent, these kinds of mortality result from poverty, and with rising incomes, fatalities are falling. But money is not the only factor. Though Kuwait has achieved a very high per capita income, its mortality rate for mothers and infants has been slow to improve (Eisler, Loye and Norgaard 1995: 45).

Devaluing women has always affected children. In most desert nations, the shaping of young children's minds was traditionally assigned to women, yet these women were commonly restricted from hands-on participation in public life. Especially in moderately wealthy urban families, mothers were supposed to be isolated from the economy, politics and public religion. Supposedly such isolation would keep mothers "pure," so they could pass such purity on to their children. And very often this succeeded (Patai 1983: 333). Many children grew up viewing their world as a vile place which can contaminate the pure soul. In such a world, many felt it religious to live out their lives in a kind of mental and social quarantine.

Perhaps the artificiality of such constraints was most exposed in times of emergency. In the anticolonial wars many women sacrificed their isolation, taking roles comparable to the first female companions of Muhammad—the *Sahabiyat*. After the 1982 Israeli invasion of Lebanon, Nadia Nouihid wrote,

> During the war I found women to be more challenging than men. They were perhaps discovering for the first time not only how much they love Lebanon but also how much they can do when they need to.... I looked at them and said to myself, "Oh my God, where were all these brilliant creatures previously?" (in Shaaban 1988: 110–11)

Women's Value in the Desert Market Economy

Over the past two centuries, the economies of the desert lands vastly diversified and expanded. The conditions of life for millions of people have been transformed. Many traditional male roles such as caravan trading, herding and raiding, often central in the old desert order, have grown ever less viable. The old roles have been increasingly replaced by jobs which women could fill as easily as men. And partly because educated women have become a new source of money, most desert nations have set up a framework for universal education. A new Arabic saying advised modern men, "The best wife is one who takes care of you at home like a mother, works in the office like a man, and makes love to you like a courtesan" (Minai 1981: 217).

In recent decades one desert nation after another has proclaimed the equality of men and women, at least before the state. In 1962, Nasser's National Charter of Egypt urged, "Women must enjoy equal rights with men. The remaining shackles that hamper their freedom of action must be removed so that they can participate constructively and profoundly in shaping the life of the country." In 1971 the Iraqi National Charter of the Ba'ath government announced "the liberation of woman from feudalist and bourgeois concepts and from conditions of bondage that rendered her a mere means of entertainment or a second-class citizen."

If such platitudes made little difference for most women, at least Bouthaina Shaaban found that her own family began to change. In 1981, Shaaban married a man unacceptable to her father. In her father's eyes, this was simply illicit sex. In his view (and according to certain schools of Islamic law), the father's permission *was* what constituted a legal marriage (Fluehr-Lobban 1993: 111–12). Yet, for this crime, Shaaban was not killed as her girlfriend was in 1968. She was merely banished from the family "for life." Several years later, Shaaban's younger sister married a man of her choice. This time her father considered the matter and offered his blessing.

In many ways, the constraints of the past seem be fading away, at least so long as the great oil age lasts and the resources for an ever-expanding economy can be found. In Iraq, women in the early 1980s comprised about a quarter of all engineers, physicists or technical experts, and over 30 percent of all doctors (Minai 1981: 210). If this continues, Soraya Altorki's prediction for her fellow Saudi Arabian women will come true: "The greater the women's contribution to the ... economy ... the greater will be their freedom of movement and participation in public life" (1986: 23).

However, if the newly discovered resources are exhausted, just as the traditional economy exhausted its grass and soil, then the wasteland may exert its grip again. We may be seeing this already in Algeria, Egypt and Iran, where a new generation of educated but largely unemployed youth has called for the return of all desert traditions—a pushing of women out of the work force, imposition of the veil, resegregation of schools and relowering the minimum age for marriage to thirteen. In such indiscriminate reclaiming of desert tradition, Algerian war hero Marie-Aimee Helie-Lucas saw the meaning of her revolution thrown into reverse:

> Education for girls was compulsory, but there were not enough schools.... Women's employment was legal, but there were no jobs.... Women could legally walk in the streets, but the men and the police harassed them: what was an honest woman doing outside her home?... In the very summer of independence [1962] the FLN army patrolled the streets of the capital, arresting men and women who were walking together in groups: "Who is married to who here? This one. Is she someone's sister or wife?" A valid marriage certificate had to be produced. (in Badran and Cooke 1990: 110)

A slowdown or collapse of the new desert economy could lead back to conditions of scarcity, with all this has traditionally meant for women. And globally, the biggest cause of scarcity is that the desert is still growing at a rate of nearly six million hectares per year (Korten 1995: 28). The effects of that migrating environmental catastrophe may be predicted in general, but not in specific. In Niger during the summer of 1992, certain clerics announced that the rains had failed because of God's anger over "indecency" among the nation's women. Soon afterwards, groups of concerned men and boys began beating and stripping women who were wearing non-traditional clothes. In response, several hundred women held a protest march in Niamey. Next, someone burned the offices of the Association des Femmes Nigeriennes (Ms. 1992: 11).

Of course, scarcity and fundamentalism go together in many women's minds too. And the traditional morality of common people involves valid protest against powerholders who lionize scarce resources. In cities where young people of the population boom are largely unemployed, popular religion often demands an apocalypse of justice. The rulers of Morocco, Libya, Saudi Arabia or Iraq may claim to rule in God's name, but these rulers commonly oppose the fundamentalists' demands for social and economic justice with any brutality they deem necessary. The old desert traditions, including those of the early Muslims, usually stressed fairness and opposed tyranny within each tribal community. The kind of fairness usually stressed, however, was fairness for men. So, in recent waves of popular traditionalism many women faithfully embraced the veil or gave up their jobs to make way for unemployed males. For many this was a free choice or a creative adaptation of traditional styles and values, but where the veil or job losses were imposed on women by threats or public policies, compliance was a matter of necessity.

Problems of the Past or the Future?

In the usual process of desertification, plants are first exploited nearly to extinction. Then the plant-eating animals suffer famine. As the impact of scarcity moves up the food chain, deprivation among humans often starts with the women and children.

Of course, on Earth we may yet achieve a new natural balance, and the process of desertification may be reversed. A more biologically productive way of life may be found that slowly returns the land to health. In a terminal stage of degradation, our planet would start to resemble Mars, but we could probably never reduce the whole Earth to that, at least not without destroying ourselves first. However, at the present rates of forest and grassland destruction, and given the possible effects of global warming, the threat of desertification now extends to every continent. And we can see what that has meant for those it disadvantaged in the past.

Chapter 4

The Devil's Environment

WHEN THE HARMATTAN WIND BLOWS SOUTH OFF THE SAHARA, IT SPREADS a veil of dust over West Africa and out into the mid-Atlantic. The harmattan is a winter wind; the dry season winds of March to May are far hotter. The winds off the desert are drying winds that seem to suck moisture from every living thing in their path. Some people in the Sahel call these winds the breath of the Devil.

In Tibet, hell is sometimes depicted as a place of terrible cold. Most desert cultures, however, describe hell as a realm of punishing, intensified heat. According to an Omani saying, the town of Muscat "gives to the panting sinner a living anticipation of his future destiny." And the coast of Oman in summer is "hell with the fires put out" (Phillips 1966: 6).

Early Muslim geographers pictured the desert as comprising one quarter of the world. Their Earth was a great ball floating half submerged in the cosmic sea. Half of the globe floated above the waterline, exposed to the air and light. Of that half, half again was the "inhabited quarter," where plants, beasts and human beings could live. The remainder was the "scorched" or "ruined quarter" (Phillips 1966: 6). These proportions may indicate that only half of the Middle East was "scorched" a thousand years ago. The concept of "ruined" implies a memory of when it was otherwise. A dichotomy of inhabited and ruined environments could be the basis for presuming that the other world is also divided between a paradise and a hell. As below, so above.

The old geographers distinguished several types of terrain within the ruined quarter. *Hamada* is exposed bedrock, the overburden of soil having been eroded away, leaving vast plates of stone baking in the sun. *Reg* is a landscape of jumbled rocks, broken into sharp fragments by extreme fluctuations of heat and cold. The third kind of terrain is of mountains, usually stripped of soil to the naked rock. These three kinds of stony terrain form about 70 percent of the Sahara today. The fourth type of ruined landscape is the *erg*, or "sand sea," with its waves of dunes. The sand is also stone, ground so fine as to resemble organic soil. At least this sand often contains clay. As Mansanubu Fukuoka (1989: 335) says, "If there is clay there is hope."

It was incorrect of early geographers to call the ruined quarter uninhabited. Actually, there are plants, insects, crustaceans, mollusks and other classes of creatures living there, ready to increase with any improvement in conditions. But, as Leslie Hazleton (1980: 86) says, "life in the desert is ... life on the brink of death. It is the outer limits of life." The old geographers cannot be blamed for failing to see the life there. It was an easy mistake. Riding through the southwest

Sahara on a camel, Geoffrey Moorhouse (1974: 118) felt, "I was a caterpillar wriggling hopefully across an eternal nothingness from which all other life had apparently been extinguished." The country through which Moorhouse rode, approaching the Ahaggar Mountains of southern Algeria, was known by Arab travelers as "the country of fear." Moorhouse discovered why:

> Thrice now, sitting in an oasis just before departure, I had experienced a deep primeval fear of the void around. Not a quaking sensation; no more than an uneasy, feathery turning of the stomach. But there, distinctly there, as a warning and as a question mark I could not ignore. (1974: 118)

The Realm of Fear

At Jiayuguan fortress on the Great Wall of China, the old Silk Road passed through an archway. For those traveling westward into the desert, a sign above read, "The Gate of the Bravest People in the World." The gate led from the world of life into a realm where ghosts and demons seemed to whisper in the wind. So the seventh-century traveler Xuan-zang (Hsuan-tsang) wrote that when the winds rise,

> both man and beast become confused and forgetful.... At times, sad and plaintive notes are heard and piteous cries, so that between the sights and sounds of the desert, men get confused and know not whither they go. Hence, there are so many who perish on the journey. But it is all the work of demons and evil spirits. (in Hopkirk 1984: 11)

"Hallucinations," Jeremy Swift explains, "are thought to be due to extreme sensory deprivation." And in the Saharan reg, these conditions occur in spades. Driving over a carpet of meteorlike stones, Swift kept thinking that a member of his party was missing. On every side mirage hills and trees floated above the shimmering ground. The pale sky faded imperceptibly into the slate-gray landscape, erasing the skyline. All sense of distance and scale was lost. Far out on the plain, Swift saw a large tree. As the party drove toward it, it turned into a stunted bush two feet high. A strangely shaped hill appeared on the horizon. As they approached, it shrank to a rock less than a foot across. Later, the feeling that an extra person was with them nagged at Swift's mind the whole time they remained in the Tanezrouft, the great reg of southeast Algeria (1978: 154–55, 166).

This desert is a land of almost constant mirage, where sense perceptions are obviously unreliable. "Phantom rain" pours from the clouds, evaporates in mid-air and never touches the ground. Illusory lakes vanish without a trace. As on the open sea, travelers may hear siren voices calling them to change course. Magnified in the retelling, grand illusions fill the desert's lore. Soldiers of the ancient Assyrian army marching through Sinai in 671 BCE claimed their ranks were depleted by large "green animals whose wings were batting" and "two-headed serpents whose attack spelled death" (Roux 1980: 302–3).

Another cause of illusion is dehydration. If the desert traveler misses a well or finds it dry, a loss of body fluids proceeds rapidly. In extreme Saharan heat, a human can easily lose several gallons of water a day (two at rest in the shade, four if walking in the sun). If one gallon is lost and not replaced, dizziness, increased pulse, labored breathing and disorientation are normal (Langewische 1996: 150).

When Geoffrey Moorhouse and his guide Ould Mohammed became lost somewhere in northern Mali, the two men entered a contest as to whose judgment was sane. Moorhouse eyed Ould Mohammed suspiciously: "He had eaten little for two or three days and he was badly dehydrated. He had also been riding all morning with his bald head exposed to the fierce sun." Repeatedly Ould Mohammed tried to turn and ride back the way they had come. Moorhouse chased him down, arguing that water must lie in the opposite direction. Suddenly Ould Mohammed flung out his arm crying, "Look, there's the town—that thing sticking up there." Moorhouse told him there was nothing to be seen. The guide cursed him furiously, kicked his camel and rode on, leaving Moorhouse alone. Every now and then, Ould Mohammed appeared in the distance, driving his camel one way or another. Moorhouse sat against the trunk of a thorn tree:

> My mind whirled with anxiety. What on Earth was my responsibility to Ould Mohammed? Somehow I ought to have prevented him from riding back suicidally into the desert, but I couldn't see how I could have stopped him physically. (Moorhouse 1974: 232–33)

Fortunately, Ould Mohammed returned from his madcap ride having found the well.

In this same region, between Taodeni and Timbuktu, a famous salt caravan of two thousand people went astray in 1805, and all hands died of thirst. These days, of course, most desert travelers use cars, buses or trucks. Fewer people die unless they get lost and run out of gas.

Though a camel can survive dehydration to the point of losing a third of its body weight, a human cannot survive losing ten to twelve percent; that is about two gallons, which can be lost in only half a day (Langewische 1996: 182, 150). If that much water is lost, the blood congeals and body temperature soars out of control. When the Swedish explorer Sven Hedin lost his way in the Chinese Taklamakan, he watched his own body wilt like a plant. After wandering several days without water, he saw a pool in the Khotan-daria riverbed, plunged his head in and drank like a camel:

> Every blood-vessel and tissue of my body sucked up the life-giving liquid like a sponge.... My hands, which had been dry, parched, and

> hard as wood, swelled out again. My skin, which had been like parchment, turned moist and elastic. (in Hopkirk 1984: 60)

Many of the desert's poets and prophets spoke of familiarity with primal fear, and of redemption at death's gate. They described visions of the next world like people used to hearing stories of near-death experiences. Even today, with all our technological safety nets, those who travel in the true desert still know that anxiety. After extensive travels in the Negev Desert of Israel, Lesley Hazleton (1980: 34) observed that we all carry a basic fear of being lost. Normally this fear lies below the threshold of awareness, but in the desert, it rises to consciousness. For anyone traveling alone in the desert, it is a nagging fear that cannot be ignored. "It seems to me," Hazleton adds, "that it is not physical death we fear in the desert so much as death of the soul" (1980: 45). Perhaps astronauts know this fear of being lost forever. And perhaps partly with that literal fear of lostness in mind, many ancient desert Muslims prayed daily, "In the name of Allah, the compassionate, the merciful, king of the day of judgment, lead us into the straight way, and not as those who have gone astray."

So the Tuareg poet Shaykh Muhammad Hamma Al-Suqi wrote to God:

> Thou art among the shepherds of creation, secretly or seen. If Thou sleepest and ignore us for an hour or for a day, then we are those who in a desert lose their way. (in Norris 1984: 317)

Of course, desert travelers did everything in their power to ensure a safe passage. They committed known paths across hundreds of miles to memory and carefully passed on what they knew. As an old Algerian man described this lore of the desert, "Yes, by the stars at night. In daylight, by local knowledge of the desert—this soil, this tree, this ruin, these tracks, these shadows before sunset. It is passed down from father to son, and spoken of among friends" (Langewische 1996: 142). Malise Ruthven (1984: 131) explains that the closest word in Arabic to the English word "good" is *Al ma'ruf*, which literally means "the known." Evil is associated with the opposite word, *Al munkar*, meaning that which is unknown, outside tribal tradition or departs from the proven path. *Al munkar* is that which would lead people astray into the void.

Most ancient people believed that threats to their survival came from the Devil. And in a killer environment, the Devil and the land seemed strangely united against humanity. In that harsh country the Devil played tricks with people's minds, conjuring illusions to deceive them and animating the landscape with his hostile spirit. Passing whirlwinds were *djinn*, or "dust-devils." Blinding sandstorms known as "black hurricanes" seemed to be all-out assaults by the Devil himself. In desert myths, the soul of all evil and the "Lord of this world" were commonly one and the same.

In lands of mirage and heat- or thirst-induced hallucinations, personal perceptions were commonly distrusted. What the whole group affirmed was more likely true than what one person "saw." In a treacherous country where one mistake could be fatal, this was common sense. What was the shortest route to water? Which leader should be trusted? Which customs must the group follow to gain God's mercy? A desert tribe was forced to gamble on such decisions at every turn. Fear of being subject to illusion made people depend on group consensus all the more. But if each member of the group was prone to deception, how did their collective knowledge become trustworthy? One common answer was that the tribe's collective wisdom came from God. Long ago the ancestors received their beliefs and traditions from the Lord. By clinging to their tried ways and proven paths, they had survived in the desert. They had been shown the way in the wilderness and saved from dangers on every side. It was a miracle.

Ancient desert people had no evidence that rain came from earthly water, evaporated from other regions. They thought moisture fell directly from the blue vault of heaven, as mercy from God on those who deserved it (Hillel 1994: 24). So many early Muslims literally interpreted their scripture: "He sends down saving rain for them after they have lost all hope and spreads abroad his mercy"(Quran 42:28). By observing certain traditions, the surviving ancestors had merited mercy through the years. Anyone who would now depart from their ways was a fool or worse. In Arabic the closest word to "innovation" is *bid'a*, and the traditional connotations of this word are almost all bad. *Bid'a* was generally presumed to involve a mistake, a betrayal of the community or a corruption of its sacred customs (Ruthven 1984: 131). Those who would stray from tradition in the desert were more likely to be seen as devils than as prophets.

Where ancient desert tribes competed for scarce resources, each tribe tended to demonize the other. In a similar way, different religious groups often viewed each other as enemy faiths, with enemy gods. During the early decades of Islam, militants of the breakaway Kharji sect demonized all other religious communities. Their own Kharji ways were of God, and all other cultures were deceptions of the great deceiver. So the Kharjis described their world as a "zone of war" where the "people of paradise" engaged the "people of hell" in mortal combat. However, Abdullah Ibn Ibad urged a more moderate view on the Kharjis. This world; he agreed, has no worth save as God's testing ground for souls, but it is not a realm of war. Rather, for every earthly creature, it is "the realm of prudent fear" (Ruthven 1984: 183–85).

Dead Earth and Holy Heaven

South of King Nebuchadnezzar's main palace in Babylon rose the great pyramid, or ziqqurat, called E-temen-an-ki—"the temple foundation of heaven and earth." Like older ziqqurats built over thousands of years, its seven levels ascended to a point supposedly halfway between the sky and the earth. This manmade holy mountain was intended to provide a meeting place where the gods of heaven and goddesses of earth could co-mingle, fostering accord and renewal for the universe. The city of Babylon itself was called "the bond between heaven and earth."

Atop the ziqqurat, according to Herodotus, was a golden chamber containing a large bed. The presiding priestess there was a "native woman chosen from all women" (Roux 1980: 363–64). She represented the goddess of the earth in the most important religious rite for ancient Mesopotamia, the annual re-enactment of creation. The fertility of the land and the order of the heavens were held to depend on the annual sacred marriage of sky king and earth queen. At least this rite was important so long as Mesopotamia remained a fertile country. Of course, even by the time of Nebuchadnezzar's rule (c. 605–562 BCE), the crop yields had been falling for many centuries. Over much of the irrigated ground a snow-white powder of salt had appeared, like a spreading leprosy on the soil (Hillel 1994: 54–55). The goddesses of the earth and the waters of the deep, it seemed, were turning against their children. As the land grew more and more unyielding, people slowly forgot the goddesses of the earth. After that, only the sky gods remained.

Later, especially from the time of the Persian empire (c. 550–328 BCE), the ziqqurats became temples for worship of the sky god, who in Babylon was Marduk. Male priests resembling the Zoroastrian magi ascended the pyramid steps to meet in prayer with the Lord above. Then they conveyed his will to the people below. These priests studied the heavens, discerning the portents and patterns of a divine order to be followed on earth. They claimed that the stars determined the course of earthly affairs—seemingly more than any direct interaction between people, plants and animals. Much earlier, in ancient Sumer (down to the 1900s BCE), astronomy was only one science among others. The Sumerian cuneiform records contain texts on the earthly topics of zoology, botany, mineralogy, medicine and mathematics. But by the time of the Persian empire, astronomy was the king of sciences and the sky gods were of supreme importance. As their land turned into a barren, sun-baked plain, Mesopotamians raised their eyes to the heavens.

In the world's many other regions where the land remained green, the goddesses of the earth lived on. So the Greeks made Aphrodite say:

> The pure sky longs passionately to pierce the Earth, and passion seizes the Earth to win her marriage. Rain falling from the bridegroom sky makes pregnant the Earth. Then she brings forth for mortals pasture of flocks and corn, Demeter's gift, and the fruitfulness of trees is brought to completion by the dew of their marriage. Of these things I am part-cause. (Aeschylus, *The Danaids*, in McLuhan 1994: 222)

The Dogon people of West Africa tell how God made the earth as a woman, then married her. Here again was a sacred marriage. But in most desert myths, God was in heaven, alone. He was a ruler, not a lover, and the earth was dead clay in his hands. The focus of religious awe was not the dusty earth but the brilliant sky. So, for the Mongols, God was Koke Mongke Tengri—Eternal Blue Heaven. The Vedic hymns of the Indo-Aryans emphasized worship of the god Indra, the source of thunder and lightning, and Varuna, the all-encompassing sky.

For the ancient Hebrews as well, God was identified with heaven. Yet something of an older myth remained. At least in the popular mind, Yahweh was of male gender, and this suggests he also once stood opposite a female goddess of the earth. John Allegro claims the holy name Yahweh is related in meaning to the Greek name Zeus. Both mean "spermatozoa," or source of life. In sky-god myths of both the ancient Semites and Indo-Europeans, "He was envisaged as a mighty penis in the heavens which in the thunderous climax of the storm, ejaculated semen upon the furrows of mother Earth, the womb of creation" (Allegro 1971: 29–30).

The old Kabbalistic *Zohar* explains that while the Temple stood at Jerusalem, the King of Heaven descended at midnight to the Temple bedchamber to make love with his wife, the Matronit. The welfare of the earth depended on that holy union, which was cut off at the Temple's destruction (Patai 1990: 142). When the Hebrews of Exodus had come to Canaan, it had been "a land of wheat and barley, vines, fig trees and pomegranates, a land of olives and honey; A land wherein thou shalt eat bread without scarceness" (Deuteronomy 8:8–9). And so long as the country remained green, the ancient Hebrews worshipped their own goddess of Zion. Then the promised land slowly dried and withered like an old grandmother. In one old tale, the prophet Jeremiah goes to Jerusalem and is disturbed to see a woman in black with disheveled hair, weeping on a hilltop. He tells her, "You are not better than my mother Zion, who has become a grazing ground for the beasts of the field" (Patai 1990: 211–12).

While the goddess of the Hebrews was increasingly forgotten, the majesty of heaven increased. The scarcity of rain seemed to underline the Lord of Heaven's absolute power to give life or take it away. The Old Testament prophets attributed virtually everything that happened to God in heaven. Isaiah quoted him as saying, "I am the Lord, there is no other; I form the light, I create darkness; I make peace and create evil; I am Jehovah that doeth all these things" (45:6–7). The increase or failure of the harvest, the rise or fall of the economy, the expansion or decline of the kingdom—all these things were in his hands. In this case, the desiccation of the land was not caused so much by war, deforestation or overgrazing, as by the Lord's displeasure over the degree of reverence offered by his creatures. In 520 BCE, soon after many Israelites returned from exile in Babylon, there was a terrible drought around Jerusalem. The prophet Haggai claimed the Lord had sent this curse

> because of my house that lies in ruins [the temple at Jerusalem], while you busy yourselves each with his own house. Therefore the heavens above you have withheld the dew, and the Earth has withheld its produce. And I have called for a drought upon the land and the hills,

> upon your grain, upon the new wine, the oil, upon what the ground gives forth. (1:9–11)

Still, many Israelites clung to an older faith. Jeremiah found that many of his contemporaries still believed that drought and other troubles came because they had abandoned worship for the goddess of the earth. Some villagers told the prophet, "But since we have left off burning incense to [the goddess] Anath and pouring out our libations to her, we have lacked everything and have been consumed by the sword and by famine." Jeremiah rebuked them:

> Your land became a desolate waste, an object of horror and ridicule, with no inhabitants, as it still is. This calamity has come upon you because you burned these sacrifices and sinned against the Lord and did not obey the Lord or conform to his statutes. (44:16–18, 22–23)

According to such logic, the groves of trees sacred to the goddess should be destroyed. The rituals of fertility and eating first fruits from the orchards should be renounced. The women's rites involving snakes, which shed their skins and symbolize the renewal of nature, should be condemned as an abomination. In other words, the values of the conquering pastoralists from the desert must root out the traditions of the gardeners.

Anthropocentric Culture in an Inanimate World

In the "Holy Land" of Israel and Arabia, the sacred places are sites of great events in human history, not places where the earth itself is holy. For most orthodox Jews, Muslims and Middle Eastern Christians, reverence for the earth itself is only a distant memory, or an error. Worship of Mother Nature is an abomination, or else a false belief in a non-existent deity. Through many older religions celebrated the marriage of heaven and earth, the writers of Genesis saw the ziqqurats of Babylon as an affront to God. The "tower of Babel" was an arrogant attempt by earthly creatures to reach towards heaven. The Lord smote them, casting the builders down and confusing their speech (Genesis 11:1–9). How dare the lowly earth try to meet heaven halfway?

The desert religious traditions show evidence of a step-by-step divorce between nature and culture. Pre-Islamic Arabia celebrated date-harvest and rainmaking festivals. These were seasonal festivals held at appropriate times in the life cycles of plants. The calendar then revolved around the seasons of nature. But the early Muslims made a new calendar of twelve equal months, cutting out the old thirteenth moon month, which had been inserted occasionally to keep the solar, lunar and earthly cycles aligned. Since the new calendar was adopted, the Islamic months such as Ramadan have gradually rotated through the seasons, like a mathematical wheel (Ruthven 1984: 47). In Europe, Christian calendars retained a connection to earthly seasons, and the greatest holy days were tied to the beginnings of winter and spring. But

Islam was more fully a religion of the desert. Its calendar was based exclusively on mathematics and the sky, as if the cycles of plants and seasons were irrelevant to God's reality.

Another aspect of the gradual divorce between nature and culture involves art. The religions from the Middle East have commonly involved "iconoclasm," or a moral aversion to "images." In very strict iconoclasm, nothing on earth should be portrayed in works of art, lest this imply worship for the clay, rather than the potter. Traditional arts depicting animals, people and plants were therefore subject to bouts of idol smashing in the old Middle East, as the aversion to depicting God was sometimes extended to all "graven images." Much of Islamic art avoided the sin of depicting nature and in many cases became a play of sheer abstraction. Raphael Patai (1983: 170) feels that this was an artistic tradition "completely divorced from nature and which was purely the product of the mind, of the artistic fiat ... an *imitatio dei* on a small and modest scale ... not copying nature ... [but creating] something out of nothing." By implication, these artists saw nature as having no redeeming value, and looked for inspiration elsewhere.

Joseph Campbell (1976a: 393) claims that all major "Western" religions have shared in making "an ontological distinction between God and his universe, creator and creature, spirit and matter." At least the orthodox versions of the great Middle Eastern faiths have involved a "separation of the reigns of nature ... and the spirit ... in such a way that neither touches the other." This Campbell describes as "mythic dissociation." That which is holy and of ultimate worth cannot be found in nature or in this world but only "out there"—outside the universe. The religions historically pervaded by this image (Judaism, Christianity, Islam and Zoroastrianism) all originated in the arid lands of southwest Asia. The universe they describe is probably the environment their founders knew.

According to this absolute division between creator and created, traditional Islamic sciences were split into two main fields of study—theology and physics. Theology was knowledge of the creator and his eternal, otherworldly truth. Physics was knowledge of earthly things, or of "all that changes" (Qadir 1988: 118). Knowledge about mundane things was important but not ultimately so. The potter was vastly more important than the clay, and knowledge of primary causes was the domain of theology.

In the first centuries of Islam, Muslim scholars were deeply concerned with society and everyday life. Islam boasted its glorious centers of learning in Persia, Baghdad, Egypt or Spain that offered the highest levels of training then available in languages, literature, medicine, law, history and theology. But century by century the higher valuation on matters of God eclipsed the other fields of study. Theology grew and physics declined. For several centuries before 1839, the only schools in the Ottoman empire were Quranic schools maintained by private donations (Mansfield 1991: 65). The greatest Middle Eastern universities such as al-Azhar in Cairo, or Qarawiyyin in Morocco, focused mainly on theology and jurisprudence. In these lands the study of human religious traditions was great, but learning about nature actually declined. This would have been like all of nineteenth-century Europe and North America being serviced by universities offering degrees in law, languages and divinity, but next to nothing in chemistry, biology, geology, agriculture, climatology, physics, forestry or any other earthly subject.

By the mid-twentieth century, institutions for natural science mushroomed across the Middle East. But as late as the 1970s, Iraqi poet Buland Al-Haydari characterized his own culture as an "oral civilization," where words, repeated correctly as in tradition, became both the means and end of social life (in Patai 1983: 59). Omar A. Farrukh, in *The Arab Genius in Science and Philosophy*, attacked the same anthropocentric focus. For some six centuries, he claimed, the Islamic world had been preoccupied with grammar, rhetoric and jurisprudence, while the scientific challenge of the natural universe had been virtually ignored (in Patai 1983: 253). All this might appear as some sort willful blindness, but perhaps the heirs of desert tradition can hardly be blamed for what they saw or didn't see. According to Campbell, for the people of the desert, "nature—Mother Nature—had little or nothing to give." In living largely without that mother, people looked all the more to their fellow human beings for sustenance and meaning in their lives (Campbell 1976a: 626–27).

In less "ruined" environments, other attitudes toward nature prevailed. Concerning Japanese culture, D.T. Suzuki wrote,

> Nature to us has never been uncharitable, it is not a kind of enemy to be brought under man's power. We of the Orient have never conceived of Nature in the form of an opposing power. On the contrary, Nature has been our constant friend and companion, who is to be absolutely trusted in spite of the frequent Earthquakes assailing this land of ours. (1994: 333–35)

Among the Mbuti pygmies of the Congo forests, Colin Turnbull found

> a communal spirit that is difficult to define.... It is, let us say, centered on a love for and devotion to their forest world, and results in their wholehearted and unquestioning identification with it. And what more powerful force toward social unity and cohesion can there be than such a deep rooted sense of identity? (1972: 155–56)

Aboriginal India venerated the living environment as holy in itself:

> [In India] the natural *is* the religious.... The river [Ganges] does not stand for, nor point toward, anything greater beyond itself; it is part of a living sacred geography. (Eck 1986: 169)

When Muslim and Christian invaders came to India, they commonly judged the native beliefs as polytheism and idolatry. They commonly took monotheism to mean that God was in heaven and nothing on earth should be worshipped. Even the radical monotheism of Isaiah or Ibn Sab'in (who said, "There is nothing but God") could be taken this way. But Hindu saints used similar words for their own view, in which, as Campbell (1976c: 12) explains, "It is not that the divine is every*where*: it is that the divine is every*thing*."

The desert habit of seeing the natural world as inanimate traveled on the back of Christianity to Europe, not as a basic tenet of Christianity, but simply as an image prevalent in the Middle East and included in its cultural baggage. Then, even after medieval Christianity lost its grip on Europe, the new philosopher-priests of the "Age of Reason" continued to explain nature as being inert matter shaped by external forces. Instead of a clay pot, nature was now a machine set in motion by its creator beyond. Still later, the agnostics of the nineteenth and twentieth centuries denied there was any divinity in the heavens, which left many people holding nothing sacred in either heaven or earth. Life was then commonly reduced to economics—how to exploit and consume dead resources at maximum speed. Only a growing minority of Westerners began turning to an opposite view, held by the "aboriginals" of the biologically rich environments all along, that both earth and heaven are holy and alive. And with this came a concern that, if the world is alive, it can be killed.

Salvation in the God-forsaken Lands

As Mesopotamia slowly changed from a riverside garden into a semi-desert, nature, which had been portrayed in local mythology as a beautiful goddess, came to be seen as a demonic hag. At some point, a new Babylonian creation story evolved and was celebrated every new year. In this myth, the universe began with the destruction of the earth goddess. The great sky god Marduk descended to fight and kill the monster-goddess Tiamat, who was no longer the mother of all creatures, but only the enemy of life.

Among the farming cultures in the Fertile Crescent, the land evolved as an ever-harsher mistress. From around 2200 BCE, a text called the "Curse of Akkad" describes fields that "produced no grain" and "heavy clouds" (possibly clouds of dust) that "did not rain." In this period, many villages in northern Mesopotamia were abandoned, and their clusters of ruins were buried in layers of dust (Wright 1998: 96–97). As is common in deforested and overgrazed regions, rain in the Tigris-Euphrates watershed produced flash floods. In that case even the nurturing rain could be a force of destruction, leaving ruined fields in its wake. For crops to grow, controlled irrigation became ever more necessary; and the more necessary it was, the more expensive. The more artificial flooding, the sooner it turned counterproductive, as underground salts floated up to accumulate in the topsoil. Under the combined effects of aridity, erosion and salinity, crop yields slowly fell, and the peasants of ancient Iraq sank toward the edge of subsistence. In their seemingly inescapable plight, they made fatalism a virtue.

People tend to believe in fate when they feel unable to influence the environment they depend on. And over much of the Middle East, resignation to fate not only flourished but was even deemed religious and spiritual. As the book of Jeremiah put it, "A man's deeds do not rest in his power; It rests in no man's power, how he moves or directs his way"(10:23). Such fatalism could become "popular" only where long experience seemed to prove it realistic. Yet, however fatalistic the Mesopotamians grew, they also felt inescapably responsible for the state of their lives. If their existence grew hard, they must have done something to deserve it. So, as Samuel Kramer explains, "The Sumerian sages believed and taught the doctrine that man's misfortunes are the result of his sins and misdeeds.... They argued there are no cases of unjust and undeserving human suffering; it is always man who is to blame, not the gods" (1990: 110).

This also was realism. The farmers, shepherds and warlords *were* partly to blame for the desiccation of their land. But in assessing blame and need for penance, they commonly ignored the direct relationship between themselves and their environment. At least their priests usually ignored it. Official religion focused on the relation between humans and gods. If the earth withheld its fruit, this was punishment for failure to give the sky gods due reverence. The priests called for yet greater devotion to the highest authorities. But whatever reverence was offered to the gods of heaven, the land of Mesopotamia tended to decline further. The gods, it seemed, were very seldom satisfied. The Mesopotamians increasingly saw themselves as creatures of dust seeking mercy from heaven. Campbell claims that the sense of human participation in a divine creation slowly dissolved. Humanity was now utterly separate from the gods, yet wholly dependent on them: "A pathos of anxiety developed in which all the nursery agonies of a child striving to gain parental favor were translated into a cosmological nightmare of mythic dependency" (Campbell 1976c: 131–32).

For many people in the Middle East, the gap between their hopes and realities seemed to grow over time. The distance between themselves and the divine seemed to widen, till it matched the physical distance between heaven and earth. That enormous gap could be crossed only after death, after leaving the earthly body and world behind. In a widespread West Asian myth, humanity was cast out from the original garden of paradise, which could never be regained in this world. This earth, then, was the fallen world. To be saved was to be taken up from it. So across the Middle East, religions of "world denial" slowly replaced the ancient cults of nature worship. The hope of joy in this life was overshadowed by hope for salvation unto another world. In one such cult, the Gnostic gospel according to Thomas taught, "Whoever has known the world has found a corpse, and whoever has found a corpse, of him the world is not worthy" (91:30–32, in Campbell 1976a: 147).

Such an anti-earthly attitude may have never prevailed among ordinary people in the Middle East, but clearly world denial became a persistent and influential theme in cultures across North Africa and West Asia. The feelings of many desert people against nature were heartfelt. But such feeling would seem a psychosis to most people of the naturally rich environments, be they Native Americans, or Polynesians on Tahiti.

A few Middle Easterners took world negation to a logical extreme. Josephus Flavius attributes the following lines to Eleazar Ben Jair, a leader of the Jewish Zealots in their mortal combat with Rome at Masada:

> For from of old, since the first dawn of intelligence, we have been continually taught by those precepts, ancestral and divine ... that life, not death, is man's misfortune. For it is death which gives liberty to the soul and permits it to depart to its own pure abode, there to be free from all calamity. But so long as it is imprisoned in a mortal body and tainted with all miseries, it is, in sober truth, dead, for association with what is mortal, ill befits that which is divine. (in Allegro 1971: 206)

Of course, orthodox Jews, Christians, Muslims or Zoroastrians would never go this far. After all, God made the world, and made escape from it through self-destruction a sin to be punished by *eternal* exile from heaven. But wherever religion seemed to project all divinity and worth to another world, there were commonly individuals who took this separation to heart with complete seriousness. The Egyptian psychoanalyst Nawal al-Saadawi reports a case from her clinic, of a pious woman living in *purdah*:

> For a year the young woman has been staring into space at night without sleeping. When she falls asleep, she sees a flood inundating the land and the Prophet Noah embarking on his ship and leaving her behind. She finds herself in life after death walking on a narrow path with an inferno below her. Her feet are bleeding and her body, off balance, is about to fall. She opens her eyes and finds herself asleep in bed under the blankets drowning in her own sweat. She reads the opening Sura of the Quran and thanks God that she hasn't yet died and has a chance to repent. She goes to the bathroom and washes five times.... After she prays she sits with God's book in her lap, reading, and asks God's forgiveness for her grave sin. There is nothing in her life except that sin. (in Badran and Cooke 1990: 201)

Campbell claims there was a point in ancient history when, for vast numbers of people, the religious aim of life was "reversed":

> Following a crucial moment that I shall term "the great reversal" ... for many in the Orient as well as the West, the sense of holiness departed from their experience both of the universe and of their own nature,

> and a yearning for release from what was thought to be an insufferable state of sin, exile, or delusion supervened. (1976c: 36)

Both Joseph Campbell and Riane Eisler suggest that this "great reversal" came along with a series of invasions, in which militarized tribes from the arid zones attacked, plundered and dominated village cultures from Europe to China. One of the greatest waves of migrants came soon after 2000 BCE. At roughly that time, China, India, Iran, Mesopotamia, Egypt and Europe all suffered serious attacks from nomadic raiders and conquerors. The Kassites hit Mesopotamia, the Hyksos rode out of the Sinai to pillage Egypt, and the chariot-riding Shang rose to power over northern China. Evidence of large-scale invasion is clear in many burned and abandoned villages in most areas bordering the arid zones. The timing of these migrations seems to complement climatic evidence of intensified aridity and desert expansion starting around 2200 BCE (Wright 1998: 96–97). The migrants from the arid zones were probably fleeing that general drought. As Langewische describes the old Tuareg invasions of Africa, "just as the crops failed, masked swordsmen would appear out of the desert like agents of hell" (1996: 228).

We have Mesopotamian and Egyptian texts dating from these early invasions: some offer praise to the conquerors for their glorious strength, others voice a nearly suicidal despair. The seemingly simple odes to joy in the earliest writings from Sumer, Egypt and China often give way to a "literature of lamentation," as in a text from around 1750 BCE by "the Babylonian Job," Tabi-utul-Enlil:

> Mine eyeballs he obscured, bolting them as with a lock:
> Mine ears he bolted, like those of one deaf.
> A king, I have been turned into a slave.
> And as a madman I am maltreated by those around me.
> The allotted time of life I had reached and passed:
> Wherever I turned there was evil upon evil.
> Misery increased, justice departed,
> I cried to my god, but he did not show his countenance:
> I prayed to my goddess, she did not raise her head.
> (in Campbell 1976c: 140)

We can gain some image of these early invasions from the better recorded descriptions of more recent Arabian, Mongolian and Ottoman conquests of the Near East. In these invasions, entire nomadic tribes swept through the Levant's farming regions, besieging towns and plundering villages. People in the towns had walls for defense, but rural villagers were at the raiders' mercy. In the wake of these invasions, whole regions were temporarily "deserted" (Bat Ye'or 1996: 107). The seemingly peaceful countryside became a place of fear, and the desert showed itself hostile in yet another way.

Campbell says that prior to the first large waves of nomadic raiders, the popular religions of Old Europe, Sumer and neolithic China were mainly concerned with celebrating nature's bounty, and with hope for the increase of earthly life. The moods deemed "religious" were joy and wonder at the Earth's beauty. But starting around 2000 BCE, a different sort of religion arose, heavily concerned with the problem of suffering. Increasingly, suffering was taken to be the most characteristic quality of life on this Earth. Soon, various paths of salvation offered escape from this sea of pain (Campbell 1976c: 139, 210). So the Persian poet Ma'arri wrote:

> This world resembles a cadaver,
> and we around it dogs that bark.
> And he who eats from it is the loser;
> he who abstains takes the better part.

With a certain pessimism, Ma'arri added, "And certain is a dawn disaster, to him unwaylaid in the dark" (in Stewart 1979: 109).

Likely both the militarization of life and the gradual desiccation of the land gave rise to such widespread cults of despair and "other-worldliness." People need to see meaning in their suffering, and many in the lands of scarcity could only suppose they had been sent into this world to be tested by the Devil.

Chapter 5

Warlords and the Politics of Scarcity

A SOMALI YOUTH LEAGUE SONG CALLED "SOMALIYEY TOSEY," GOES IN PART:

> Somalia, awake!
> Unite the warring tribes.
> Give help unto the poor,
> And strengthen the weak.
> If one of your camels is stolen,
> To save it you risk your lives.
> But for our whole lost land,
> No man even raises a stick. (in Laurence 1963: 37)

In the 1950s this was an anticolonial song. The post–World War II generation hoped to unite Somalia against foreign rule and then move ahead to greater things. Since the vast majority of Somalis were of one blended ethnic group and one religion, there was already broad social unity. The divisions of the past were mainly practical ones, such as rival claims to pastures and wells. But those divisions were hardly small or petty. Margaret Laurence wondered how they could ever be overcome in place "where membership in a tribe was a nomad's only protection in a harsh environment" (1963: 87–88). In times of drought the Somali herdspeople banded together in small, tight groups. No allegiance was "higher" than their loyalty to each other, and no cause was "greater" than their mutual survival. In the most severe dry seasons, so little grass remained that dead sheep and goats lay strewn about the landscape. Hyenas reportedly did not eat them: the carcasses held so little protein, they were not worth chewing (Jordan 1982: 751).

In many other environments of the world, nature is usually gentle and generous. In the tropical forests or coral reefs, galaxies of creatures thrive in webs of mutual benefit. On their tropical coastlines, pre-modern Thais found an easy abundance of fish in the water and rice on the land. Life in old Malaysia or the Amazon basin could be relaxed and largely concerned with the quality of friendships. But in northern Somalia, nature taught crueler lessons. To survive there, people learned to juxtapose extreme self-sacrifice with ruthless self-interest.

Virtues Appropriate to Scarcity

Famine years in the desert were traditionally times for raiding, as those who ran short of supplies tried to steal from others. The practice was hard to condemn when so many survivors owed their lives to it. To go raiding for one's people was a deed of valor. And in really hard years, the numbers of raiders in old Somalia or Arabia could outnumber the raided. The Umayyad-period Arabian poet al-Qutami explained, "Our business is to make raids on the enemy, on our neighbour and on our brother, in case we find none to raid but a brother" (in Phillips 1966: 151). As a last resort, it could come to that.

These traditional raids and counter-raids were more like small-scale wars than crimes, and by the simple absence of anything to live on, whole tribes could find themselves engaged in conflict. Irregular rains in the desert caused grass to appear in different areas each year. Traditional pastoralists, however, were seldom free to follow the grass and rain. Each tribe generally held a delimited territory, usually set by trucelines from past land disputes. If a tribe needed to cross its boundaries, the lines established by fighting had to be reset, often by more fighting. With a meager and uneven dispensation from nature, the competition for pasture and water could be fierce. And any sign of weakness in defending tribal territory could be exploited by neighbouring groups (Walton 1969: 138). A traditional prayer of Kenya's Kikuyu people (who live south of the Kenyan deserts but have commonly faced drought and attack by nomadic tribes) expresses the sentiments of such conflict with a frankness recalling the Old Testament:

> O Mwene-Nyaga [God], the Greatest Elder, you will give us knowledge to make strong bows and arrows that shoot well and to the mark, so we may keep our enemies at bay, who seek to take our cattle and therefore starve our women and children and make the tribe weak. O Mwene-Nyaga, who dwelleth on Kere-Nyago [Mount Kenya], you will give us strength so that when our enemies come to close quarters we will be able with your guidance and our strong muscles to drive our spears right through their hearts, and prevent them from depriving us of the gifts which you the Lord of Nature have bestowed upon us. (Kenyatta 1966: 23–24)

In the late 1960s, Colin Turnbull described a large group of Turkana nomads who were "invading" the hills of northeast Uganda in order to flee a drought in their own Kenyan lowlands. The Turkana camped in Uganda for months, grazing their cows and taking what they needed to live. Finally, the Ugandan government issued an ultimatum: the Turkana were to get out or face military attack. Turnbull visited the Turkana elders, thinking they might not realize their danger, and described the Ugandan army's arsenal of heavy guns and tanks. The Turkana shrugged and said they would take their chances. If they went back to Kenya before the rains returned, they would die anyway. Turnbull walked to the ridge of Mt. Meraniang, which overlooks the plains of northern Kenya. During the past season, the countryside below had changed colour, from the yellow of dead grass to the red of bare earth (Turnbull 1972: 107).

A similar problem arose in northeastern Kenya, as the Somalian desert grew southward. By 1981, one and a half million of the approximately five million Somali people were environmental refugees. As usual when the deserts advance or intensify, masses of hungry and armed people were on the move. They pushed eastward into the hills of Ethiopia and southward into Kenya. Their clashes with local officials and other groups there escalated into low-scale wars.

In old Arabia, the Bedouin men were well schooled in the virtues appropriate to scarcity. On the one hand, they were prepared to sacrifice themselves in defending their people's means of life. On the other hand, they were ready to plunder a stranger. In their morality, sacrifice and cruelty were two sides of the same blade. The generous open hand was for insiders and guests; the face of ruthlessness was for outsiders. There was not enough for all, but a real man would find a way to provide for his own strictly delimited circle. The line between "my people" and others was closely drawn and very hard to cross. In the competition between Arabian tribes, some, such as the Shammar, Rwala or Mutayr, emerged as militarily and socially superior. The losers in past wars, such as the Rashayda, Hirshan or Soluba, became vassal tribes who paid tribute to superior tribes and lived in a state of customary inferiority. It was almost inconceivable for members of superior and inferior tribes to intermarry (Patai 1962: 251–55). Pride in one's strength and contempt for the weak were two sides of the same virtue.

The desert tribes often took their rivalries with them as they migrated to greener regions. In eighteenth-century Palestine, the bordering towns of Bethlehem and Hebron were rival communities, each mainly descended from a different Arabian tribe. Each group guarded its fields against incursions by the other. As the best defense was an offense, they indulged in constant posturing to intimidate each other and kept hundreds of men armed and ready for combat (Patai 1962: 218–19). Out of such practical rivalries over meager resources, the politics of Middle Eastern nations evolved. So in recent years an officer of the Israeli "Green Patrol" explained the hard facts of life to Leslie Hazleton, who seemed to him a bleeding-heart Western liberal:

> Listen. Get one thing straight. There's no such thing as justice in this world. One man's justice is another man's getting screwed. That's it. Now there's one thing I'll fight for to the ends of the earth. And that's that this land remain ours.... Land that we bought in the Negev with blood and money is ours. It was the Bedouin's before? Fine. It's ours now. (Hazleton 1980: 193)

The politicians and citizens of richer lands might wish that every person's needs and aspirations could be somehow fulfilled. But in lands of scarcity, a crueler politics has usually been accepted as realism.

The Receding Agricultural Frontier

The Nile River's annual floods used to cover a wider area, flowing out through portions of the western desert to the Fayum depression. In ancient times, most of the Fayum depression was underwater; *pa-yum* meant "lake" in ancient Egyptian. But during the time of the Pharaohs, soon after the Saharan savanna turned to desert, the flood corridor to Fayum slowly closed. Perhaps it silted over, as the Nile floods slowly reduced in volume. Fayum was cut off as a shrinking oasis in the western desert. Eventually its lake shriveled into a tiny, brackish dead sea, 50 meters below sea level. But first the Pharaohs and local people fought back. For centuries they strove to keep the floods flowing, digging out the old watercourse from the Nile (Claiborne 1970: 152). This may illustrate how irrigation was born: irrigation probably did not burst upon the world as somebody's brilliant idea for increased production. It probably first arose as a defensive measure. The first farmers planted where crops could grow naturally. Then, when the water supply dwindled, they carried water or moved the earth to keep their land alive.

In northern Mesopotamia, many ancient village sites dating back to 5500 BCE are located where rain-fed crops cannot grow today. The same was true in ancient Pakistan. The rain was probably adequate at first but later declined. Villagers may have carried water to their gardens in pots till the local streams and wells went dry. Around 2000 BCE, many villages in northern Iraq were completely abandoned. Walton writes of this period, "All the archaeological and geomorphological evidence points to the aridity of Bronze Age times and the withdrawal of animal and human populations to the water points afforded by oases and river valleys" (1969: 44).

The arable zone of ancient Iraq shrank eastward toward the Zagros Mountains and narrowed around the river valleys. The villages left behind by the rain found themselves "overpopulated" compared to the food they could grow. Many of the "surplus" people migrated to the banks of the lower Tigris and Euphrates; there the soil was good, but almost nothing would grow without water carried or channeled from the rivers.

As land surrounding the rivers grew more desolate, the competition for riverside land increased. This was probably the sort of rivalry over shrinking resources that transformed raiding into war as we know it. According to cuneiform records from about 2550 BCE, King Eannatum of Lagash ordered an attack on the city of Umma. He demanded from Umma an area of irrigated land called Gu-eden, or "the field beloved of [the Goddess] Ningirsu." According to a scribe of Lagash,

> [King Eannatum] hurled a great net upon them and heaped up piles of their bodies on the plain.... The survivors turned to Eannatum, they prostrated themselves for life, they wept.... Eannatum decreed that since the residents of Umma had dared to oppose his will, they must pay a heavy tax in barley. Then he raised a new boundary

stone claiming the field of Gu-eden for Lagash.

The stone was carved with images of vultures eating the bodies of the vanquished (Roux 1980: 137).

War over arable land was still new and shocking in those last centuries of ancient Sumer. In the older Ubaid culture, people probably lived without such organized murder. As in other premilitary cultures across Eurasia, Ubaid houses and graves were all of similar size. The religious shrines were modest, like those of many villages in India. No significant signs of war, such as defensive walls, caches of weapons or mutilated skeletons were found. Only later did the shrines of rulers mushroom into monuments, and the homes of leaders swell into palaces (Starhawk 1987). Perhaps the Ubaid people were more spiritual, less greedy and less violent than all generations to come. More likely they had enough water and land for their needs, and landlessness had not yet been invented. The Ubaid people lived in a more plentiful time, before the ownership of any particular plot of land became a matter of life and death.

Civilization as we know it was attained when irrigated lands and royal granaries became objects of war. The losers in these wars might be used as workers or milked for rent and protection money. Otherwise the conquerors might slaughter the local men and enslave the women. If the defeated villagers ran away, there might be nowhere for them to go except the surrounding wastelands. Perhaps many of the pastoral desert people were children of such refugees, wandering in the deserts and inheriting dreams of vengeance against "civilization."

If history is read as a march of technical progress, then civilization seems to flow from the practical inventions of irrigation, writing, metallurgy and accounting. From a "gross national product" viewpoint, both swords and plowshares were tools for increased income. The ancient accountants, like industrial-age economists, counted wealth which was "in hand," not what was left in the bush. Their measure of progress was the amount of wealth transferred from the environment into their treasury storehouses. With improved tax collection, full-time guards were needed to protect the rulers' storehouses. Next the guards became units of professional soldiers. Finally, imperial armies were needed to keep the land and its produce from others. The whole rise of sedentary civilizations and nomadic empires in the Middle East can be viewed as a concentration of shrinking resources in the hands of competing warlords.

The Barbarians beyond the Walls

The first pastoralists and farmers were probably relatives, specializing in different crafts but living together. In Syria, Jordan and northern Arabia, where rains came in the winter, the ancient pastoralists and gardeners formed a cycle of cooperation. The farmers planted in the fall, and the pastoralists then left the village, taking their animals away to winter pastures while the crops grew. After the harvest, the herds returned to graze the crop stubble and leave their droppings as fertilizer (Walton 1969: 137).

In parts of Saharan Africa, the pastoralists were only semi-nomadic and owned oasis farms to which they returned for part of each year. But as desiccation advanced over time, the pastoralists' animals required ever wider treks for pasture. These longer journeys edged the semi-nomadic people toward a full-time life in the wilderness. Meanwhile, in the watered areas, rulers fought each other for shrinking parcels of productive land, and the crops were ever more closely protected against grazing animals. So, in gradual steps, the followers of nomadic and settled lifestyles were slowly wedged apart.

The people of the watered lands and the arid country came to see each other as separate communities with conflicting interests. Where mutual hostility forced an end to intermarriage, these communities soon saw each other as different races, or even different species of humanity. So an ancient Sumerian poem characterized the nomadic Amorites:

> The weapon is [his] companion
> Who knows no submission,
> Who eats uncooked flesh,
> Who has no house in his lifetime,
> Who does not bury his dead companion. (Patai 1962: 81)

Where desertification shrank the croplands, it damaged the surrounding pastures even more. With even a modest decline in conditions, the pastoralists often found their means of life stripped away. Even in the best years the nomads usually needed grain and vegetables from the farming areas. In the worst years they came to the watered regions en masse, as when the Hebrews went down to Egypt as refugees, offering to work for food. Other nomadic people, such as the Libyans, Hyksos or Arabs, came to the Nile as conquerors. As in the biblical legend of Cain and Abel, blood was spilled between farmers and shepherds.

To counter the demands of nomads, those who held the productive land began to erect walls. In response to being excluded, the nomads attacked. So, from the seventh and sixth centuries BCE, bronze arrowheads of the Scythian type are found embedded in the walls of towns across Mesopotamia, Syria and Egypt. The nomadic people were scattered over vast regions; but being highly mobile, they could come together for overwhelming local superiority against a village. Ruby Rohrlich says, "The impulse to urbanization seems to have occurred when the neolithic villages and towns drew together … in defense against the periodic raids by nomadic pastoralists" (1980: 80).

Many people now see the old raiding nomads as the inventors of militarism and patriarchy. But if these migrants were so patriarchal and militaristic, how did they get that way? Riane Eisler says,

> At first, it was like the proverbial cloud "no bigger than a man's

> hand"—the activities of seemingly insignificant nomadic bands roaming the less desirable fringe areas of our globe seeking grass for their herds. Over millennia they were apparently out there in the harsh, unwanted, colder [she might have added "hotter"], sparser territories on the edges of the earth. (1987: 44)

The environments Eisler speaks of were deserts and arid steppes. In describing the migrants from those lands, she clearly identifies with the settled people, who feared the depredations of the barbarian. "We have," Eisler continues, "nothing to go by but speculation on how these nomadic bands grew in number and in ferocity and over what span of time" (1987: 44).

It is true the early raider-nomads left few records by which we might understand them. But not all is speculation. For example, out in Central Asia, a Turkic kaghan called Bilga died in 734 CE, leaving a barbarian view of his wars written on a funeral stele:

> I did not reign over a people that was rich; I reigned over a people weak and frightened, a people that had no food in their bellies and no cloth on their backs.... When I became kaghan, the people who had dispersed in different countries returned, at the point of death, on foot, and naked. To re-establish the nation I led twenty-two campaigns.... Then by the grace of Heaven ... I brought back to life the dying people, the naked people I clothed, and I made the few many. (Sinor 1990b: 312–13)

In the view of agrarian cultures, such nomads were a lesser breed, probably godless, and certainly without the goddess. They were described in terms of what they lacked, or what they wanted from the settled lands. The *Tso-chuan*, written in third-century CE China, says, "The Barbarians of the west and the north are ravenous wolves who cannot be satiated." The Huns, according to Ammianus Marcellinus, "burn with an infinite thirst for gold." John of Plano said the Mongols were "most grasping and avaricious, exacting in their demands, most tenacious in holding what they have, and most niggardly in giving" (Sinor 1990a: 4–5). The barbarians were typical dispossessed people.

Near the eastern end of the arid lands, the tribes of Xinjiang (Sinkiang) and Mongolia raided China for thousands of years. Sechin Jagchid claims that many Chinese rulers "failed to discover that poverty and famine caused the nomads to invade China to supply their needs by force" (1970: 40). Sometimes the so-called barbarians merely threatened to attack unless they were given favorable terms of trade (meaning trade of what little they had for what they needed). In the 800s CE, the Yugurs (Uighurs) made such a trade agreement with the Tang dynasty. Qiu Tangshu (Chiu T'ang-shu) described the terms:

> Usually they came every year, trading one horse for forty pieces of silk.

> Every time they brought several tens of thousands of horses.... The Barbarians acquired silk insatiably and we were given useless horses. The court found it extremely galling. (in MacKerras 1990: 338)

If the Chinese rejected such trade agreements, the nomads would have to resume raiding. So the desert people came to rely on war as a primary means of survival. Like professional predators, they honed their arts of battle with increasing success. At times, nomadic confederations such as the Xiungnu (Hsiung-nu) forced China to pay them tribute. Later, in the tenth to twelfth century CE, the Qidan (Kitans) and Nuzhen (Jurchens) actually conquered most of northern China and lived as lords off tax revenues. Then the Mongol hordes (of the 1200s to1300s) and the Manchu banner units (of the 1600s to 1900s) overran all of China.

To the western end of the steppes, serious nomadic raiding started with "Kurgan Wave Number One" around 4300 BCE. The archaeological record suggests the early villages of eastern Europe were looted and destroyed by raiders who knew nothing of farming. Marija Gimbutas reports, "Millennial traditions were truncated, towns and villages disintegrated, magnificent painted pottery vanished; as did shrines, frescoes, sculptures, symbols and script" (1977: 281). Europe's eastern frontier of agriculture was pushed back towards the west, not by advancing deserts but by the force of raiders from the arid grasslands.

During the Roman empire, more steppe migrants poured into Europe in a series of invasions known as the *Volkerwanderung*. The farmers of the eastern plains were driven into central Europe, partly by the so-called Huns. And of these latter-day Kurgan invaders, Denis Sinor says,

> Their sole productive activity and at the same time their only marketable skill was military action, in which they excelled.... As a result of either a conscious decision ... or, more likely, of a short sighted policy aimed at short-term advantages, the Hun economy became almost monocultural, with booty and ransoms as its main products. (1990c: 204)

In the see-saw battles between settled and nomadic lords, military control of production often became more important than production itself. The actual producers—the farmers, craftspeople or shepherds—were often viewed as less important economically than the warriors who looted them. In response to nomadic attacks, farming cultures bordering the arid zone increasingly devoted their surplus production to defense. Then, whether they won or lost their anti-barbarian wars, their societies grew into military states. A rationale arose in which a farming country's king was not a leader of farmers but the head of the military guard. Sinor explains, "To combat him [the barbarian] is the foremost duty of the ruler, in fact it may be the justification of his power over his own people" (1990a: 17).

The term "barbarian" portrays the pastoralists as subhuman. The arrogance of pastoral warlords towards farmers was reverse discrimination. So with amazed (and certainly mutual) disgust, a thirteenth-century Persian described some prisoners of war from the Mongol horde:

> Their eyes were so narrow and piercing that they might have bored a hole in a brazen vessel, and their stench was more horrible.... Their heads were set on their bodies as if they had no necks, and their cheeks resembled leather bottles full of wrinkles and knots.... Their chests ... were covered with lice which looked like sesame seed growing in bad soil. Their bodies, indeed, were covered with these insects, and their skins were as rough grained as shagreen leather, fit only to be converted into shoes. (in Moorhouse 1990: 87)

Geoffrey Moorhouse visited the site of Merv in Turkmenistan, a city destroyed by the Mongol horde. Only isolated bits of masonry protrude through the steppe, marking an area where hundreds of thousands of people lived until 1221. Moorhouse thought he could understand why the Mongol empire faded away. But why, he wondered, had it risen in the first place? What had driven these nomads to launch such an all-out murderous attack: "Was it simply in obedience to the urge of one man who, bereft of father as a boy and betrayed by his clan, had sworn to revenge himself on the world?" (Moorhouse 1990: 100).

Between the nomadic people and the settled civilizations there was a grossly unequal division of the Earth. Warlords rose to defend or overthrow it. The farmland warlords claimed to fight for civilization against the forces of darkness. As Sinor puts their high-minded fear and loathing:

> The aim of the Civilized cannot but be the banishment of the Barbarian beyond the borders [and] the prevention of further incursions. This was the spirit which prompted, for instance, Alexander the Great of legend to shut out, beyond iron gates, set into impenetrable mountains, the "impure people" of Gog and Magog, mythical embodiment of the quintessential Barbarian. Yet there can never be certainty that he will not emerge from his northern lairs, if not earlier, then at doomsday, when the hosts of Gog and Magog will bring universal devastation to a world from which they have been excluded. (1990a: 18)

"Amazons" and the Male Elite

In Europe until World War II, the "Indo-Europeans" were popularly credited as the source of all higher civilization in the world. Supposedly the Aryan conquerors of India, the Persians of Cyrus the Great and the German master race all had a common source among the virile warriors of Inner Asia. Many feminist scholars have reversed this theory. They view the Kurgans, Indo-Europeans and other barbarian warriors much as the ancient Europeans saw them—as destroyers of all that nurtures life. Eisler points out that some raider-nomads literally worshipped their weapons as symbols of holy power, in contrast to those who worshipped the head of grain, the fertile bull or the mother with child.

Eisler's view of history casts the steppe and desert peoples in the role of aggressors, which they often were. The early agrarian cultures she casts as victims brutalized by their attackers, which was also commonly true. But this polarization of roles arouses suspicion. First, it bears too great a resemblance to the old Eurocentric view, in which the world beyond Europe's pale was a realm of uncivilized tribes. Second, if the nomadic raiders are blamed as the source of militarism, then understanding the roots of militarism inside sedentary cultures may be neglected. Third, if the nomads are labeled as the original patriarchs, then the role of their men is emphasized, possibly at the expense of understanding the female half of those tribes. The Kurgan waves were, after all, half female.

Herodotus described the female nomads of the Sarmatian tribes north of the Black Sea in this way:

> The women of the Sauromatae have kept to their old ways, riding to the hunt sometimes with, sometimes without their menfolk, taking part in war and wearing the same clothes as men. The language of these people is Scythian, but it has always been a corrupt form of it because the Amazons were never able to learn to speak it properly. (in Sitwell 1984: 53)

Among Sarmatian female graves, about 25 percent contain battle weapons. Among the Scythians, who ruled the steppe several centuries earlier, nearly 37 percent of discovered female graves hold weapons (Melyukova 1990: 106, 111–12). It seems that Herodotus didn't call them "Amazons" for nothing.

The ancient Greeks and Romans were apparently amazed by the barbarian fighting women of the steppes and of Libya. But limited military roles for women were actually quite common around the world. The usual pattern was that women fought in self-defense only. A.I. Melukova doubts if the Sarmatian women made full careers of war and plunder. Rather, he suspects they carried arms to guard their camps, especially when the men were off raiding other camps. They were mothers who knew how to defend themselves. In a military economy, they had considerable need to do so.

Theoretically, men and women could be equal as warriors in a military economy. But war required losers, and raiding involved hunting human beings for loot and slaves. Women could supposedly join this hunt, and some did. Perhaps they even helped take female slaves for their own menfolk. But the acquisition of male slaves by women was seldom recorded. Defeated captives from foreign tribes were possibly not most women's top choices as helpmates, lovers and fathers for their children.

In practice, the military standard did not make all people equal but it separated them into dominant and subordinate classes according to fighting prowess. And though all stories of the ancient "Amazons" stress their valor in battle, it must have always seemed senseless to risk mothers of small children in knifefights. The early nomadic women may have fought fire with fire to defend themselves. But the art of war slowly evolved far beyond the dimensions of personal defense or individual combat. Over time, units of professional warriors honed their science of raiding. The Mongols would eventually field armies of one hundred thousand horsemen, capable of raining thunderstorms of arrows on any target within 250 meters. Against such units of professional raiders, resistance by women with children grew increasingly pointless.

But if women dropped out of the military rat race, what became of their status in a raiding economy? In pre-Islamic Arabia, according to Fatima Mernissi, "[Women] were regarded as second class citizens, because they did not take part in war" (1991: 122). In that society, raiding was a basic means of survival, along with herding, trading and inheritance. All these vocations or privileges were reserved mainly for men.

On the Eurasian steppes, ancient moundlike "chieftain tombs" commonly contained several women buried along with their lords. At least some of these "wives" were prizes of battle. Apparently this was the sort of concubinage awaiting any Amazon captured by another tribe's warriors. The rising steppe warlords both gained their extra wives and provided for them by raiding. In medieval Mongolia women were also commonly acquired as men's battle loot. Batu Khan had twenty-six women, and Genghis Khan (Chingis Khan) took forty with him to the grave. So across the arid belt the best raiders tended to dominate or enslave the non-fighting classes. Farming areas near the deserts became danger zones, where village women commonly lived in fear (Bat Ye'or 1996: 67). Perhaps the widespread fear of the "evil eye" rose from the fear of raiders. In a land plagued by marauding horsemen, a child or woman who let her good looks be admired could seem to be advertising herself as a prize worth plundering. The costs of being a non-warrior were high. Some Arabian women felt they were unacceptably high.

At least some Arab women of Muhammad's day were ready for drastic remedies. In Medina, Umm Salama called on the Prophet: "During the pre-Islamic period, men excluded women and children from inheriting, because they said, they did not go on raids and did not share in booty." And, "Messenger of God, why do men make war and we do not?" (Tasfir Al-Tabiri, in Mernissi: 1991: 132). Umm Salama believed that all Muslims were equal and the Quranic recitations on women's rights should be enforced. She also accepted that women would never be equal in a military economy except as warriors. She was politician enough to make her demand for equal "rights" in war sound religious, claiming women should have the right of "sacrificing oneself for God." The right was there, because some Muslim women took it and became fighters, martyrs and even generals. But Al-Tabari records that most women in Medina responded to Umm Salama's arguments with resignation: "'It is too bad,' the others said, 'that we are not men: if we were, we could go to war and gain wealth like them'" (in Mernissi 1991: 134).

According to Mernissi, the role of Arabian women in war became the single biggest contradiction for any Islamic vision of equality among believers. The community at Medina was unusually dependent on warriors during its early days of conflict with Mecca. The Islamic community was then living under siege and trade blockade. For the faith to survive, Medina had to win the war. But the same warriors on whom Muhammad depended were accustomed to stealing women in battle, and the women they took became female slaves within the Islamic community. Many captured women, however, claimed conversion to the faith and demanded full rights as Muslims. Some even called for the right to bear arms, a right unheard of for a captive or slave—because if captives had weapons, they would not be subject to their master's will.

Obviously Muhammad faced an important and complicated problem. If female slavery was banned within Islam, the soldiers would no longer have their traditional incentive to fight: "Female slavery was a source of sexual gratification, of domestic labor, and of reproduction of that labor force." The demand for female equality was "threatening a huge reduction in the wealth that a man could gain by raids" (Mernissi 1991: 132).

While Mecca threatened to exterminate his community, Muhammad could not oppose his soldier's interests. Yet, according to his faith, if an enemy embraced Islam, he or she should be received, not plundered. A convert should be freed and not enslaved by fellow Muslims. Muhammad knew that if this principle was applied, he would have to ask his men to fight without the right of plunder or slave-taking. But these were the soldiers' only sources of pay. The enemy was fighting eagerly to plunder Muslims and to take Muslim women; but if the Muslims won a battle, the enemy could simply announce conversion, and neither their property nor their women could be touched. The Muslims would then be the only army on Earth fighting for nothing but souls (Mernissi 1991: 131–32).

Only after Mecca's defeat did Muhammad try to force the issue. At Ta'if, the next town south of Mecca, the Muslim army approached and demanded the ruler's submission. Speaking outside the walls, Muhammad promised freedom for all slaves who defected to the Muslims. And some did defect. Clearly, the rulers of Ta'if would fight to retain their slaves, and the slaves, be they male or female, had reason for hope.

The Muslims were victorious, capturing the town along with six thousand women and children. At this point, the Ta'ifan commander came forward offering the town's submission to Islam. Muhammad announced that the women and children should not be enslaved. He tried to lead by example, freeing his own share of the captives and urging his troops to do likewise. But the soldiers were unwilling to give up their rewards for risking life and limb in battle and argued against Muhammad's lofty plea. At the next Friday prayer meeting, Muhammad offered his own wealth—six sheep to ransom each captive. The troops considered this, and at least a large majority refused. They said the captives were worth far more than six sheep. Perhaps Muhammad realized he could not yet take this step. Later, when the war was over, it might be done. But for now, without spoils and female captives, he could not field an army against the traditionalist rulers who were anxious to destroy Islam (Mernissi 1991: 137–38).

Not long after the battle of Ta'if, Muhammad died. The institution of female slavery and the capture of women in war lived on. Islam managed only to modify the practice. Forcing female slaves to give sexual service was condemned (Quran, Sura 24:33). Also, slaves could only be taken from among the unbelievers. But if unbelieving women could still be legally captured, then Muslim women had to signal their identity by wearing the veil; otherwise they could be taken as fair game. As in a war, a female Muslim had to wear a uniform designating whose side she was on. If she was not dressed as a pious Muslim, she could be taken for an immoral infidel and treated with appropriate immorality.

Such methods of distinguishing abusable women were nothing new. Reference to veiling is found in the warlord state of ancient Assyria, where an inscription from around 1500 BCE mentions veiling as a way for men to show custody of women. In that society of raiders, veils would show which women were already "taken" (Phillips 1966: 144). The *Encyclopedia of Islam* holds that the tradition of veiling women came from roughly that area:

> This custom [the *hijab*, or veil], which appears to have been unknown to the early inhabitants of the Hijaz [Muhammad's region of Arabia], seems to have been introduced into Islam by the Umayyads, probably under the influence of Sassanid [Persian] civilization.

As Islam expanded across almost all of the Afro-Asian arid zone, there remained two main classes of women in Islamic communities—local family women, and women captured in war. With this as a legalized reality, the empire became an intercontinental market for female slaves. Rich men could buy exotic enslaved women from Persia, the Byzantine lands or Buddhist Central Asia. Soldiers could take their female captives home (Zaydan, in Mernissi 1991: 195). The Prophet had reportedly said, "The worst of men is the seller of men." But at the height of the Islamic empire, Muslim merchants "ransacked the whole known world for choice young females" (Phillips 1966: 87).

Many twentieth-century Middle Eastern women grew up surrounded by the memory of that traditional raiding economy. Mernissi says the capture of concubines in Morocco ended only in her mother's generation: "My grandmother was kidnapped in Chaouia plain, sold in Fez, and bore my mother to a member of the landowning bourgeoisie.... This group was the main buyer of female slaves for decades after the French occupation in 1912" (1987: 48). The soldiers and raiders of richer environments such as Europe or China were not morally better. They commonly raped "enemy" or "bystander" women with impunity. And they might massacre civilians indiscriminately and go unpunished. But it was generally held unacceptable in Europe or China for soldiers to capture foreign women and bring them home at swordpoint to serve as additional wives. That right came to soldiers only where looting was more central to the economy and women were seen as the possessions of males.

Rulers of the Arid Zone

Absolute monarchy started in the Middle East, but not among the desert nomads. For the nomads who ranged in small groups across vast spaces, power was never an impersonal, centralized institution—it was a matter of trust and loyalty among tribe members. Raphael Patai says, "The authority of the tribal shaykh rests not on force, which as a rule does not stand at his disposal, but on the esteem, renown and prestige he enjoys" (in Phillips 1966: 149). Such prestige was never simply inherited. A son might try to follow his father as shaykh or khan, but if he failed to personally earn the tribe's admiration, the people would follow someone else. Whom the nomads followed, where they went, and whom they raided were determined by what Robert Lacey calls "the boundaries in men's hearts" (1981: 164–66). So the great military confederations of nomads, of Attila the Hun or Tamerlane, were forged of personal allegiance and seldom survived the leader's death. When a great man died, the tribes swirled into another configuration of heart boundaries.

In the farmlands of the Middle East, kings and emperors were also unknown at first. The earliest rulers where mainly religious figures. For early Sumerian kings, "godship of the land" involved sacred marriage to the goddess for ritual insurance of the earth's fertility. The Sumerian city states were theocracies "owned," not by the kings but by their patron deities. For example, one-third of the land in Sumerian Lagash belonged to the temple of the goddess Ningirsu.

With the spread of warfare over irrigated land, and with increasing raids by nomadic people, the role of king evolved under military pressure. By about 2600 BCE, records in Sharuppak, Iraq, indicate the local king kept six or seven hundred bodyguards (Roux 1980: 129). The lands of temples and private farmers in Lagash started to shrink as the ruling palace took more land. Where the king had been the high priest, he now began to focus increasingly on military matters, leaving religious functions to others. Georges Roux says, "Already in about 2400 BC, Entemena, *ensi* (king) of Lagash, was no longer high priest of that city ... because a silver vase he dedicated to Ningirsu says the priest of Ningirsu is Dudu" (1980: 133).

If the king was no longer synonymous with the high priest, at least he was not yet synonymous with the general. In Uruk, a city assembly of prominent men and women appointed special leaders in case of war. At first, these military commanders were dismissed once a period of fighting was over, but around 2600 BCE, during the reign of a certain King Gilgamesh, the Uruk assembly excluded women. Then, to telescope a gradual political revolution, the office of war leader was made permanent, and the war leader became the king. In the *Epic of Gilgamesh*, which looks back to this period with assumptions from a later age, the king is presented as a divinely appointed absolute ruler, whose chief role is war (Starhawk 1987: 39–40).

The period of about 2600 to 2334 BCE was for Mesopotamia somewhat like the "Warring States" period of Chinese history. The land was divided into city states which were constantly seeking to swallow each other. This arms race culminated with a warlord from the fringes of the arable area conquering the whole chessboard. King Saragon of Agade (who ruled c. 2371–2316 BCE), of the Semitic-speaking Akkadians from north of Sumer, overran the chief island of agricultural civilization in West Asia. His conquest laid down a pattern which repeated itself for the next four thousand years. The historical cycles of Middle Eastern history as described by Ibn Khaldun had begun (see Stewart 1979: 130).

In this Khaldunian pattern, various groups of arid country nomads vie with each other in raiding the watered zones. Slowly, one of the raiding tribes grows stronger and bolder than the others. At some point this group moves in, not just to raid but to actually overthrow the king of the farmers. Then the lord of the raiders assumes rule over the farmlands, not as an administrator but as an absentee landlord. Being a semi-nomadic general, he takes the agricultural area as his tax farm. With the proceeds, he builds an army beyond the dreams of mere subsistence raiders and supplements his treasury with further conquests abroad. But living amid so much stolen and inherited wealth, the heirs of the conqueror slowly forget the martial virtues by which their fathers rode to power. Living in luxury, they go soft. A rot of corruption slowly undoes the empire's strength. Finally, another tough nomadic leader rises up to exploit the empire's weakness. Then a new warlord moves in to repeat the game. Saragon set all this in motion. He mounted sedentary civilization and rode it like a warhorse.

In his prime, Saragon pushed his empire outward in every direction. He personally led his troops into Armenia, the Mediterranean, Arabia and Iran. "Now," he boasted, "any king who wants to call himself my equal, wherever I went, let him go!" (Roux 1980: 148). This king was no roleplayer in a religious drama, praying for rain and crops. Saragon better resembled the legendary Gilgamesh:

> His arrogance knows no bounds by day or night. No son is left with his father, for Gilgamesh takes them all, even the children; yet the king should be shepherd to his people. His lust leaves no virgin to her lover, neither the warrior's daughter nor the wife of the noble; yet this is the shepherd of the city. (Starhawk 1987: 49)

Under the Akkadians, the irrigation systems of Mesopotamia decayed. Revenues from farming were invested in military adventures or court splendor, leaving little to maintain the canals and dikes. The old Sumerian religion was also neglected. The temples now came under the king, as "landlords among other landlords, tax payers among other tax payers" (in Roux 1980: 170–71).

With the Akkadian and Assyrian kings, politics transcended the tribe or the city state. The kings of all-Mesopotamia were no longer just heads of tribes or ethnic communities; they ruled over many tribes of farmers and nomads. Power no longer came from a council of clan heads but from success in generalship. The warlord king's subjects were simply all those whom he was able to plunder or force to pay tribute, regardless of race or creed.

The Assyrian kings rose possibly from Amorite migrants, became warlords in an often-conquered area of northern Iraq and took the Akkadians' position in Mesopotamia between the 1200s and 609 BCE. In almost every one of those six hundred years, they led military expeditions to plunder new areas or intimidate conquered regions into paying more tribute. In one such raid, King Tiglath Pileser III forced Tyre to pay 150 talents of gold. In another, Saragon II plundered Musasir for five tons of silver and more than a ton of gold (O'Connell 1995: 154). For poor rural areas, a preserved document shows King Ashurnasirpal's imposition of yearly tribute on a district chief: it lists one thousand sheep, four thousand bushels of grain, two minas of gold (a mina equals eighteen ounces), and thirteen minas of silver (Luckenbill 1926–27, in Roux 1980: 265–66). Failure to pay in full and on time would bring military assault, with indiscriminate slaughter of the local population. Basically, the Assyrians treated their empire as a huge hunting ground, "a geographical area through which they could raid without encountering effective opposition" (Hogarth 1950, in Roux 1980: 264). And as their region was depleted by raiding, they were forced to go farther and farther afield, like nomads chasing ever scarcer pastures.

In China, the emperor was also a chief of the armed forces. But in most native Chinese dynasties, that role was minor compared to the emperor's many other responsibilities. Besides fighting the desert barbarians, he was to establish roads or canals, open the planting seasons, perform rituals for celestial harmony, review the performance of officials and repair dikes against floods. In the Chinese environment, scarcity and famine were assumed to be unnatural. Such problems were seen to result, not from nature itself or from the people's sins, but from poor administration. If the people had not enough food to survive after taxes, many Chinese believed that heaven would withdraw its mandate from the ruler.

A similar attitude was common across Europe, as seen in the cycle of Grail myths. The fertility of the land somehow depended on a good king or queen. If the ruler was not in harmony with the people and with nature, the ominous appearance of a wasteland would make God's displeasure known.

However, in much of West Asia, the wasteland had already arrived, and the rulers commonly took it for granted. The problem was less how to prevent a wasteland than how to endure or escape it. In desiccated country, the creation of food or wealth was often so constrained that it seemed the inadequate level of production could hardly be increased. Since there was not enough for all, the distribution of wealth had to be decided politically or militarily. Every warlord wanted the lion's share, and only one in each region could have it. Meanwhile, producing farmers, pastoralists, craftspeople and merchants were squeezed between a weakening environment and the arbitrary demands of competing warlords.

During the Persian empire (of the 500s to 328 BCE), King Darius milked the Middle East for about 14,560 talents worth of money and produce per year. A talent was worth only around $2,000 U.S. in 1990s terms, but its ancient purchasing power was astronomical (Collins 1972: 151–53). This income went mainly for upkeep of the royal court and the Persian army, which enforced collection. In times of war, rulers squeezed harder. During the first Punic war between Carthage and Rome, Carthage doubled its taxes on the Berber farmers from 25 percent to 50 percent of the crop, in order to pay mercenaries. About the same time, the Ptolemies in Egypt took one-half the people's produce, partly to pay troops to keep down rebellion against the level of taxation.

There were environmental costs associated with pushing the people so hard, and these costs were especially high in fragile, semi-arid farmlands. To meet imposed annual production quotas, farmers often had to sacrifice fallow years for their fields. The land was then forced into constant use without letup. Between the requirements of the military and the survival needs of the farmer's families, any reinvestment in the land was often squeezed out.

It was not the local farmers who were so shortsighted. The peasants continued to worry about the fertility of their land and how their grandchildren would live on it. Generally they did what they could as families and villages by, for example, digging the underground irrigation channels of Iran. But the kings often counted their wealth in money diverted from such purposes. So ancient water conservation systems tended to fall apart: the massive series of dams at Marib in northern Yemen collapsed for lack of maintenance. The irrigation network of Sumer silted over. There was money in Baghdad for larger armies, wonderful palaces and patronage of temples, mosques and the arts, but seriously redoing the irrigation system had to wait nearly four thousand years. So, desiccation begot warlords, who begot desiccation.

The great monarchies of the Middle East arose in islands of watered land such as Egypt and Mesopotamia, and on the coastal peripheries of North Africa, Turkey, Arabia and Iran. But the rulers of these productive islands were commonly raiders from the arid hinterlands. The resulting civilizations were not truly agrarian or nomadic, but centaurlike hybrids with semi-nomadic warlords on top and farmers on the bottom. Many of these rulers looked on their farmers much as nomads looked on their sheep or cows. Not only did the conquerors claim a heavy portion of the farmer's yield, but they also claimed to own the land by right of conquest. So down to the nineteenth century all land in the Ottoman empire was ultimately owned by the sultan, who could confiscate it if he desired. In Persia, the nineteenth-century shahs could, and did, seize individual land holdings whenever they needed extra cash (Mansfield 1991: 142).

This legal tradition of despotic control over all significant resources has continued down to recent or present times. In modern Egypt, for example, the central government holds veto power over all appointments to administrative boards for the nation's approximately twenty thousand charitable, social, cultural, religious, scientific, artistic and environmental associations. As Fawzy Mansour explains, "These are all channels of access to social power and prestige and are very well controlled and manipulated in such a way that only those who are conforming to the line of the government are given access" (1997: 165–67). Such control of power and resources requires a means of withholding access. So, in the modern nations of North Africa and West Asia, the military remains a major political force. So, as rains failed across the African Sahel in the 1970s, military coups toppled the governments of Niger, Chad and Ethiopia. The armed forces did not meekly wait in line for their share of supplies in difficult times.

The Hebrews of the Old Testament, while oppressed by ancient Assyria and Babylon, called to their god for a day of judgment against their enemies. And sure enough, that day finally came for the Assyrians. One day in 609 BCE, after the Assyrians had plundered the same vast area for hundreds of years, deporting whole populations to uproot resistance till the Middle East was an ethnic mish-mash, their enemies closed in like a pack of wolves. As an Assyrian civil servant had written to his king, "The king knows that all lands hate us" (O'Connell 1995: 157). Joseph Campbell suggests the Assyrians had built up a lot of bad karma. Their capital Ninevah was sacked with a powerful vengeance. The avengers wiped the Assyrian ruling class off the face of the earth and burned the city to a field of rubble. The flames fired the royal library's clay tablets so hard that we can still read them today. That could happen to warlords. Transitions of power were seldom easy in the arid lands. The step between ruler and subject was usually too vast to take voluntarily.

Chapter 6

Tribes without Nature in Common

IN SOME TRADITIONAL CHINESE LANDSCAPE PAINTINGS, THE ARTIST SEEMS to be standing on a mountain looking down. Far below, tiny humans scurry about like insects, overshadowed by vast forests, mountains and rivers, in a land of mystery. This terrain shows the artist's sense of proportion. Especially in the wet and mountainous country of southeast China, this was how important the plants and animals, the earth and living rock, seemed in relation to humanity.

For North American natives also, the place of humans in nature seemed small. Native myths are mainly populated, not by founding fathers or conquering heroes, but by the spirits of corn, buffalo, salmon or caribou. These plant and animal powers are seen as ancestral to humanity and are treated with filial respect.

In the Afro-Asian desert, however, humanity stood in the foreground of attention. In a vast and barren landscape, the human figure stood out starkly. The eyes focused on a traveler approaching in the distance. Old Arabia had perhaps the grandest traditions of hospitality on earth. A fellow human being in that desert was received with a joy or concern which would surpass belief in a modern city. Almost as if greeting a fellow countryman met by chance on another planet, the poet Omar ibn al-Farid opens his heart to a guest;

> Welcome to him whose approach I am all unworthy
> Welcome to the voice announcing joy after lonely melancholy
> Good tiding thine; off with the robes of sadness; for know
> Thou art accepted, and I myself will take on me whatever grieves thee.
> (in Phillips 1966: 116)

Relatedness to other humans was the central and almost exclusive concern for people in the desert. Perhaps only sailors on the sea felt themselves so greatly dependent on their fellow travelers. The worst possible calamity for a Bedouin was expulsion from the tribe. To be cut off from the group in that wasteland would be like floating alone in the ocean (Hitti 1960: 14).

The Community as Everything

Among traditional nomads, life was an endless camping trip. The fellow campers saw each other day and night, year after year. A certain intense intimacy was almost inescapable. Inside a Mongol yurt, or mobile tent, very little could be hidden from others. Outside on the open plains, everything was also exposed to view. In that culture, privacy was neither expected nor particularly valued (Jagchid and Hyer 1979: 70). All this could be shocking for those unaccustomed to nomadic life. Amid the various hardships of crossing the western Sahara by camel, Geoffrey Moorhouse felt,

> Most trying of all was the lack of privacy. Apart from the delicate matter of *le cabinet*, everything one did was with half a dozen onlookers sitting within a few feet, scrutinizing, commenting, implicitly participating. (1974: 55)

This is how life was for traditional desert people almost every day of their lives. Their ways of eating, washing and relating were constantly on display. Moorhouse felt that from such conditions a deep concern arose to regulate all bodily functions in certain approved ways. The lessons were reinforced constantly from childhood and internalized far more deeply than in the privatized Western world. Over time, each desert tribe tended to develop its own complete code of behaviour covering almost every detail of daily life. By the nuances of such codes, an outsider could be easily discerned.

According to a line of foreign and native sociologists, the traditional Middle East was a "shame society." Shame in this case means regard for the judgment of others, as distinguished from guilt, which is an individual's judgment on her- or himself. A shame society is one where shame prevails over guilt, so that the main concern is to belong and conform. As Wai'il Kheir explains, "In a shame culture it does not really matter how you feel. The most important thing is that you should not violate the standard behaviour of the community" (in Butt 1977: 234–35). Any deviance from communal standards is usually hidden, repressed or practiced in secret; as the old Arab saying goes, "Hypocrisy is the homage which vice pays to virtue." On the negative side, shame is taught by shaming techniques, such as critically comparing one child to another (Hamady 1960: 34–39). On the positive side, success in conformity is a source of pride.

This labeling of the old Middle East as a "shame society" would be a racist generalization if it was just a description of "ethnic characteristics." But what if such sentiments have a cause beyond ethnicity? Speaking of the whole Mediterranean world in the time of Jesus, John Dominic Crossan suggests, "Honour and shame are the constant preoccupation of individuals in small scale, exclusive societies, where face to face personal … relations are of paramount importance and where the social personality of the actor is as significant as his office" (1991: 10).

Among Old World pastoral tribes, deviance from group standards was a serious concern, and not just a matter of personal taste. One person's violation of custom could bring excruciating shame on the entire extended family, and in many Middle Eastern societies, this was a calamity to be avoided at all costs. A rejected individual might find no place to live, either inside or outside the tribal economy. A non-conforming woman might "shame" her male relatives so deeply that they would feel they must kill her. Unless the cause of their shame was cut out and destroyed, how could they ever hold up their heads in public? So the anxiety to avoid shame could overshadow every other consideration. A family head would commonly seek to control all aspects of his women's and children's lives, lest their independent actions shame him. A tribal patriarch could feel compelled to sternly enforce a thousand correct customs among "his people." As for the rights of individuals, Fatima Mernissi says,

> Our traditional identity hardly acknowledged the individual, whom it abhorred as a disturber of the collective harmony.... The idea of the individual in a state of nature ... is nonexistent. Traditional society [in Morocco] produced Muslims who were literally submissive to the will of the group.... Individuality in such a system is discouraged; any private initiative is *bid'a* (innovation), which necessarily constitutes errant behaviour. (1991: 22)

Such deep concern for communal harmony often rose from real devotion to others. Wealth was shared; hoarding was shameful. To be deemed generous by the community was often more important than to have possessions. The ethic of self-enrichment was slow in coming to Arabia. Muhammad himself was a merchant, but he preached that the prosperous would be judged by how they shared their wealth. The paying of the Muslim *zakat* tithe (of 2.5 percent of annual income) was named as one of the five pillars of the faith. This "tithe" was not intended as a tax for support of any government or clerical office. It was a direct sharing of wealth with more needy individuals, as an expression of covenant among all members of the community.

As anyone from a traditional village or tribe knows, intimate community always has its good and bad sides. Millions of moderns have fled from that intimate codependence, preferring the "freedom" of relative anonymity in cities and towns. That old codependence was probably more intense in the desert than anywhere else. The ancient Europeans may have said, "When in Rome, do as the Romans do." But most desert tribes took their traditions far more seriously than that. Their customs tended to transcend the status of "social habits." Their ways of eating, washing, speaking or loving were held to come from the tribe's founding fathers and prophets, who received them from God. Ethnic culture and true religion were therefore one and the same. As Raphael Patai generalizes for the whole Middle East:

> All custom and tradition are basically religious; for whatever is old and customary and traditional is hallowed by religion. Religious practice itself is mainly tradition and custom, so that practically every activity is either in conformity with or contrary to religion. These observations hold equally good for Islam ... for the Eastern Christian churches, and for Judaism in its Middle Eastern form. (1962: 288)

In this context, if a man abandoned his people's traditions, he would lose his soul. If other desert tribes clung to different customs, then their ways must be wrong in the sight of God. Joseph Campbell says that for each Semitic desert tribe, God was "made known, not in the sun, the moon, the cosmic order, but in the local laws and customs—which differ, of course, from group to group" (1976b: 431). For ancient desert dwellers, the human community was virtually everything. The tribe was almost deified as the only holy thing on earth. The right ways of acting were determined long ago by tribal patriarchs. After that, each generation of young people found their various options in life pre-labeled as right or wrong. Each person had an acknowledged freedom to choose *between* the right and the wrong, but not the acknowledged freedom to question *what* was right or wrong.

Communities without Places

In the environmentally rich lands of Europe or China, villagers commonly lived in one place all their lives. From childhood to old age they watched the same trees grow. Home for them was a place on the face of the earth. The locals there spoke of "fatherland" and "motherland," and citizenship was often determined by place of birth, regardless of ethnic lineage. In Guinea, the Sousou people buried each child's placenta and navel cord beside its home so that, as Prince Modupe explains, "When I go home I shall stand on the spot where the waiting ones stood that night [of my birth] … and I shall speak these words: 'My belly is this day reunited with the belly of my Great Mother Earth!'" (1958: 7).

Nomadic people could have no such special relation to a particular place. They had to move about like scavengers of wild grass. They could not plant a tree and watch it grow. As a herdsman in Mali explained:

> When [we] nomads plant trees we can look after them as long as we stay in that place, but as soon as we move on, stray animals destroy them. So long as we are not settled, we have no means of protecting the trees. (in Cross and Barker 1991: 78)

Where aridity imposed a nomadic life, groups of wanderers became communities without places. Their identity was defined by blood and social custom, not by the land over which they traveled. So, for example, "feudalism" in old Mongolia was not a set of relations between people and their land. It was an order of ranking among members of a "nation on horseback" (Jagchid and Hyer 1979: 267).

The Sahara's southern "shore," where the desert meets the arable land, has a long history of collisions between the cultures of farming and desert people. Around Lake Chad, farmers from many tribes and languages settled close together, while warlike nomads such as the Tuareg or Zaghawi often ruled over them. The farming people tended to gradually bridge their differences and to form a regional culture based on common experience in a shared environment.

For these villagers, the kingdom of Bornu (around Lake Chad) was a homeland. But the Zaghawi rulers of Bornu saw the region in a different way. For them, the kingdom of Bornu was the royal family and its possessions. Membership in their community came through birth, not through neighbourly friendship (Trimingham 1962: 107–8). The Zaghawi remained aloof from the local culture of their subjects, and their loyalty to extended family remained stronger than any devotion to "country." In the twentieth century this attitude continued, as the governing officials of Chad and Niger commonly displayed loyalty to kin first. But in a modern territorial nation, this kind of loyalty has been labeled "corruption."

For many modern pastoralists, blood relations and tribal customs still form the bedrock of personal identity. Even long-settled descendants of nomadic tribes often inherit such values. In Uzbekistan, where most of the Uzbek nomads settled as villagers centuries ago, a saying goes that, "Anyone parted from his land will weep seven years; whoever is parted from his tribe will weep until he dies."

Neighbours without Common Ground

In ancient Europe, the tribal lore of Celtic or Germanic people was largely a knowledge of nature. Traditional wisdom concerned the uses of flowers or mushrooms, the ways of sea animals, patterns of stars, or sources of sacred water. Each tribe had its own stores of knowledge and myth, but the objects of knowledge were similar across tribal lines. The religions of Old Europe had differing ways of relating to the same powers of nature. The gods and goddesses were not primarily tribal deities but the holy spirits of thunder, sea or grain. Zeus was more the lord of all storms than he was a patron of the Greeks. Zeus was called Jupiter in Rome, and Jupiter was obviously analogous to Thor in Germany (Campbell 1976b: 431). When the Romans made this comparison, the Germans did not seem offended. But most desert tribes would recognize no such correspondence between Yahweh and Allah, much less Yahweh and Zeus. When the ancient Greeks ruled Israel, King Antiochus IV (who ruled c. 175–163 BCE) attempted to relate the religions of Jews and Greeks by erecting a statue of Zeus in the main temple at Jerusalem. This was the "Abomination of Desolation" against which Israel exploded in revolt. Yet John Allegro claims the relationship of Zeus to Yahweh was valid: "This association of gods was, in fact, perfectly historical and legitimate. Zeus was, indeed, Yahweh in origin: Both names meant the same, 'seed of life,' … and both had their common origin in the underlying fertility religion of the ancient Near East" (1971: 79).

In more ancient times, the Middle East contained many popular cults of nature featuring goddesses of the earth. These goddesses were popularly known to have many names in many languages. But by classical times, perhaps, the earth was so depleted as to constitute no important common ground between cultures. Each tribe increasingly focused, not on relating to nature but on its own social traditions. Each group increasingly saw its own customs as the only "human" ones. The Berbers of Mauritania revolted against mixing their culture with that of the Greeks and overthrew their king, Juba II, who was one of the ancient world's greatest scholars, and author of over fifty works in Greek. For these rebel Berber traditionalists, the Greek cosmopolitan dream (of a cosmos-city) was nothing but a corruption of their tribal heritage.

So, in arid countries, the spotlight of religious concern tended to fall on the tribe's social conventions, ethics and visions for itself in the future. Outside of that, what else on Earth was worthy of inquiry, respect or devotion? Campbell explains that for a valid act of Jewish Orthodox worship, no fewer than ten males over the age of 13 must be present. By implication, "The individual has no relation to God, save by way of this community, or consensus. God—the only God there is—is apart, and the body of his chosen people is the only holy thing on earth. The individual apart from that is null" (1976b: 138–39). Campbell contrasts this form of worship with the solitary quests of Indian yogis, or sages in China. In these eastern Asian traditions, seekers of profound religious insight commonly went alone into the forest. Enlightenment for them involved a realization of identity with the natural universe. But according to the covenant of Moses, "God, who is transcendent, is neither within nor in nature, but in the group—this group alone, with its laws, which are the only facts of real moment to be known" (Campbell 1976b: 138–39).

If one's code of tribal laws were the holiest thing on earth, then to mix customs with other tribes would be a sacrilege, signalling treason against the true community. And in the desert, where dependence on the community rather than nature was almost complete, the mixing of cultures was commonly taken as an abomination against God.

The True Social Contract

Ancient pastoral people often depended on each other for their lives. As if climbing mountains, the group members often hung suspended by ropes of trust. In war, drought or storm, the vows they made to each other had to be reliable (Jagchid and Hyer 1979: 148). They exchanged solemn oaths promising safe passage, rights to grazing, betrothal, marriage, alliance and mutual help. If any of these sacred trusts were betrayed, the consequences could be fatal. The wronged parties would fall back on other pledges, not only for survival, but for justice.

Mary Boyce explains that the proto-Indo-Aryans before Zoroaster had a great concern with pledges and how they become binding. Two sorts of pledges were recognized. The first kind was called "Varuna." The term probably stems from the Indo-European root word "ver," meaning "bind" or "tie." By this sort of oath, a man bound himself to do—or not to do—some specific act. The second kind of oath was called "Mithra." This word likely comes from the Indo-European root word "mei," which means "exchange." In a Mithra pledge, two parties made some agreement, and both promised to uphold it. For these old Indo-Europeans, taking an oath was a religious act. It involved calling divine powers to witness and enforce the agreement. Later, if one party accused the other of betrayal, the gods Varuna or Mithra would pass judgment through a trial by ordeal. For an accused breach of personal oath, the trial was by water. The accused was held underwater while an arrow was shot, and a runner sent to fetch it. If the accused was still alive when the runner returned, Varuna had shown him innocent of disloyalty. For violation of a mutual covenant, the trial was by fire, either running between bonfires, or having molten metal poured on the body. Mithra, the witness of all covenants, associated with the all-seeing eye of the sun, would protect the faithful from grievous harm and burn the guilty with a vengeance (Boyce 1979: 8–9).

In that society-centered world, violation of a social covenant was an apostasy deserving death. Witches were people who called on the deities of other tribes and cast spells to harm their own people. They were tried by ordeal. If found faithless toward the cult of the tribe's leaders, they were cast into the flames. Without such a drastic punishment for disloyalty, many feared that bad faith might spread like gangrene through the social body, rotting or cutting the ropes of trust and leaving the tribe members disunited before their enemies.

Zoroaster was perhaps the first tribal priest to cast the social contract in truly cosmic dimensions. According to his preaching, not only would each person be held accountable to whatever oaths or covenants he or she might make, but in addition all people would be judged by a single great covenant, imposed by an almighty god. A day of universal judgment would come, in which the test of loyalty to the Zoroastrian compact would resemble the Mithraic trial by fire: "For him who is righteous it will seem like warm milk, and for him who is wicked, it will seem as if he is walking in the flesh through molten metal" (in Boyce 1979: 28). Zoroaster proposed that all humanity must be judged by a single standard of behaviour. That universal standard, however, contained elements of social custom from Zoroaster's own tribe.

This idea of a day of judgment later took hold across the Middle East. On that day, all would be tested for loyalty to the covenants of their lives. Apocalyptic Zoroastrians, Jews, Christians, Manichaeans and Muslims all preached their visions of one socio-ethical standard, one Lord and one judgment. In popular imagination, the heavenly rewards for loyalty grew lavish—some Muslim poets described harems of forty eternal virgins for each faithful man. And the punishments for disloyalty grew ever more terrible—Campbell (1976a: 50) claims the early Christians conceived of the ultimate punishment, namely eternal damnation in a pit of fire.

If one tribe's social contract was the true cosmic principle by which all must be judged, then the domain of that contract should be expanded over the Earth. The natural desire of each tribe to expand then became a matter of manifest destiny. The ultimate conclusion would be one covenant made universal, and this was the dream of many tribal prophets. One day, a mighty king would arise to unite the world, putting the warring tribes under one law. A new world order would be established, perhaps with justice and equality for all. How could this glorious vision be achieved? By what steps could a prophet's followers actually move from their limited group covenant towards a universal one? One way was to simply proclaim that one tribe's customs *were* the universally valid ones of God, and that the followers of all other traditions would be condemned on judgment day. For some, this seemed to solve the problem of covenants in collision—through the power of moral judgment alone.

More practical people tried to take covenant-building one step at a time. Obviously, each tribe needed to develop its own code of mutual obligations. Each had leaders who wanted to instill fears of breaking the social contract. In relation to other tribes with different loyalties, agreements had to be worked out and honoured. By some gradual process of socialization, commerce, diplomacy, forgiveness and mystical insight, a trans-tribal society would slowly emerge. Unfortunately, such a gradual path of self-transcendence was not what some Middle Eastern rulers desired. They wanted the world united, but only if it was united under themselves.

The King of Kings

The imperialistic shortcut to one world order was tempting. If a rival tribe could be conquered, the victors could impose their own social contract on the vanquished. A giant step would then be taken towards the universal rule of truth. Many tribes believed that expanding at the expense of their neighbours was divine will. But few tribes came anywhere near conquering the world. The Persians were the first to seriously attempt it.

Before the Persian empire (of the 500s to 328 BCE), each aggressive tribe had been able to stretch its soldiers only a limited distance over the Earth. But when the Persian armies marched east to India, north to Turkmenistan and west to Greece and Egypt, they left local administrations largely intact. Rather than replace these leaders with his own people, the Persian king oversaw other tribal kings. He became the "King of Kings," ruling the entire central section of the Afro-Asian arid belt. This was manifest destiny almost fully realized. Campbell says, "The Persian [Zoroastrian] answer to sorrow … was the building of a soundly governed progressive world empire under God" (1976c: 245).

Smaller nations like the Hebrews were certainly influenced by the King of Kings. His grandeur, however, did not make them believers in Persian spiritual leadership for the world. Instead, the Hebrews dreamed of their own kings of kings. The Israelites commonly held that their own law, not that of Zoroaster, was the true cosmic principle by which all would be judged. But could tiny Israel actually aspire to world leadership and domination? Some of its prophets and fanatics, such as the Essenes, believed the Lord had promised them an earthly victory. As Isaiah had proclaimed,

> Enlarge the place of your tent, and let the curtains of your habitations be stretched out; hold not back, lengthen your cords and strengthen your stakes. For you will spread abroad to the right and to the left, and

> your descendants will possess the nations and will people the desolate cities. (54:2–3)

And Second Isaiah continued more grandly:

> Kings shall be your foster fathers and their queens your nursing mothers. With their faces to the ground they shall bow down to you, and lick the dust of your feet. (49:22–23)

But over the course of Middle Eastern history, no ethnic group ever managed to eclipse all others. Even the Arabs failed to unite the whole world. Most Jews were content to believe that the upholders of the true covenant would be vindicated in the end, when God came to claim his own. As successive Persian empires failed, most Zoroastrians settled for a similar view. They accepted that though their society and race were based on the one true covenant, their fate until judgment day was to be surrounded by infidels.

The Chosen People

For thousands of years, both Palestinian and Jewish settlers lived together in the same environment. For extremists of each group, however, this was only a problem to be corrected. The zealots of each ethnic group made exclusive claims to both the land and the truth. Only in the forgotten past had all these neighbours been one people. As T.J. Meek confirms,

> The contention that Yahweh was of Arabian origin is clearly in accord with the Old Testament records, which connect him with the Negeb and with southern sanctuaries like Sinai-Horeb and Kadesh.…The most probable [origin of the name] in our opinion is … from the Arabic root "hwy," "to blow." (in Campbell 1976b: 132–33)

The Bible says that the Hebrew people came from the Canaanites: "You shall solemnly recite before the Lord your God: 'My father was a homeless Aramaean who went down to Egypt with a small company'"(Deuteronomy 26:5, in Bloom 1990: 3). Ezekiel told the people of Jerusalem: "Canaan is the land of your ancestry and there you were born; an Amorite was your father and a Hittite your mother" (16:2–3) The Hebrew language largely evolved from Old Canaanite. But since that time, the Palestinians and Jews have tended to emphasize their differences. They have claimed separate lines of patrilineage and unique God-given traditions.

During times of drought, arid-land tribes, such as the Tuareg, Kurds, Pathans and Hebrews, all fought their neighbours over land and water. In the lore of each people, these were epic wars between good and evil, as in the classic stories of the Old Testament. Over time and repeated conflict, the battle lines hardened into walls of ethnic exclusivity. On returning from exile in Babylon, for example, the Jews around Jerusalem faced several years of drought in a row. Their homecoming turned into a trial by starvation. Allegro describes the deliberations of anguished Jews as they gathered at the temple:

> What unatoned sin barred the Chosen People from the favor of their God? Had not their prophet proclaimed their penance paid in full, even to double the extent of their iniquity? ... From the turmoil of their disillusionment and heart searching came fresh accusations of their shortcomings, and pious directives to Israel's restitution to grace. She had compromised her religious purity by accepting foreigners with their syncretistic worship in her midst. Only by purging the community anew of these alien elements could her cultic innocence be restored. No sacrifice was too great to achieve this end. Loving husbands must banish their foreign wives and children.... Jewish women must leave their Gentile husbands and forsake their offspring. Only when the nation had purged itself racially would Yahweh fulfill his promises. (1971: 14)

Just as modern racism rears its head in times of high unemployment, so the greater scarcity of drought seemed to force a delimited self-interest. Leaders like Nehemiah rose up to rebuild the sectarian walls and attribute xenophobia to God:

> In those days also I saw the Jews who had married women of Ashod, Ammon, and Moab; and half their children spoke the language of Ashod, and they could not speak the language of Judah but the language of each people. And I contended with them and beat some of them and pulled their hair; and I made them take an oath in the name of God saying, "You shall not give your daughters to their sons, or take their daughters for your sons or yourselves." Did not Solomon King of Israel sin on account of such women? ... Foreign women made even him sin. Shall we then listen to you and do all this great evil and act treacherously against our God by marrying foreign women? (Nehemiah 13:23–27)

So the zealots for cultural purity pruned back the growth of love and friendship across tribal lines. They labeled inter-tribal relationships as high treason. To stamp out the fires of forbidden love, the drive for exclusivity had to be overwhelming. An iron-clad morality was required and supplied. The tribal elders proposed a firm, closed circle of logic, in which all hardships were punishments from God, to chastise people for departing from their ancestral customs. The good times were rewards for clinging still tighter to tribal tradition. So a zealot could draw the same lesson from every experience. Negation of other tribes could become an article of faith. The divisions between

Middle Eastern cultures could actually deepen with each passing century. Eventually, Jewish and Palestinian neighbours would learn to see each other as utterly different in culture, religion and even race.

At their most extreme, the desert-style tribal religions could evolve into complete non-recognition of other cultures. So numerous "Muslim" fanatics during the old Islamic empires repeatedly targeted non-Muslim people for extortion, vandalism and even murder, on the assumption that other cultures had no right to exist. Likewise, many modern Israelis have taken their Palestinian neighbours as enemies of God. Miriam Levinger, a leading Israeli settler in Hebron (on the West Bank), told visitors in the 1980s:

> Who are the Palestinians? They have no history books, they have nothing like the Bible, they have no heritage. Who was their last Prime Minister? Who was their last king? They have no credentials. (in Smith 1988: 63)

In other lands such as North America, it would be normal for Jews, Christians and Muslims to live together on the same streets as legally equal citizens. But in much of the Middle East, such sharing of an environment would be close to unthinkable. In its exile across the world, Judaism flowered into an international society of learning and spiritual exploration, vitally linked in dialogue with other cultures. But the Middle Eastern environment did not seem to produce a common society. On the West Bank of modern Palestine, the descendants of Pharisees and Samaritans lived scattered amid the dusty hills and olive groves, feuding over security or water, and often feeling they had nothing in common.

The Lord of Lords

If the communal order transcended nature in importance, and its laws were the standard by which God would judge the world, then surely he would send a messiah warning and calling all people to embrace his ways before the judgment day. So the Zoroastrian savior Saoshyant would herald the final confrontation between God's followers and the people of the lie. He would lead his people into the battle of Armageddon to destroy the forces of darkness, and the faithful would emerge victorious over their foes. Their savior would reign as a king of kings and lord of lords. His people would then bask in God's favor forever. This was the greatest glory any arid-country tribe could imagine for itself. And this myth, in many variations, percolated through all the major cultures of the Middle East.

The Christians claimed their slain leader was this expected messiah. But in the eyes of most Jews, neither Jesus nor his followers fit the part. The Christians were not noted for Jewish patriotism or for leadership in the independence struggle with Rome. As far as most people of Judah could see, no king of Israel had risen up to triumph over the nation's foes. Besides, many early

Christians showed a disturbing lack of concern for that previously primary object of religion, the communal order. Instead of urging a rededication to ancestral laws and the temple-state, Jesus showed a dismissive indifference toward both Jewish and Roman authorities. He and his followers seemed willing to bend tradition for a supposedly higher cause. Even the Romans, who were heirs to the Greek cosmopolitan dream, were often scandalized by Christian indifference toward the civic order. Both Orthodox Jews and Roman officials were right to point it out—the early Christians' main concerns lay elsewhere.

Like some sects of Buddhists or of Hindu *bhakti* followers, early Christians displayed concern, not for communal law and order, but for creative experiment and personal fulfillment. What if people lived as if ordinary compassion mattered more than obedience to authority or tradition? Jesus seemed to invite everyone to try it. To a group of specialists in the holy law, he reportedly said, "I tell you there is something greater than the temple here. If you had known what that text means, 'I require mercy, not sacrifice,' you would not have condemned the innocent" (Matthew 12:6–7) .

According to some texts, Jesus spoke of the kingdom of God as a state of inner awareness. In the gospel of Thomas he says, "The Kingdom of God is spread upon the earth and men do not see it" (99:16–18) The gospel of Philip urges:

> Let each one of us dig down after that root of evil which is within one, and let one pluck it out of one's own heart from the root. It will be plucked out if we recognize it. But if we are ignorant of it, it takes root in us ... it is powerful because we do not recognize it. (83:18–29)

Such concern was akin to introspective psychology. It would seem to encourage self-exploration more than group conformity. Where a traditional cult of the tribe foresaw "salvation" through a national hero-king, Jesus was presented as a village healer of souls. Such religion evidently had popular appeal. As when the Sufi brotherhoods bloomed across of the Islamic world, Middle Eastern society was showing a hunger for neglected personal dimensions of life.

Still, the Christian movement was divided over its own departure from Jewish tradition. On the one hand, Paul of Tarsus claimed his own Jewish people were the true retainers of God's rules; on the other, he proclaimed that the Christian message applied to all souls, regardless of tribal identity. Perhaps in the Christian paradise according to Paul, people would arrive in all their ethnic diversity. Their souls would stand transcendent of all delimited identity, whether of culture or even gender.

Such a movement of soul seekers was worse than useless to most tribal leaders and empire builders. It met none of their dreams, in which the people would sacrifice themselves for their ruler's aspirations. Yet both Christians and royalist zealots seemed to speak in the same religious vocabulary. Both spoke of the Messiah, the kingdom of God and the new world order, but the meaning of these words depended on who was saying them. In the mouths of zealots, these words referred to the rise of ethnic kings, the defeat of tribal enemies, and ultimate victory for a chosen people. For some early Christians, the extinction of such ambitions would be a better gateway to God—"Everyone who glorifies himself will be humbled" (Matthew 23:12). But which meaning of these common religious symbols would prevail? Would the Christians transform the traditional cult of the tribe? Or would they gradually conform to the world into which they were born?

For over three hundred years the Christian movement was a multi-ethnic society, cut loose from the mold of one tribal-orthodoxy versus another. Each group of Christians seemed to pursue their own line of thought and experience. "Gnostic" Christians were fascinated with the world of the psyche. What, they pondered, were the inner roots of violence and greed? Could those compulsions be understood and mastered through insight? Could one person's spiritual needs be different from those of another? Less intellectual Christians asked how they might live together in a more compassionate way. But such questions could imply disrespect for rank and privilege. Valentinus of Alexandria argued around 140 CE that the true church was that part of humanity which recognizes its divine origin. Society, he said, creates its own forms of religion, and the church is not an institution set down from on high, but a human means of self-discovery (Pagels 1989: 122–23).

However, for many church leaders, such inwardness and questioning of society were deeply disturbing. To "church fathers" like Tertullian (c. 160 to c. 225 CE, from Carthage) it seemed that the Gnostic Christians were guilty of a heinous sin: they took their own desires or values as more important than the consensus of God's community. Was it not dangerous to dabble in the dark forces of personal desire? Surely it was the duty of religious leaders to determine the standards of behaviour for society and enforce them.

Already, as the "official" gospels were set down, many leading Christians grew concerned over the conditions of membership in their community. Membership and conformity to group norms became requirements for entrance to the kingdom of God. The image of Jesus began shifting "from that of a … teacher who espoused a way of life that any might try, to that of an authoritative model, judge, and keeper of accounts" (Mack 1993: 137). In the hands of the emerging church leadership, Jesus began to resemble a communal patriarch. According to many Middle Eastern bishops, Jesus would hold his flock accountable to every letter of Jewish law. Anyone who deviated from tradition would be damned. Jesus would return in a second coming, and this time he would meet all the old expectations. According to the second-century writers of Thessalonians, Jesus would return as a destroyer: "The Lord Jesus is revealed from heaven with his mighty angels in flaming fire, inflicting vengeance … on those who do not obey the gospel of our Lord Jesus. These will suffer the punishment of eternal damnation" (2 Thessalonians 1:7–9). Christ had come into the world to teach "obedience unto death" (Mack 1995: 124). The revised message of Jesus seemed to be: "You must obey the correct authorities, or else you will be punished."

John Chrysostham (347–407 CE) was one of the last bishops of early Christianity to actually argue for an anti-authoritarian church: "We do not have 'authority' over your faith, beloved, nor do we command these things as your lords and masters … the counselor speaks his own opinions, not forcing his listener, but leaving him to his choice. (Mack 1993: 137). But many church leaders felt that Chrysostham was unqualified for spiritual authority as they knew it. He was chosen bishop of Constantinople largely because of his popularity among the laity. But in the Byzantine corridors of power, Chrysostham was a misfit from the primitive church. He was dismissed from office.

While church doctrine remained fluid, views like those of Valentinus or Chrysostham could be aired and discussed in public. But early church councils, such as that at Nicaea in 325 CE, operated on an assumption that their task was to bring dialogue to an end. At these state-sponsored councils, appointed regional bishops met like clan heads to decide the meaning of Christ's life. By the end of the council, they were to reach final and binding decisions concerning the official books, creeds and doctrines of the faith. After that, further debate would be heresy against God.

The intent of the councils was to unite all Christians in a single set of beliefs and customs. But this effort to impose uniformity was just what split the church into basically ethnic blocks. Where an international movement had flourished at first, now the national churches of Egypt, Syria, Armenia and Ethiopia, the Nestorians of Persia, the Donatists of North Africa, the Roman Catholics and the Greek Orthodox all broke from each other. From a Eurocentric point of view, Rome excommunicated all these other ethnic churches from its true community. Actually, the various churches excommunicated each other, taking each other's cultural differences as deviations from God's customs. The word "excommunicate" expresses perfectly how the doors between the ethnic churches were shut.

At the time of Christ, three major kinds of religion flourished in the Middle East. First were the civic religions of family and communal virtues practiced by each ethnic community. Second were the diverse mystery cults or ancient nature religions. Third, and last in popularity, were the philosophical schools, mainly from classical Greece (Norris and Walker 1985: 1–7). At first, Christianity was a mystery cult of the Madonna, her son and the renewal of life through death. The early Christians also engaged the classical schools of philosophy in vigorous debate. But the first kind of religion soon prevailed. Mainstream Christianity reverted to a Levantine-style communal faith. The church patriarchs formally banned mystery cults and philosophical schools. Orthodox Christianity was largely reduced to a set of correct social mores to be obeyed under threat of punishment.

Of course, the mystery cults and philosophical schools were never fully stamped out. But for the next thousand years they flowed as underground streams. Meanwhile, the mystics and "free thinkers" of Christendom lived by a compromise like that of the Sufi saint Al-Ghazali (1058–1111 CE). Al-Ghazali proposed to avoid all conflict between the social order and the private life of self-realization, saying that the self-explorer "makes the Law his outer garment and the mystic Way his inner garment" (in Campbell 1976b: 451).

The Tribe of Tribes

Before Muhammad, Arabia was a semi-continent of tribes at war, divided between the hostile superpowers of Byzantine Rome and Sassanian Persia. These competing empires bought, lured or forced the leading Arabian families into alliances against each other. This imperial rivalry was clothed as a religious confrontation of Christianity versus Zoroastrianism. The troops in this holy war generally fought without mercy for the civilian population, and often on Arabian soil. During the fourth and sixth centuries CE, the Persians invaded Arabia in force, destroying rebellious villages down the Gulf coast to Oman. On search and destroy missions, they filled in wells, cut down date palms, tore up irrigation works and drove "disloyal" Arab farmers into the desert. Then the Persians set up the Lakhmid clan as their puppet rulers over the region. The Byzantine Romans countered by supporting the rival Quda'ah dynasty, and the Ghassan tribe Christians. The Romans also helped the Ethiopian Christians invade southern Arabia in the 520s, because King Dhu Nuwas of Himyar had made his country a Jewish kingdom and was said to persecute Christians.

Each superpower imposed heavy taxes on the Arabs, making them pay the expenses of their own occupation. As the foreign armies drove farmers from their oases into the desert, Arabia's "people of the palm" commonly became "people of the camel." From those wars until the nineteenth century, the nomadic way of life predominated in Arabia. Persia and Byzantium would soon have reason to regret the increased numbers of Arabian nomads.

The town of Mecca was a trade center partly because it had been a holy place for the old nature religions. The town's Quraysh clan overlords were custodians of a "haram," or holy site, where a truce prevailed between warring tribes. Trade in that place was secure, and Meccan merchants could operate as middlemen, indirectly linking tribes in conflict. But Mecca's importance as a trade center partly depended on the instability of surrounding regions. Most Meccan merchants wondered how they might profit from existing conditions of sporadic war. One merchant proposed to end those conditions, even if it undermined Mecca's advantage. Muhammad's revolt was a break with tribalism itself. He would subordinate the tribal chiefs to a greater loyalty and then turn them all against those who used the Arab people as pawns in their conflicts. To the Christianity or Zoroastrianism of the foreign powers, Muhammad countered with a native revelation. He would unite the warring tribes into one greater tribe of faith. Then not only the valley of Mecca but perhaps the whole world would be a realm of peace.

The Islamic basis of unity between tribes was partly shared faith and partly common social custom. But this was an ingeniously delimited core of cultural unity. Instead of attempting to standardize whole galaxies of tribal traditions, just five pillars of faith of were initially established to bind the tribes: confession of faith in Allah, daily prayers, payment of the *zakat* tax for common welfare, the month of fasting at Ramadan, and the pilgrimage to Mecca. All upholders of these five articles would consider themselves as brothers and sisters.

Apart from accepting the five pillars and the Quran (which establishes only a few principles of social law), the early believers from diverse tribes continued to follow their own local customs. The Quran encouraged healthy respect for diversity: "For each We have appointed a [different] divine law and a traced-out way. Had Allah willed He could have made you one community.... Vie then with one another in doing good works" (Sura 5:48). So the believers were to respect each other's customs, confident that Allah had shaped each local culture for his own reasons.

Muhammad's "constitution of Medina" involved a pact much like a traditional *hilf*, or pact of tribal loyalty (Ruthven 1984: 73). But this was a pact for the tribe of tribes: loyalty to the *umma* (Islamic community) was to override every lesser loyalty. So Muhammad defended fellow Muslims against his own Quraishi clan relatives. He also held that members of the *umma* could not enslave each another. The Moslem *jihad* sometimes involved wholesale liberation of slaves. Islam therefore threatened the traditional tribal heads with a reduction of sovereignty, not to mention the loss of their slaves.

At first the *umma* tried to live in peace with other communities. Only tribal heads who sought to suppress Islam faced the wrath of the tribe of tribes. The Quran advised: "Fight in the way of Allah against those who fight against you, but begin not hostilities. Lo, Allah loveth not aggressors" (2:190). As for other desert religions, it is said Muhammad once told Umar, "He who oppresses a Christian shall have him for an accuser on the Day of Judgment," and "He who injures a Christian, injures me" (in Phillips 1966: 26). The points of difference between Zoroastrians, Christians and Jews had been cause for holy war, with Arabia as the battlefield. Now Muhammad held that all these "people of the book" (people also desiring one universal social covenant) should be affirmed and encompassed within the crowning unity of Islam.

In its youth, the tribe of tribes was a loose, creative alliance. For well over a hundred years the Islamic empire was perhaps the most culturally diverse, intellectually open society in the world. By accepting Islam, Boyce claims, a Zoroastrian could be freed from a burden of priestly obligations accumulated over centuries: "for the thinking man there was the further attraction that as a new religion, Islam had yet to create its own hard shell of scholastic dogmatism, and so laid few constraints on independent thought" (1979: 148–49).

True, Islam's early front-line troops were often book burners, but they were mainly illiterates, probably trying to destroy any magic of the powers against them. They were soldiers and nomads of the ancient world, not like the highly literate dogmatists Islam would later know.

After Arabia was united, it stunned the world by overthrowing the powers of Persia and Byzantium. These empires would have held Arabia in subjection, so revolt against them was a war of liberation. The early caliphs used the desert nomads as Islam's shock troops. As Caliph Ma'mun said, "I will seek out the [Bedouin] Arabs. I will bring them from their deserts and settle them in all the towns that I will conquer, until I attack Constantinople" (Bat Ye'or 1996: 131). But Arabia's Bedouin warriors were both Muslims and nomadic raiders, traditionally rapacious toward settled communities. When these soldiers seized the wealth of Persia or Egypt, their interests in religion and plunder were combined.

After each battle the raiders commonly wished to divide the booty of land, goods and slaves among themselves. Caliph Umar demanded the troops leave at least a fifth of the plunder to the whole Islamic community. More than that, he urged that the conquered people would be better left in peace than plundered, because then they would be prosperous, loyal taxpayers. What would the soldiers do, Umar asked, if after confiscating the productivity of villages across the Middle East, the villagers could pay nothing more for the upkeep of the conqueror's families? What would happen when "God will harden the hearts of the tributaries, who will refuse to pay the tax on their properties" (Bat Ye'or 1996: 125). The interests of Islam and its nomadic warriors could conflict deeply. In many localities the soldiers did not curb their greed, and the empire's revenues suffered accordingly.

The first face of Islam in many regions was that of the Bedouin troops. These nomadic warriors displayed the ancient virtues of manliness, fierceness, pride and courage more than the Islamic virtues of humility, kindness and brotherhood. Yet these fighters often proclaimed themselves the greatest Muslims, and those they conquered often believed them. As Ibn Khaldun (1332–1406) explained, "The vanquished always want to imitate the victor in his distinctive mark, his dress, his occupation, and his other conditions and customs" (in Patai 1983: 20). So in the great wave of conquest, Islam's message and the ways of its bearers were often confused. Perhaps inevitably, Islamization was confused with Arabization, and Arabization was confused with Bedouinization. How could the conquered people distinguish these things? How could the conquerors themselves?

At first the Islamic community made a clear distinction between the Quran as Allah's revealed word and Muhammad as its human receiver. But the Quran leaves many practical questions unanswered, making only a handful of rulings on social issues, mainly concerning fairness in marriage, divorce and inheritance. As this is an insufficient base for an entire body of public and international law, a demand arose for further divine guidance. Several decades after Muhammad died, his personal acts and words were taken down from second, third or seventh-hand reports and collated into the volumes of the Sunna, or the sayings and deeds of the Prophet. With that, the daily life of Muhammad the man became a secondary source of revelation. The jurists began to presume that all Muslims, in addition to observing the five pillars of the faith, should conduct all their practical affairs in conformity to Muhammad's personal example. As Karen Armstrong explains,

> Because Muhammad was believed to have surrendered perfectly to God, Muslims were to imitate him in their daily lives. Thus by imitating the way Muhammad spoke, loved, ate, washed and worshipped, [they might] ... live a life that was open to the divine. (1993: 160)

In that case, the local customs which Muhammad the man practiced became more than the incidental trappings of the message bearer. The particular local ways of this seventh-century Arabian messenger were increasingly taken as the message itself.

Imam al-Shafi'i (767–820 CE) was the first scholar to elevate the sayings of Muhammad to scriptural status. This greatly increased the body of religious material out of which a more complete legal system might be formed. Yet even the authentic sayings of the Prophet left many questions unanswered. For example, Muhammad made no rulings about contraception or abortion. Concerning government and succession to power, there is little consensus on what he said, if anything. Possibly he did express convictions on these issues, and these were later purged from the record. At any rate, the Quran and Sunna did not offer complete guidance in all things, as the tribal cultures of "the time of ignorance" had done. Possibly the omissions were deliberate—leaving room for local diversity and future creativity. But most clerics saw the holes as gaps to be filled—by omnipotent, universal rulings made to last for all time.

Where every act was to be justified by appeal to the Prophet's authority, the trade in quotes alleged to Muhammad expanded with the empire. No believer could argue with an act endorsed by the Prophet; nor could they easily find out if Muhammad had really endorsed it. Did Muhammad actually say that women are mentally unfit to inherit wealth, or that those who entrust their affairs to a woman will never know prosperity? Was it Muhammad who said that slaves who converted only after their capture were to remain as slaves, and that their Muslim children must also be slaves? Obviously, the opportunities for abusing the sacred were many. Muhammad Abu Zahra entitled a chapter in his history book, "The Increase in Lying Concerning the Prophet and the Schisms and Divisions in the Ranks of the Fuqaha." Mernissi explains the trend as if it were a law of sociology: "The greater the number of people affected and excluded, the more violent were the dissentions within the governing elites, and the more pressing the need to manipulate the sacred" (1991: 46).

The jurists who codified Islam into thousands of rulings on the conduct of everyday life started out trying to draw the line on a flood of rumor. Perhaps the greatest of these was Muhammad ibn Ismail al-Bukhari, commonly known as

Al-Bukhari—the man from Bukhara. Al-Bukhari (810–870 CE) was a believer, in spite of knowing Arab imperialism at its worst. By the time the *jihad* reached his oasis in Uzbekistan, Islam already involved a full-blown cult of Arab ethnic supremacy. In his history of Bukhara, Narshakhi, describes the forced Arabization:

> After the [Muslim] conquest of Bukhara, it is recorded, the Arab commander, Qutaiba, thrice converted its citizens to Islam, but they (repeatedly) apostized and became infidels. The fourth time he made war and seized the city and established Islam there after much difficulty. He instilled Islam in their hearts and made (their religion) difficult for them in every way.... Qutaiba thought it proper to order the people of Bukhara to give one-half of their homes to the Arabs so that the Arabs might be with them and informed of their sentiments.... He built mosques and eradicated traces of unbelief and the precepts of the fire worshippers. (in Boyce 1979: 147)

This was the common lot of conquered communities. The Arabian troops often commandeered half the houses and properties for themselves. They often made war on a community repeatedly if the locals "apostized," usually by failing to pay taxes in full. In Bukhara the Arab commanders demanded that all local Jews must identify themselves by wearing black clothing only. Jewish shop owners had to work on their knees (or stand below ground level), so their heads would be lower than those of Muslim customers. The Quran, of course, contained no such rules (Bat Ye'or 1996: 93, 99).

As the Prophet lived in an age without mass media, the memory of his words and deeds seemed to flow in family lines, like an inheritance from parent to child. Those born into families close to the companions of the Prophet could claim the most direct and authoritative inheritance of memory, and these families were from western Arabia. The Arab conquerors of Bukhara, Sind or Morocco could presume themselves better guided in all things than the newly converted. If these invaders justified their every action by a quotation attributed to the Prophet, on what basis could the conquered non-Arabians protest?

Al-Bukhari undertook to draw the line on lying concerning the Prophet. To do it, he catalogued every quotation then in circulation, finding, it is said, about 600,000. Then he set out to distinguish the counterfeit spiritual currency. His method was to trace the quotations to their sources, documenting just *who* quoted *whom*, back to the Prophet. Then he researched the lives of the transmitters. If Al-Bukhari found record that a self-proclaimed authority had lied on other occasions (for example, bringing a charge of adultery against a woman without evidence) then every saying of the Prophet reported by that authority was held doubtful. If it was said that "A" heard a saying from "B", but Al-Bukhari found that "B" died before "A" was born, then that saying too was tossed on the rubbish heap. In this way, Al-Bukhari, the human computer, sifted his 600,000 quotations, ruling only about 8,200 (omitting repetitions) to be authentic. It was a lifetime labor of love and a service to the faith. It was also perhaps the greatest intellectual feat of spiritual self-defense in recorded history.

The codifiers of Islamic law such as Al-Bukhari, Imam al-Shafi'i, Malik ibn Anas or Muslim ibn al-Hajjaj led a kind of peaceful cultural revolt against their political rulers. In effect, these jurists took control of the faith from the hands of politicians and generals. After the major tomes of the Sunna were set down, the sultans, amirs or caliphs could no longer manipulate the sacred with impunity. Now the clerics proposed to guard Islam, much as rabbis guarded the law in Judaism. After all, the "Islamic" rulers were heirs to the grasping warlords of the arid lands whom Muhammad had revolted against. Al-Bukhari himself was so contemptuous of such rulers that he refused an invitation to read his works before the amir of Bukhara. "Go tell your master," he replied, "that I hold knowledge in high esteem, and I refuse to drag it into the antechambers of sultans" (in Mernissi 1991: 45).

Al-Bukhari helped block control of the faith by desert warlords. But in doing so, he conceded one central point to the Arab imperialists. He refused to be bound by any Arab tradition unless it was convincingly attributed to the Prophet. But by this same argument, he bound himself to accept all Arabian customs the Prophet *did* practice. So the door was opened for a new cleric-led Arabization of the tribe of tribes. As Muhammad the man was an Arab, the imitation of his personal ways could not help but involve a wholesale canonization of Arabian seventh-century culture. Already in the late 600s CE, Caliph Abd al-Malik (who ruled 685–705 CE) required that Arabic be the only language of administration for all peoples of the Islamic empire. For this caliph, the spiritual growth of Muslims and their assimilation to Arabian culture were one divine process. As Amir Feisal, the King of Syria, would put it (in 1919), "We are Arabs before being Muslims, and Muhammad is an Arab before being a Prophet" (Mansfield 1991: 228). Al-Bukhari was set to lose the next round of his fight.

As questions unanswered in the Quran could be resolved by the example of Muhammad the man, so questions unanswered by a saying of the Prophet could be resolved by referring to Arabian tradition. And as Arabia in its heroic age was largely a nomadic warrior society, the Arab Bedouin were commonly seen as the purest representatives of the Islamic way. Of course, Muhammad himself was not a Bedouin, and he stood squarely opposed to many desert tribal values. Against the Bedouin virtue of clan loyalty, Muhammad claimed loyalty to an international brotherhood. Against the old law of blood revenge, Muhammad urged forgiveness. Instead of fierce pride and combative manliness (*muruwwa*), the Prophet honoured those who were neither proud nor boastful (Patai 1983: 99–100). But with the passage of time, even these Bedouin values could become quintessentially "Islamic." Muslim jurists often decided fine points of Islamic law by referring to Bedouin tradition. The same went for Arabic philologists who ruled on correct grammar in accordance with "pure" desert Bedouin usage, rather than the more cosmopolitan Arabic of the towns (Patai 1983: 74). So in many ways, Islamic orthodoxy came to favor Arabian desert traditions over those of any other culture (Patai 1983: 140). The Bedouin, after all, were Islam's greatest heroes. They were the ones who had conquered the entire Afro-Asian desert and beyond.

So, from the Quran, the sayings of the Prophet and the traditions of Arabia the clerics fashioned complex codes of Islamic law. They set out to make Islam "more" than a religion, faith or spiritual practice. They would make it "a complete way of life," with guidance for every situation. Furthermore, they hoped to impose this massive body of law on all Muslims, be they Persians, Africans, Indians or Spaniards. By the 900s CE, the main work of the legal jurists was nearly done. Almost every conceivable human action was assigned to one of the following moral categories:

1. the commanded (such as the five pillars of the faith),
2. the recommended,
3. the legally indifferent (such as drinking coffee or smoking hashish),
4. the reprobated and
5. that which is forbidden by God himself.

Melise Ruthven (1984: 163) says the accumulation of legalistic rights and wrongs grew so detailed that observance could become all-consuming. Like Jewish doctors of the law, the Islamic mullahs built up a mountain of requirements so particularistic as to raise a huge barrier against the very universalism Muhammad sought.

The Modern Theocratic State

In 1908, the "Young Turk" leaders of the Ottoman empire tried to revive Islam in the face of Western aggression. They hoped to foster pan-Islamic feeling, inspire loyalty to the caliph in Istanbul and unite all Muslims behind the Ottoman state. This effort to build Islamic unity, however, assumed Turkish rather than Arabian leadership. The Young Turks upheld both Islam as a basis for religious fellowship, and Turkey as the basis for political unity. The difference between these goals grew apparent with a decree that all schools and government offices in the empire must operate in Turkish. The Muslims of Arabia now experienced the same cultural imperialism their ancestors had forced on others. Once more, Middle Eastern rulers tried to make religion a banner for military and ethnic supremacy. The dream of a tribe of tribes was taken to involve one ethnic nation ruling and imposing its culture on the rest.

For General Zia-ul-Haq, the ethnic diversity of modern Pakistan was a problem he hoped to solve by establishing a "pure" Islamic code for the nation. As an initial step, he authorized a Commission for Islamic Ideology to advise the government on its path of return to the true law. Obviously the "Islamization" of Pakistan would involve finding ways to uphold the five pillars of the faith, including the *zakat* tax. Zia's government acted on a commission proposal to establish a particular form of *zakat* tax, which would be a mandatory tax collected and administered by the government. But when this religious tax bill was passed, leaders of the large Shi'ite minority objected that, in their tradition, *zakat* was a voluntary tithe and not something imposed, collected and disposed of by people outside their religious community. In the resulting controversy, many Sunni Muslims saw the Shi'ites as blocking the path to unity under God's law. The Shi'ites were finally granted their exemption from the official tax, but only after a round of bloody riots reopened old communal wounds (Faruki 1987: 74).

The Commission for Islamic Ideology made a series of other recommendations. One proposal was to treat murder as a private offense to be settled between the families concerned, as it was in seventh-century Arabia. Another proposal concerned restoring the death penalty for blasphemy against the Prophet. When the elected Pakistani legislature balked at implementing these suggestions, many fundamentalists demanded a constitutional amendment, to give a committee of clerics veto power over all laws passed by the people's assembly (Faruki 1987: 67–68).

In Iran, the traditionalist clerics did assume veto power over the legislature. The November 1979 constitution established the principle of "rule by the jurisconsultant." In that case, defending one standard of culture from ancient Arabia was enshrined as the most important function of the state. The ruling jurisconsultants proposed greater conformity to ancient Arabia's social code as the main solution for all the nation's political, ecological or economic issues. But that ancient code originally concerned a tribal, face-to-face society that was basically a patchwork of personal relations rather than an institutional tangle of public and private corporations. Now the clerics insisted that the problems of a largely corporate society should to be solved in the old way, by conventionalizing all personal relationships between young and old, male and female. The committees for public virtue set forth into the streets, initially filled with zeal to correct the behaviour of their fellow citizens, especially the females. Meanwhile, other issues were often neglected. As Shahrough Akhavi observed in the early 1980s, "The danger for the regime is that it has no coherent industrial and investment policy—not to mention an agricultural one—so that the temptation is to rely on oil revenues to import necessities to service the needs of the people" (1987: 44).

In practice, the establishment of Islamic law was every bit as complex as an establishment of "Christian law" would be in Europe or the United States. Whose Christian law? Which sect would be recognized by the state as possessing the one true code of life? The Anglicans? The Mormons? But the formation of ethno-theocratic states was widely accepted as normal in the arid-belt countries. There, to speak of faith, blood and the state in one sentence was not fascism but simply orthodoxy, whether in Israel, Morocco or Iran.

Eternalization of Desert Tradition

In desert cultures such as those of the ancient Hebrews or Arabs, the words for "holy" and "sacred" took on special meanings. These words meant not only that which inspires awe and reverence but also that which is absolute, eternal and unchanging. Each tribe's core of sacred traditions evolved over time, as the Old Testament records the development of Jewish ritual and law. But those who took their customs with ultimate seriousness often claimed them eternal and unchangeable—either from the beginning or from a certain point forward. For traditional Mongolians, this happened in their moment of greatest glory under Genghis Khan. The great khan left a codified set of customs known as the Yassak; later khans could give new laws so long as they did not contradict it. The decrees of later khans could be repealed, but the Yassak could never be altered (Lessner 1955: 130).

For the Arabs, the heroic age was the early Islamic empire. At that time, Islamic civilization was rapidly evolving and as plastic as an erupting volcano. But as the empire started to cool and subdivide, its customs slowly hardened into stone. From that time, basically from the 800s, it was increasingly accepted that, in Nabih Amin Faris's words, "Piety and virtue lie in obedience and conformity, while nothing is more repugnant than change and innovation" (in Patai 1983: 280).

Perhaps desert environments offer natural tendencies toward changeless culture. The desert seems to change far less than a more obviously living landscape. When change comes, it is commonly bad news—involving starvation, decay or death, rather than growth and progress. The real deserts of Afro-Asia did not offer their ancient peoples hope that life springs eternal, or that change is a renewal of life. Desert people commonly took a more pessimistic view, that all which changes dies.

When people in the green countries wished to identify with something undying, they could relate to the ever-returning cycles of nature. They might say that the soul sheds its coil like a snake to be born again, or that the earth's children return to their mother's womb in the soil and come again like the return of spring. But the true deserts had no season of splendourous earthly renewal. The face of the earth there resembled a corpse or a wasted old woman in all seasons. In this wasteland, where might people look to see life transcending death? Without Mother Nature, two things remained which seemed to never die: the eternal order of the heavens above, and the tribe's code of sacred traditions as handed down to future generations. The true code, according to most desert cultures, had been received from heaven. If mere earthly mortals had composed it, it would not be of divine origin. If mortals ever changed it, it would not be eternal. In that case, to change religious tradition is a heresy. So, an idea grew prevalent in the Middle East that the holy is recognized, not by its capacity for evolution or renewal, but by its eternal fixedness. By that standard, a rock of ages could seem more godly than a flower, a flowing stream or a mother with child.

Where the social code was "eternalized," child rearing became a matter of teaching conformity and obedience, rather than encouraging children to explore their own capabilities. Observing traditional schools in the Middle East, Raphael Patai (1983: 37) felt that education was an experience of constant correction rather than recognition for risks and achievements. The standards of correct behaviour were beyond questioning by the students. Conformity, it seemed, was next to godliness.

Heresy was seldom a capital offense in religions from outside the Afro-Asian desert. Hindu yogis, Yoruba healers, Native American shamans or Thai villagers (part-Buddhist, part-Confucian and part-Hindu) generally had no great religious orthodoxy to defend unto death. In pre-Christian Europe also, F.M. Cornford says,

> Greek theology was not formulated by priests nor even by prophets, but by artists, poets and philosophers.... There was no priestly class guarding from innovating influence a sacred tradition enshrined in a sacred book. There were no divines who could successfully claim to dictate the terms of belief from an inexpugnable fortress of authority. (1923: xv–xvi)

For artists, sages or villagers of the green lands, the sacred was living and fluid—not frozen in unalterable form. Those who composed new versions of classic legends were not excommunicated; they spoke as human beings about human experience, not as mouthpieces for an omnipotent god. But where the community and its orthodoxy seemed the most important thing in the world, heretics were commonly executed. Otherwise, disloyalty to the social contract could spread. It was treason to ridicule an eternal tradition, and treason to change it; either way, the sacred status of the social contract would be threatened. Thus the Sudanese government condemned its opposition critic Mahmoud Taha to death in 1985 for saying that elements of the Shari'a law were backward. In this "blasphemy," Taha seemed to both ridicule the eternal laws and imply they should be changed. When Taha was executed, the presiding judge explained that, "The Prophet said that those who change religion would be punished, and those who come with false ideas should be killed, and killed, and killed" (Wright 1985: 205).

In the modern Middle East, popular aversion to change often comes from hard experience. Where rulers and politicians manipulate the rules to receive the lion's share of scarce resources, the common people often try to hold both their rulers and themselves accountable to a God-given moral code. Many people would sooner place their trust in an impartial religious law than in the "process" whereby modern politicians and lawyers tamper with society's rules. So, Mawlavi Yonus Khalis convincingly claimed that Islamic law would restore justice to modern Afghanistan:

> Acquisition of [state] power by the force of capital or by the sword is against Islam. Islam is opposed to the practice of giving jobs to the unfit people on the basis of kinship, connections or corruption.... The faithful should not allow that the prevalence of such practices bring about the destruction of Islamic society. (in Ghani 1987: 91)

The eternal traditions of poor people and of rulers are often vastly different. Throughout the modern Middle East, people gather in their places of worship, earnestly discussing how to apply their values. Their holy books offer inspiration, showing how their ancestors dealt with scarcity, tyranny, communal hatred, and massive inequality. The same issues remain today and may be growing more complex with the rise of "global" Western culture, mass unemployment and the exhaustion of natural resources such as groundwater. Many "people of the book" feel they face a moral crisis requiring drastic action. Like local people everywhere, they need to articulate and defend popular morality against vested interests. Only they can decide how to uphold the values that matter most to them. But the "fundamentalist" logic of defending an eternal tradition against all external influence tends to lead this discussion toward an indiscriminate response. The complexity of issues may then be reduced to one basic choice between good and evil. For those who accept this context, upholding the good may seem to require a simple rejection of all persons and practices which depart from the true community's eternal standards. In the desiccated lands, that indiscriminate defense of ethnic tradition has probably passed for God's will more commonly than anywhere else in the world.

The Day of Judgment

In the starkest versions of fundamentalism, only one tribal tradition among all those of the world is true. The upholders of that tradition must either suppress all others or isolate themselves to remain pure. Where such beliefs influence state policy, concern for cultural purity demands a return to an ethno-theocratic state. So in modern Iraq, Iran, Israel and the Sudan, those who do not belong to the true and dominant group have been commonly treated as ethnic pollutants. In the 1980s, the Kurds, Zoroastrians, Baha'is, Jews and Sunni Muslims of Iran came under varying degrees of persecution, though their communities were still represented in the legislature. Apparently, the task of managing the national environment in the best interests of all was not seen as a sufficient basis for social unity. Instead, unity must rest on a broad cultural homogeneity. The zealots for cultural purity believed that their nation would be judged by God according to its degree of conformity to a single cultural standard. And the judgment day could arrive at any time. Then, in the final battle of Armageddon, God would glorify those who kept his ways and damn the rest.

Every major, desert-born religion across the Middle East adopted belief in an apocalyptic day of judgment and described it in similar terms. This was not a day of universal mercy, when the unity of all life would be revealed (as if it is not already revealed). This was a day for settling accounts with a vengeance. All enemies of the true tribe will be condemned on account of their false ways. Even members of God's chosen people would be damned if they departed from the customs of their fathers. As the Christian book of Revelation foresaw it,

> And out of His mouth goeth a sharp sword, that with it He should smite the nations: and He shall rule them with a rod of iron: and he treadeth the winepress of fierceness and the wrath of Almighty God. And he hath on his vesture and on his thigh a name written, KING OF KINGS AND LORD OF LORDS. (19:15–16)

The day of judgment is often portrayed as the ultimate battle. God and his prophets will appear as omnipotent warriors, smiting their enemies in vengeance for all the wrongs of Middle Eastern history. It is hardly surprising that many believers have taken this image to mean that the struggles of history must be solved in a final battle of ethnic cleansing. So the "Pious Ones" of the ancient Maccabee revolt rose up in their hour of fury, turning on any fellow Jews who "saw nothing amiss in integrating their religion and culture with that of their pagan neighbours" (Allegro 1971: 81–82). The book of Maccabees records the battle in the streets as these zealots "joined their forces, and smote sinful men in their wrath, but the rest fled to the heathen for succor" (I Maccabees 2: 44).

The Common Earth in Other Lands

Over the vast environment of China, stretching between the eastern seas, the northwest deserts and the southwest mountains, the local languages of ancient times were as diverse as those of the Middle East. China's cultural and religious diversity was probably greater than that of the Middle East or Europe. Yet for over two thousand years this semi-continent was seen as a single realm of "Chinese civilization." Most likely, the greatest factors in that unity were not emperors, political parties or official religions, but the biological continuity of fruitful earth and farming villages. In Chinese classical art, that living landscape was painted in all its guises, like an outpouring of portraits of a beloved woman. If words were to accompany these paintings, the lines of master Zhangzai (Chang Tsai, 1020–1077 CE) might serve:

> Heaven is my father and earth is my mother, and even such a small being as I finds an intimate place in their midst. Therefore, that which fills the universe I regard as my nature. All people are my brothers and sisters, and all things are my companions. (in McLuhan 1994: 17).

It seems the more denuded lands of the Afro-Asian deserts had no such power to unite their peoples.

Chapter 7

Desertification and Culture in Ancient Egypt

SOUTH AND WEST OF THE SAHARA, VARIOUS AFRICAN CULTURES CLAIM roots in ancient Egypt. A line of Yoruban kings reportedly came to Nigeria from Egypt, and Wolof elders in Senegal make the same kind of claim. Some West Africans say their languages are related to ancient Egyptian. Science, of course, tends to dismiss all traditional "information" pending harder evidence. According to British scholar P.L. Shinnie, these myths have "no more truth ... than the eighteenth-century histories that sought to bring the British from Troy" (1971: 435). He dismisses arguments for the kinship of Yoruban and ancient Egyptian languages and asks why such wishful theories keep appearing. He suspects an emotional attachment "to a diffusionist theory which demands that Egypt must be the center of cultural dispersion" (1971: 437).

If the Saharan Desert is taken as a permanent feature in history, it seems nearly impossible that the widespread myths of cultural linkage between peoples north, south, east and west of the desert could have any basis in fact. But if certain cultures formed *in* the Saharan region *before* the modern desert and dispersed from there, then the linkages would be possible. Perhaps this is why Chiekh Anta Diop can follow certain Wolof and ancient Egyptian verbs through all conjugations, showing hardly a letter of difference. He states that, "These phonetic correspondences are not ascribable either to elementary affinity or to the general laws of the human mind, for they are regular correspondences on outstanding points extending through an entire system" (1989: 25).

Did the Wolof language emanate from the Egyptians and travel to the Atlantic coast, or the other way around? From a Green Saharan viewpoint, a certain Egypt-centric bias appears in Shinnie's admission that: "It can be seen that, here and there, there are strong resemblances to Egyptian objects and to Egyptian culture scattered throughout Africa." These objects include the small harps of southern Sudan and Uganda; the sandals, wooden headrests and ostrich-feather fans of West Africa; the whips, crooks and flails of kingly regalia in old Nigeria; and the Yoruban goat-god Shango, who is held to resemble the Egyptian god Amon (Shinnie 1971: 447–48). All these items or ideas could have been freely shared in the prehistoric Green Sahara, rather than carried thousands of miles over searing sand and rock more than a thousand years before the domesticated camel came to Africa.

The Saharan Diaspora

Just as linguists in Eurasia theorize a common "Indo-European" source for most languages in their part of the world, so linguists of Africa, such as Diop or J.H. Greenberg, see a common "Afroasiatic" source for five major language groups. Counter-clockwise around the Sahara, these are Berber (north and west of the desert), Chadian (to the south), Cushitic (to the southeast), Egyptian (which is dead or transformed into "Colloquial Arabic") and Arabic (Diop 1989: 25; Davidson 1989: 47). If all these languages have certain roots in common, the similarity would probably stem from a Green Saharan meeting ground. As harder evidence, Italian archaeologist Elena Garcea reports: "It is a fact that the Saharan sites with the earliest pottery and the first domestic fauna are earlier that those of the ... Nile Valley" (1993: 19–20).

But the recorded history of the Saharan zone starts only after 3000 BCE. It begins with the background of a growing desert in the heart of Africa, pushing wave after wave of migrants outward in all directions. So, according to pre-Islamic legend, an ancient Hausa queen led her people out of a deadly wilderness into their promised land of flowing water by the Niger River (French 1985: 53). And the pre-dynastic Egyptians were a jumbled assortment of tribes, many of them recent arrivals from the deserts.

The Emergence of Egypt from the Sea

In the 1970s, deep drilling operations in the Mediterranean discovered massive deposits of sea salt, beginning around 150 meters below the sea floor and extending downward for a mile. The deposits contain anhyrite crystals, which form when salt water evaporates at temperatures above 35° C. Such salts are found on the banks of the Great Salt Lake in Utah and the Dead Sea of Palestine. According to one theory, these deposits formed when the continents of Africa and Eurasia moved into contact, temporarily closing off the Mediterranean at both ends. For perhaps a million years, probably between 6.5 to 5.5 million years ago, the Mediterranean was a dead sea. In time it evaporated, becoming a vast bed of crusted salt gleaming in the sun, like a vision of hell beside the Riviera. For such deep beds of salt to form, the basin must have received constant minor inflows of ocean water which rapidly evaporated. Over a million years, the salt accumulated to become over one and a half kilometers thick. Then, probably due to massive earthquakes, the western land barrier collapsed, and the Atlantic thundered in through the Straits of Gibraltar, submerging the land of salt (Attenborough 1987: 9–15).

This story is relevant to Egypt, because during the time of the dry Mediterranean, the Nile River poured its waters into a basin thousands of meters below the present sea level. The total vertical drop in the river from its source near the Ugandan Mountains of the Moon to the seabed of the dry Mediterranean was perhaps three kilometers greater than it is now. The gravity-induced velocity of the river must have been enormous. Like the Colorado River with its Grand Canyon, the Nile carved a basin downward through the rock. Soundings into the soil detect the rocky walls of that old canyon more than two kilometers beneath the streets of Cairo (Attenborough 1987: 12).

When the Mediterranean refilled, the sea must have intruded up the vast Nile canyon, perhaps all the way to Aswan. Only slowly would river sediments have refilled the valley. H.W. Fairman estimates that the lower Nile Valley became suitable for farming only after 8000 BCE, as solid land emerged from the bays and marshes (in Davidson 1989: 13–14). After studying under ancient Egyptian historians, Herodotus (c. 485–425 BCE) reported that at the founding of the Old Kingdom (c. 2685 BCE),

> the whole country, except the district around Thebes [later called Luxor, north of Aswan], was marsh; none of the land below Lake Moeris [the Fayum depression, southwest of Cairo], was then showing above water. To this place from the sea is [now] seven days passage up the river. (in Chandler 1989: 120)

This could explain why the civilization of Egypt seems to emerge quite suddenly, with settlements in the northern valley and delta appearing only after 4000 BCE (Gautier 1987: 175). If the early dynasties were built on relatively new ground, there would be little evidence of older cultures on the same sites. The predecessor cultures would lie mainly to the south. And this is what hurried excavations behind the Aswan dam uncovered in the 1960s. The series of thirty-three royal tombs unearthed near Qustul, deep in Egyptian Nubia, seem to date back to 3300 BCE (Van Sertima 1989: 86).

The Shrinking Oasis

Since the lower valley emerged from the sea, the story of Egypt concerns a gradually narrowing strip of watered land. As the Nile Valley dried into solid ground, the regions around it dried as well. The great valley slowly turned from a swamp surrounded by savannas into the world's longest oasis surrounded by deserts. Tributary rivers from the old Green Sahara vanished, leaving the Nile as a single artery of rain from tropical Africa. Still, for a few more thousand years, the oasis was wide and green enough to make Egypt the lushest, richest and most envied land in the Middle East.

The paintings on ancient Egyptian tomb walls depict fond memories of the best times in life. Most commonly, they show loving couples and the joys of hunting or fishing on the waterways. According to these pictures, the Nile was bordered by vast wetlands of tall reeds. The reeds and shallows were sanctuaries for millions of birds, which rose in flocks so dense that a hunter's sling could hardly miss. Fish were also abundant, and various records suggest that the ancients got much, or even most, of their food from sporting in the marshes. Here was a world of tranquil waterways, endlessly fascinating for children. Perhaps no words can express what the gradual loss of these wetlands has meant for rural Egypt's quality of life.

Egypt—The World's Longest Oasis

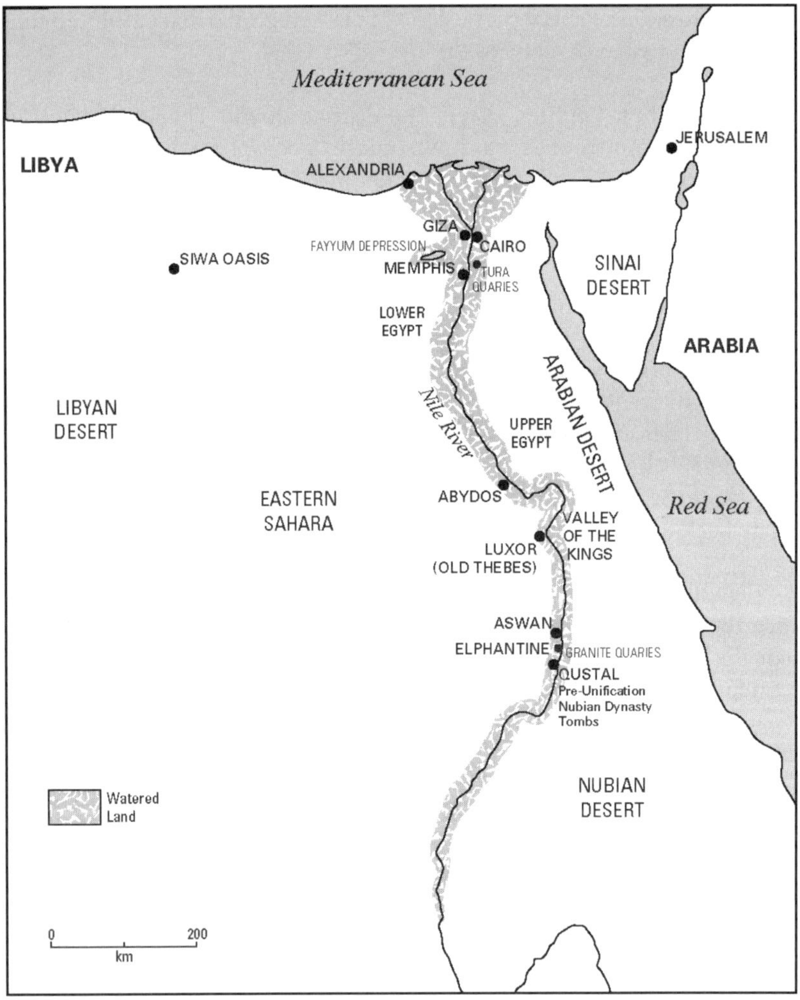

As the lands beyond the Nile basin turned to barren wastes, refugees trickled into the river valley. Many of the earliest migrants were displaced Saharan farmers and hunters from the Green Sahara, such as the Tasian and Badarian people (Davidson 1991: 15). As the population rose, the green valley tended to shrink toward the riverbanks. Over time the volume of water in the Nile diminished, as rains declined over the length of its watershed. Farmers began digging channels to extend the floods. Outlying villages such as Marimde, near the delta, or Tasa, off the right bank, were swallowed in the desert and remained extremely well preserved. Within the valley, dust and sand from the surrounding desert became a constant irritant. At some point, Egyptian women took to lining their eyes with a black oil-and-malachite paint, partly to catch the dust (Watterson 1991: 115). Airborne sand fell into the grain and was baked into every loaf of bread. Sand in their food ground down the Egyptians' teeth.

The desert was said to isolate and protect Egypt. But over the next five thousand years, invasions from the North African or West Asian deserts slowly grew more common. Some of the early myths imply conflict between valley farmers and outsiders. The valley people symbolized themselves as the forces of life and labeled their foes as forces of death and dryness. Plutarch wrote that the Egyptians "give the name of Osiris to the whole source and faculty of creative moisture, believing this to be the cause of generation and the substance of life-producing seed; and the name of Seth [the mythic murderer of Osiris] ... they give to all that is dry, fiery and arid, in general antagonistic to moisture" (in Metzner 1997: 265). Osiris was a green-skinned god of vegetation who was killed but rose again as the ever-returning force of life from death. He was the lover of Isis, goddess of the earth. As in the story of Cain and Abel, Osiris had a brother, and this brother was identified with the arid country, perhaps a pastoralist rather than a farmer. Seth was a jealous, violent man who murdered his brother and mutilated his body. Ralph Metzner (1997: 265) reads the myth as a reflection of early clashes between green-valley farmers and pastoral warriors from the desert.

Another legend says that the first ruler of all Egypt, Menes (c. 3100 BCE), was a leader of troops defending the northern valley against light-skinned invaders. This victorious general then consolidated military power over his own people, giving rise to the lofty but basically parasitical order of pharaohs. So the Egyptians countered force with force, sacrificing much of the peaceful life they hoped to protect.

Through the Old and Middle Kingdoms (2685–2180 and 2060–1785 BCE), Egypt was generally left in peace. Its towns and cities usually had no walls, and the standing army was small. As in ancient Crete, the old artwork shows almost nothing of war or military glory (O'Connell 1995: 140). But over the following centuries, invasions and migrations from the desert grew more frequent. Some, like the Hebrews, arrived as refugees from famine; others came as organized attackers. Before the end of the New Kingdom (1567–1085 BCE), Egypt faced major assaults from the Libyans, the Hyksos and the "People of the Sea." After the Libyan conquest in 935 BCE, invaders ruled the great oasis for almost all of the next 2,800 years, and Egypt's farmers served as tax mules for one foreign-based empire after another. In time, the Assyrians, Persians, Greeks, Romans, Arabs, Mamluks, Turks and modern Europeans would all come after the natural wealth of Egypt. After Arabian migrants overran the country, much of the culture and language of ancient Egypt was gradually lost to living memory. So Egypt slowly changed from one of the ancient world's greatest farming civilizations into a colonial oasis overrun by cultures from the surrounding desert.

Timetable of Egypt's Foreign Invaders

c. 4000–3000 BCE	Migrations into the Nile Valley from the drying Sahara
c. 3300–3100BCE	Pre-unification Nubian kings
c. 3100–2685 BCE	Egypt united, early dynasties
2685–2180 BCE	Old Kingdom
2180–2060 BCE	Period of disunity and chaos
2060–1785 BCE	Middle Kingdom
1670–1570 BCE	Invasion of the Hyksos from (probably) West Asia
1567–1085 BCE	New Kingdom
935–730 BCE	Libyan rule
671–664 BCE	Assyrian conquest
525–404, 341–333 BCE	Two periods of Persian rule
332–30 BCE	Macedonian Greek invasion and Greek Ptolemic dynasty
270–279 CE	Egypt briefly falls under control of the Palmyran empire (of Syria), then the Blemmyes of Sudan
30 BCE–641 CE	Roman and Byzantine control, Egypt officially Christianized
641 CE	Arab invasion, Egypt annexed into the Islamic empire
969–1171	Rule of Fatimid Shi'ites from North Africa
1169–1252	Saladin founds Ayyubid dynasty, Sunni Islam as official religion
1250–1517	Administration by Mamluk generals
1516–1882	Rule by appointees of the Ottoman empire

Women of Old Egypt

Of all cultural differences between ancient Egypt and the rest of the old Middle East, perhaps the status of women is most striking. Jacquetta Hawkes says the artwork of Egypt presents a "marvelously varied gallery of noble, companionable, graceful, merry, loving, and shamelessly seductive women" (1973: 399). All this seems to have met the approval and admiration of the officials who commissioned these works. In tomb paintings, husbands and wives are generally the same size, which was the artistic convention to show equality of importance (Watterson 1991: 62).

Until the Arab conquest in the 600s CE, Egyptian families, including the royal families, commonly traced descent through the female line. Children belonged to their mothers, and none were deemed illegitimate. Women normally owned their homes. The word for wife was *nebt-per*, meaning "ruler of the house" (French 1985: 55). The old Egyptian houses had "women's quarters," but these were places from which women could exclude men, not rooms in which men confined women; when women went out, they did so unveiled (Watterson 1991: 24–25).

Such status and freedom for women were not remarkable by ancient world standards. Only in contrast with less abundant lands of the Middle East do they seem unusual. However, the powers of Egyptian women slowly diminished over time. Under the pharaonic dynasties, women played minor roles in public life, and only four female pharaohs ruled in a span of 1,500 years. Barbara Watterson says that "no woman could aspire to be a king's scribe, an army general, a governor, or an ambassador to foreign lands." The tomb paintings show women weaving, housekeeping, hosting, weeding, harvesting, grinding grain and cooking. An eleventh dynasty tomb of Pharaoh Amun contained a wooden model of a weaving shop, showing women as workers and supervisors. The other professions open to women were midwifery, dance, music, mourning and the priesthood (Watterson 1991: 35–38, 53).

The female dancers, musicians and mourners were often related to temples, or cults of a goddess. In early Egypt these were professions for a relatively few women. Gradually, however, such arts became communal traditions rather than specialized vocations. According to Watterson, so many women were involved in religious performances during the New Kingdom that "it almost seems as though every female in the land, from the highest to lowest, took part in a cult" (1991: 41).

The temple priestesses may have been influential socially but seem to have held little institutional power. Their religious roles seldom involved controlling major properties or enforcing moral rules upon the public. Their main functions were to impersonate goddesses such as Isis, Nephys, Neith, Sekhmet or Hathor in temple rituals (Watterson 1991: 40). Their seasonal rites thanked and evoked the divinities of nature. In the month of Hathor, as the flood began, the priestesses led women in making clay phalluses and throwing them into the Nile as it "turned to blood." To these people, it was obvious that the goddess of the earth was menstruating and giving forth her red flood of fertility (Walker 1996: 752). The roles of Egyptian goddesses, and the priestesses who impersonated them, reflected popular reverence for the "female" powers of life, and this reverence was reflected to some extent in honour for ordinary women. An old papyrus written to praise Isis says, "You have made a power for the women equal to that of the men" (in Watterson 1991: 27).

From birth, girls in the Nile Valley had better chances of surviving than girls in the harsher lands surrounding Egypt. In a country where families could normally feed all their children, parents rarely killed "unaffordable" female babies. When a young woman married, it was generally enough for her and her lover to move in together. Giving a dowry was an optional custom, which suggests that brides did not consider themselves bought or owned. A married woman did not change her name, and both men and women identified themselves by their mother's names, as in "Ahmose, the son of (his mother) Abana." Married women kept property in their own names, and they could inherit, own or sell as freely as a man. Apparently, female peasants were deemed roughly as productive as male ones. Only starting with the Greek Ptolemaic dynasty (in the 200s BCE) were women made legal wards of male guardians. And only later still, after West Asian warlords had ruled Egypt for hundreds of years, were women commonly regarded as the unimportant and unproductive sex (Watterson 1991: 23, 30, 31, 59, 62, 122).

The Nature Observatory

The farmers of the Nile were keen observers of the seasons, the river and the cycles of stars. For example, an ivory tablet dated to the Old Kingdom (2685–2180 BCE) portrays the star Sirius in alignment with the rising sun, as the flood of the Nile begins on Egypt's new year's day. The inscription reads, "Sirius: Opener of the Year; Inundation One" (Lamy 1981: 6). It might be presumed that this tablet portrays an annual coincidence of stellar, solar and earthly cycles, easily observed and recorded by ancient villagers. But this alignment of the sun at dawn, the star Sirius, the Egyptian new year's day and the start of the flood season actually occurs only once every 1,460 years. Such a conjunction came in the year 2780 BCE.

When this ivory tablet refers to "the year," it suggests the "great year" of the old Egyptian calendar cycle, also called the "sothic year" of Sirius. This was the time it took Sirius to come full circle around in the sky and rise in alignment with the sun on new year's day. As the ancient calendar was calculated to 365 days with no leap year, it lost a quarter of a day per year. This seemed to cause a gradual movement of the point where Sirius rose each new year's morning. The star came full circle around the sky only after 1,460 leftover quarter days (from 1,460 years) had added up to an even year of 365 days (Finch 1989: 328). Cheikh Anta Diop (1989) suggests that this ivory tablet shows recognition of a recurring cycle. So, the Egyptians must have known about the preceding conjunction—otherwise how would they know it was a cycle? The preceding conjunction had occurred in 4240 BCE (Chandler 1989: 120). From such scraps of clay or bone come signs of a remarkable capacity to learn from nature, gather information over generations and fuse it together in leaps of understanding.

"Science" in Egypt involved a long accumulation of folk wisdom about nature. The Nile Valley was a land of unusually regular natural patterns. The residents needed to predict the cycles of their world, and in this environment they found that they could. They learned to observe nature's regularities and use them to advantage. In nature they saw, not so much the will of a great pharaoh in the sky, as a set of patterns by which the Earth seemed to live. And just as this planet had provided a nourishing cradle for life to evolve, so the richest environment in the Middle East provided a setting for the evolution of folk science.

Life was less predictable in the Sahara or South-west Asia, where sandstorms, droughts or attacks by rival tribes came and went in sudden and mysterious ways. Behind these turns of fate seemed to lie some omnipotent and unknowable will. In many old Hebrew folk tales, those who complain of suffering or injustice often receive a vision from God, but what the vision reveals is their own utter inadequacy to ever know or judge the Lord's plan. In the later Jewish, Christian and Muslim theologies of the Middle East, the almighty could overturn any natural law with a word. As Theophile Obenga says, deities such as Enki the Ordainer (of Mesopotamia), Yahweh and Allah were portrayed as unconstrained by any natural law, "independent of their creation, above and before all of their work" (in Toomer 1971: 292). Expressed as a mathematical theorem, this religious common sense would read, "A + B = B, IF God wills."

The ancient Egyptians wrote constantly of nature's "laws." Their term for "the law" was *Maat*, meaning "divine order." But Maat was an order of both nature and society, not just a sacred social code. For the Egyptians, both nature and society were holy, as can be seen by their "polygamous" worship of animals, goddesses, kings, fertile earth and primal water. The classical Greeks reported Egypt as a land of vast learning, including knowledge of medicine, botany, geometry and astronomy. The rich land and the knowledge went together, because a living environment tends to stimulate wonder and curiosity. But if the Egyptians learned so much, what exactly did they know?

Study of the Human Body

The human body is the part of nature closest to home, and studying it carefully might seem the normal thing to do. However, if people are unaccustomed to finding predictable patterns in the outside world, they often neglect to look for patterns in themselves as well. The ancient Egyptians not only looked, but also led the world in studies of anatomy, medicine and surgery. Many Greek "fathers" of medical science, including Galen, Pythagoras, Herophilus, Thales of Miletus and Alcmaeon of Croton, took their training in Egypt (Harris 1971: 112). Fortunately, we have some of the actual Egyptian medical texts and can compare their observations with those of modern science.

Among the surviving ancient medical texts is a treatise called the "Edwin Smith papyrus," which is a fifteen-foot scroll describing anatomy and treatment in apparently actual cases of injury to the head and neck. The complete text probably went on to describe the whole body (Finch 1989: 329). Dr. Charles Finch records some of the case titles as follows:

- Case #6 — A Gaping Wound in the Head with Compound Comminuted Fracture of the Skull, and Rupture of the Meningeal Membranes
- Case #7 — Compound Fracture of the Skull Displaying No Visible External Injury
- Case #22 — A Wound to the Temple Causing Total Aphasia [inability to speak due to brain damage]
- Case #31 — Dislocation of a Cervical Vertebra. (1989: 342–48)

Finch points out that cases 29 through 33 deal with various types of vertebral dislocations. And since modern physicians find it difficult to identify such injuries without X-rays, it is unclear how the ancients diagnosed and described these problems with such accuracy (1989: 349). Case 6 refers to a rupture of the meningeal membrane surrounding the brain, and this also is an observation not matched until recent centuries in the West. The descriptions of symptoms from brain injuries, as in case 22, suggest some understanding of which parts of the brain control certain bodily functions. In Western medicine, the area linked to speech was "discovered" by nineteenth-century French doctors and named "Broca's area."

The Edwin Smith papyrus is written by a physician with almost modern abilities to observe and interpret the body, but with primitive skills in treatment. Many cases are described as "an injury not to be treated." But at least the physician is sensitive to whether or not available treatments will do more harm than good. The papyrus which the text is written on dates from around 1200 BCE. But according to James Breasted, the writing style resembles the language of Old Kingdom inscriptions from around 2400 BCE (in Finch 1989: 329). The papyrus refers to yet older medical works called *The Book of What Pertains to the Embalmer* and *The Book of What Pertains to Wounds* (Finch 1989: 335).

Other surviving ancient texts show uses of incantations and spells in healing. It seems that no clear distinction was made between a physician and a priest. But even this contrasts favorably with medicine in the rest of the ancient Middle East. J.R. Harris (1971: 114) says that Egyptian healing "was never wholly dependent upon a belief in the equation of sin, possession, and disease, such as vitiated the medical methods of Babylonia and Assyria" or, we might add, the priestly medicine of ancient Israel.

Egyptian medicine was heavily concerned with women's health. The Kahun papyrus, for example, contains thirty-four prescriptions for problems of gynecology or pregnancy. Some of the ancient tests for pregnancy or infertility involved watching the effect of a woman's urine on certain plants. This, depending on plants used and interpretation of results, has potential as an accurate method. Some tests reportedly distinguished whether the man or woman of a childless couple was infertile (Harris 1971: 122–24). In more recent times, after Southwest Asian culture came to prevail over Egypt, infertility would generally be blamed on women only. Many men would rather divorce their wives for barrenness than submit to a test of their own sexual potency.

For well over five hundred years, generation after generation of Greek physicians came to study in Egyptian medical schools (Ghalioungui 1973: 31, in Finch 1989: 327). Herophilus, for example, is commonly credited with discovering diagnostic pulse-taking. But he learned his craft in Egypt, and this sort of pulse-taking is mentioned in both the Embers and Edwin Smith papyri (Finch 1989: 327). Hippocrates also drew heavily on Egyptian medical texts. Harris says a "typically Egyptian ... feeling is apparent in the corpus [of Hippocratic writing] and notably in the disinclination to attempt treatment of the incurable." The Hippocratic uses of drugs also seem to parallel uses in older Egyptian herbal texts (Harris 1971: 118).

Like other civilizations with rich environments, the range of Egypt's herbal lore was vast. Old Egyptian texts mention hundreds of herbs, but many have not been identified. Those we can identify comprise at least a third of all medical plants known today (Watterson 1991). And this plant lore, when conveyed to Europe through Greek writers such as Galen, Hippocrates and Pliny, became a major source for the European *Materia Medica* (Harris 1971: 115).

Of course, not all Egyptian medical theories have stood the test of time. But even though these ideas later proved to be wrong, major Greek authors tended to pass them on as received truth. Hippocrates faithfully records old African ideas that the child's bones come from its father, and the flesh from its mother; that the penis is an extension of the backbone, and the sperm descends from spinal fluid originating in the brain (Harris 1971: 124). Of course, Greek medicine grew ever more innovative as Egypt "declined" under centuries of foreign rule, but we can understand why many of the greatest Greek physicians looked to Egypt as the world's most ancient center of natural science.

Using the "Laws" of Space, Mass and Form

The claim that geometry was a "gift of the Nile" arose because farming in the Nile Valley required constant engineering of earthworks. The flood could not be left to flow wherever it might, or the silt to pile up anywhere it fell. The farmers needed to direct the ebb and flow of water over numerous irregular units of land. Surviving texts show the Egyptians' rather cumbersome ways of calculating the area of odd-shaped plots, and the volumes of water required to fill them. The discovered texts contain no theorem-like abstract formulas or mathematical proofs. But their way of calculating the area of a circle—"Subtract from the diameter its one-ninth part, and square the remainder. This is the area" (Rhind Mathematical Papyrus, in Palter 1996: 230)—gave a *pi* value of 3.16. G.J. Toomer points out that this was more accurate than many old Mesopotamian math tables. He claims, however, that the Egyptian accuracy "is not mathematically significant ... [because] it is the product, not of deductive theoretical reasoning, but merely of careful observation and measurement" (1971: 35–36). Such careful observation is characteristic of people who learn directly from nature. The Egyptians may have been proud of arriving at their answers through actual measurement rather than theoretical reasoning and might have taken Toomer's criticism as a compliment.

The papyrus records contain instructions for many practical math calculations. On one papyrus, a scribe ridicules a colleague for errors in computing the number of bricks required for a ramp, or the food supplies needed for a company of soldiers. Toomer concludes, "Egyptian mathematics remained at too low a level to be able to contribute anything of value" (1971: 37, 45). Yet, however little their calculations can be documented, the Egyptians constructed the largest buildings known in the ancient world, dwarfing those of Mesopotamia, Greece and Rome. To do this, what must they have known about geometry, physics and applied engineering?

According to popular theory, the Egyptians built their pyramids, the Labyrinth and other massive constructions by making gangs of men push huge stones up long ramps. Since this seemed a sufficient explanation, the theory remained untested until relatively recently, when several research teams tried to do what the ancients did. Mark Lehner's American archeological team managed to raise a nine-meter-high pyramid using 2.3-tonne blocks of limestone. The work crews of ten or fifteen men found they could push the blocks up a ramp using wet clay for lubricant (Roberts 1995: 26–28). Perhaps the work went faster with draft animals. This, however, hardly deals with the whole task. Before the blocks even arrived at the construction sites, they had to be carved from solid rock, transported to the river and shipped over the Nile. How were these feats done?

In 1978, a Nippon Corporation research team demonstrated the enormous difficulties of major construction using the technologies believed available in the Old Kingdom. First, the team members went to the old limestone quarries near the east bank of the Nile opposite Giza and strove to hack out blocks of rock. Using stone hammers and chisels, they pounded out rows of holes, outlining the blocks to be cut. Then they rammed wooden stakes into the holes, and soaked the wood in water. The expanding wet wood was to force cracks in the stone. The process was far more difficult than expected, and weeks passed with little progress. Finally, the team used air-jackhammers to carve enough blocks to carry the experiment forward. Similar problems arose at each step. The team employed local strongmen to help haul the blocks overland to the river. But the skids kept plowing downward into the soil, till the stones had to be dug out with bulldozers. At the Nile, the team brought a thirty-meter papyrus barge, like those found in Old Kingdom paintings and models. This barge simply flipped over every time a stone was loaded aboard, so the team finally contracted a steamboat for the job. More problems appeared in unloading the stones from the boat. Finally, the team managed to push their stones into a rough line, like a sample row of pyramid blocks. But the stones by this time were badly chipped and some had their corners knocked off—quite unlike the ancients' fine, square casting stones, which were often set to within 1/1000 of a centimeter from perfect alignment (Van Sertime 1989: 87).

What was learned from this experiment? Some would like to claim that Nippon Corporation showed that the pyramids were built by UFOs or by levitating stone through the air. But the team found no direct evidence for such theories. What their experiment clearly showed was that building with stone on such a scale with ancient tools required a science of engineering more advanced than the fragmentary written record shows. The Old Kingdom builders solved problems that would be challenging for twentieth-century engineers, including the use of enormous slabs of granite for inner chambers. Granite was the prized "rock of a million years," so hard that it served as a symbol of eternity. But the main quarry for granite was 800 kilometers south of Giza, near Aswan. From there, blocks of five tonnes, and sometimes up to fifty tonnes, were cut from solid rock and carried the length of Egypt (Hawkes 1973: 327). The builders were obviously able to learn engineering principles from nature and to apply what they had learned to ever larger accomplishments.

The Repository of Ancient Learning

When Alexander the Great came to Egypt, he arrived as a conquering pilgrim. As soon as the fighting ceased, he set out for the ancient oracle at Siwa Oasis and came away proclaiming himself a true son of the Egyptian god Amon (Bernal 1987: 115). Alexander's followers placed a high priority on collecting Egyptian knowledge. Largely for that purpose they built the library of Alexandria, which reputedly held 700,000 volumes. This library was partly a collection of translated Egyptian texts, and Greek commentaries upon them. If it was purely a repository of Greek thought, then why would this, the greatest library in the ancient Greek world, have been located outside Greece?

As few Greeks mastered Egyptian writing, Egyptian teachers such as the historian Manetho bridged the language gap. According to Finch, "The scientific renaissance of the Alexandrian period was ... due to the Greek world's unlimited access to the learning of ancient Egypt" (1989: 327). The Greeks, Romans and all later civilizations possessed their own originality, but Egyptian learning served as a foundation for the Mediterranean world, at least till the end of the "Classical Age." After that time, the conquerors of Egypt usually had less use for native Egyptian thought.

Creation and Enslavement of Agricultural Egypt

Because the Nile floods came naturally, one might assume that the land received its water and fertilizer without human intervention. But irrigating the valley properly required constant, large-scale cooperation. Teams from each "basin" (of 16,000 hectares or so) dug ditches to lead the flood beyond where it would otherwise reach (Hawkes 1973: 327). When the river retreated, they captured the water behind dikes, so more would soak into the fields. In some places the workers dug extra drainage channels to prevent waterlogging. So the whole face of the valley was landscaped into a vast gravity-flow network. All the dikes and ditches had to be partly rebuilt each year, because the floods would wash away the previous year's work (James 1984: 112–17).

The Nile's Native Administrators

This vast system of food production, plus the riverside marshes rich in waterfowl and fish, yielded the relative abundance for which Egypt was famous. Many foreigners referred jealously to "the flesh-pots of Egypt." However, when this great ecosystem was neglected or abused, the famine that followed could be terrible. The collapse of the Old Kingdom around 2180 BCE involved serious droughts, reduced floods and a temporary collapse of the farming system. Famine and disease raged like fires in the villages, and political strife nearly completed the damage. The sage Ipuwer described the resulting chaos in his *Admonitions*:

> The wrongdoer is everywhere. There is no man of yesterday. A man takes his shield when he goes to plow. A man smites his brother, his mother's son. Men sit in the bushes until the benighted traveler comes, in order to plunder his load. (in Hawkes 1973: 298)

Decent management usually averted such disasters, and the district governors often praised their own diligence in irrigation work. Governor Akhtoy, who presided over Upper Egyptian Nome XIII in the difficult years after the fall of the Old Kingdom, claims he made the flood reach the parched areas with new dikes and canals: "I nourished my city: I made the ordinary workman into someone who ate barley." He then makes what could be his proudest boast of all: "I made spittle possible in the middle of the day" (in James 1984: 116).

Egypt's native rulers usually taxed the villagers in moderation. Their harvest tax was progressive—up to half the crop for large estates, but less than a third for small farms. In bad crop years, official precepts advised the collectors to remit two-thirds of assessed grain back to the peasants (Hawkes 1973: 393, 341). The ancient administrators commonly prided themselves on the general health of their realm. They trusted that increased prosperity for all meant increased revenue for the rulers. But if the general health of the land was viewed as just a means to increase revenue, then taxation practices could be easily corrupted. In that case, the wealth extracted would *define* the health of the realm.

Some district heads combined ambitious tax collection with cost-cutting by neglecting irrigation works. The state's profit was maximized, but an erosion of social and ecological foundations ensued. Probably some such combination of mismanagement and drought helped bring down the Middle Kingdom around 1785 BCE. This could be the period described in Genesis, when Joseph reportedly presided over seven years of famine. The Genesis account suggests a catastrophe with ominous implications. Representatives of the villagers came to Joseph (who administered the state granaries) ready to bargain for their survival:

> "My lord, we cannot conceal it from you: our silver is all gone and our heads of cattle are yours. Nothing is left for your lordship but our bodies and our lands. Why should we perish before your eyes, we and our land as well? Take us and our land in payment for bread, and we and our land alike will be in bondage to Pharaoh. Give us seed-corn to keep us alive, or we shall die and our land will become a desert." So Joseph bought all the land in Egypt for Pharaoh, because the Egyptians sold all their fields, so severe was the famine; the land became Pharaoh's. As for the people, Pharaoh set them to work as slaves from one end of the territory of Egypt to the other. (Genesis 47:13–22)

Could this reflect a legend of how the rulers took ultimate ownership of the land and people? Joseph answers the peasants:

> Listen; I have today bought you and your land for Pharaoh. Here is seed-corn for you. Sow the land, and give one-fifth of the crop to Pharaoh. Four-fifths shall be yours. (Genesis 47:23–24)

By the later standards of most empires, that was a light tax. The Bible presents it as an innovation, arising as a debt payment for the means of life in a drought. Was this how landlords rose to be masters, even in the absence of war? Perhaps drought alone was sufficient to make normally self-reliant villagers feel "obligated for their lives" to parasitical rulers.

The pharaohs of the New Kingdom (1567–1085 BCE) displayed unprecedented power to acquire and divert Egypt's wealth. When West Asians and Libyans attacked the kingdom, these pharaohs retaliated with wars of expansion. Egypt's revenues were poured into armies and campaigns, and taxes rose for the war effort. In one papyrus letter from the New Kingdom, a scribe overseeing field workers describes his importance in extracting the bottom line: "I am carrying out every order given me by my lord with excellent attention and the hardness of copper. I shall not let my lord be dissatisfied with me" (British Museum papyrus, in James 1984: 129). A country in which the old cities had had no walls against invaders or peasant rebels now began to resemble a military state. Farmers were conscripted and foreign mercenaries hired to fill the ranks. Slaves from the wars began to replace free Egyptians in the ruler's estates or weaving shops (Barber 1996: 206).

By the reign of Rameses III (1198–1166 BCE), officials complained of peasants fleeing the land to escape harsh tax collectors, and of royal granaries falling close to empty (Hawkes 1973: 393). Rameses held off the attacking Libyans from the west, and the "People of the Sea" from the northeast, but maximum extraction with minimum investment undercut the country from within. The New Kingdom fell apart in 1080 BCE—more than a century before the Libyans conquered the country in 935.

The Libyans and other foreign rulers to follow generally had even less restraint than native pharaohs in milking the Nile peasants. When Assyrian king Esarhaddon overran Memphis in 671 BCE, he ordered his army to carry away all movable wealth, including "cattle beyond counting." Then, having drastically reduced the region's capacity to work the land, he imposed a crushing grain tax, "to be paid without ceasing" (Zenjirli Stele, in Roux 1980: 303). After that, the Persians arrived (525–404 and 341–333 BCE), with less devastating results. It may have helped that an entire Persian army vanished after it marched off to conquer the Sahara Desert.

The Nile as a Greco-Roman Tax Farm

The Greeks after Alexander may have revered Egyptian learning, but they also loved to enrich themselves at the expense of the peasants. Alexander was, after all, a horse-army officer from Macedonia. His men were Indo-European warriors, perhaps little different in cultural background from the Assyrians or Hyksos. C.H. Roberts says, "Countless [Greek] papyri [in Egypt] show how the principle that the subject existed for the benefit of the state was ruthlessly applied in every department of life" (Roberts 1971: 381). A petition to the Greek prefect of Egypt in 280 BCE is endorsed with the reply, "With a view to what is expedient for the revenues ... his Excellency ... shall sift the matter with utmost equity" (Roberts 1971: 383). After due consideration, Egypt was taken as a breadbasket for the eroded land of Greece.

The early Roman empire took about a third of its grain supply from Egypt and North Africa, and Rome's reliance on African farmers steadily grew as Italic crop yields fell. By 330 CE, Rome lived almost entirely on grain from Africa. Egypt was so vital a resource that no senator was allowed to visit there without imperial permission. The Egyptian farmers on whom so much depended were reduced to tax-mules. Greek records suggest that the country supported about seven million people when Alexander conquered Egypt, and the Roman records indicate roughly the same population for the next five hundred years. Evidently the peasants were taxed so heavily that population growth was zero (Praux 1971: 349–50).

In the late Roman empire, the flight of Egyptian peasants from the land became epidemic. Villages were assessed taxes as a whole, and so when villagers ran away, the tax burden simply increased upon the rest. Surviving papyrus petitions describe villages where only a few people remained, yet the tax burden was as high as ever. The petitions beg for forgiveness of tax "debt." Those who ran from their productive duties were threatened with beating and torture. Perhaps this concern to control African farmers was the main reason the late empire banned its subjects from switching occupations or giving up the work of their fathers. As Claire Praux explains, once farmland had been abandoned by peasants, it could produce nothing for the state: "It was therefore necessary to bind the peasants to the soil, to create for them a statute of enforced residence, and this in fact was what was done" (1971: 350).

The Emerging Feudal Order

When Egyptian peasants ran away from their farms, some joined the rapidly growing number of Christian monastics in the desert. Still more fled to tax-exempt plantations of foreign rulers. There they received wages, instead of paying the tax quotas that had left them hungry or in debt. Ruling Romans had taken these estates to invest the wealth they gained through collecting taxes in office. But soon the expanding number of private estates began to undercut the whole taxation system. As the rulers' estates swallowed taxable land, the flow of villagers to them became a flood. As Egypt's producers departed for the domain of the rulers' private profit, so the strength of the Roman state slowly withered away. At the same time, the ancient villages suffered social erosion. The young people drifted away to jobs on the private plantations, or to the towns. Uprooted from their ancient communities, many sought community in a new series of urban religious cults, such as Christianity. The population was ripe for the non-localized and seemingly universal faiths of uprooted Southwest Asian immigrants.

So, as the state revenue fell, the officials' personal wealth increased. Some in the ruling classes protested the trend but, with no decisive reform, the economic center of gravity began to shift. The ship of state grew top-heavy, with a weakened base and a bloated summit. Finally, so much revenue and land was being diverted to the Roman private sector that the ship overturned. A feudal-style order arose in which peasants worked like serfs on virtually autonomous estates. After that, the Roman empire existed mainly as a figure of speech (Praux 1971: 351). At least the new order was less centralized. The land's lords now lived closer to the land and depended more directly on the results of their own management decisions.

The feudalized order may have been more economically viable than the old imperial state, but it had one big weakness: the landlords could no longer pool enough military resources to defend their plantations. As the whole empire fell into a growing number of fiefdoms, warrior armies from the Eurasian steppes were relatively free to enter Europe as mercenaries and claim their own landed estates. Still, Egypt remained under Roman landlords of the Byzantine empire until 639 CE. In that year an Arabian army rode into the Nile Valley and seized it easily. The peasants either received them with indifference or greeted them as liberators. Many Egyptians promptly embraced Islam and joined the Arabs in looting the former Roman estates.

Egypt under the West Asian Empires

The new Arabian invaders were no more nor less predatory than the Romans had been. But most were Bedouin warriors from the true desert who ruled with a different style. After each of their victories in Egypt, the Arab soldiers divided the spoils as in Bedouin tradition. Their division of captured land, slaves and movable wealth was "Islamic," in that portions were reserved for the caliph (or for the whole Islamic community). After the fighting, many Arab veterans were rewarded with administrative positions and estates in Egypt. But since the Bedouin usually had no experience in managing farmland, they tended to act as military-class absentee landlords. Because most of the invading Arabs were traditional pastoralists, whole tribes arrived in the Nile Valley bringing their animal herds to feed on the village greenery.

The Arabs imposed their own system of taxes, which included special taxes on non-Muslims. The eighth-century jurist Abu Yusuf claimed that such taxes should not involve oppression:

> In order to collect poll tax [on unbelievers], one must not beat the taxpayers, nor expose them to the sun nor resort to other such methods, or inflict upon them repulsive physical torments. They must be shown gentleness, or imprisoned in order to extract payment from them for what they owe, and they are not to be released until they have paid in full. (in Bat Ye'or 1996: 322)

The annual rates of taxation on unbelievers were set by law, but collection was enforced by Bedouin soldiers. Pseudo-Dionysus (d. 845) describes the common result as he saw it in newly conquered Iraq:

> When collecting the poll tax and many others, they [the governor's assistants] demanded several times the amount. Not only did they exact the tax which was due in a place, but the same tax several times. There was neither beginning, nor middle, nor end.... When they obtained the sum by violence, they began to exact it again. (in Bat Ye'or 1996: 311)

So, by both the carrot of Islamic teaching and the stick of punitive taxation, the majority of Egyptians were won to the invader's religion. The conversion, however, offered relief from only one category of taxes.

For several centuries, Egypt's provincial governors were appointed by the caliphs in Damascus or Baghdad. Some of these governors, such as Ibn Tulun and his successor (866–96), invested in irrigation and public welfare to improve their revenues in a positive way. Others were accused of milking the cow till nothing remained for the calves. As a group, these governors were neither wiser nor more short-sighted than those of any other empire. But their armies were largely composed of foreign troops, because masses of armed Egyptians were not trusted to serve a foreign master's interests. The governors claimed to rule in the name of God, but most of them worshipped power and believed in Arab ethnic supremacy. Their Egyptian subjects learned of Islamic spirituality through common mullahs, local saints and Sufi brotherhoods—not from the ruling warlords.

For the next twelve hundred years, Egypt remained a magnet of natural wealth, attracting invaders from the surrounding wastelands. One Middle Eastern empire after another—the Umayyads, Abbasids, Fatimids, Ayyubids, Mamluks and Ottomans—all came to milk the Nile Valley. The Arabian rulers especially invited wave after wave of desert migrants into the country.

Erasure of Heretical Learning

The civilization of the Nile had already been submerged under a general Middle Eastern culture even before the rise of Islam. After the New Kingdom dispersed (1080 BCE), the long series of foreign rulers prized knowledge of war (and of their own foreign tribal traditions) more than any knowledge of nature possessed by the Egyptians. The old learning about agriculture, art and healing became concerns for the non-ruling classes. Egypt's native religions, including the cults of Isis and Osiris, were slowly replaced by cults imported by the new rulers and their migrant supporters.

Several rumors remain in circulation concerning the vanished library at Alexandria. And whatever legend is believed, it did disappear. Possibly it happened in the 270s CE, in a street war between Jewish rebels and the Roman army. At that time, Rome was engaged in an all-out war with Zoroastrian Persia. Alexandria was home to a large population of exiled Jews, displaced by the earlier Jewish-Roman wars in Palestine. Some Jewish leaders believed that Persia would deliver them from the Romans, as it had when Cyrus the Great freed the Jews from exile in Babylon. Some local Jewish radicals organized a revolt to hasten their deliverance, and the Roman army responded with a pogrom worthy of the Nazis. It may have been during this mass destruction that the great library vanished. If so, then the main repository of ancient Egyptian wisdom was destroyed by sparks from Middle Eastern religious wars.

The Establishment of West Asian Culture

Less than a hundred years after the Jewish revolt in Alexandria, the rebels' aims were basically achieved. A Middle Eastern monotheistic religion, Christianity, with its partly Jewish and partly Zoroastrian roots, became the state religion of both Egypt and Rome. It was as if the Roman army had held its own militarily against Zoroastrian Persia but lost culturally. In an empire where numerous ancient nature religions once flourished, a single official cult was now proclaimed—with growing intolerance for any other creed. In 391 CE, a large company of righteous Christian monks took it upon themselves to destroy the pagan temple of Serapis (Osiris-Apis) in Egypt's Brucheium district. In the process they destroyed Alexandria's "daughter library," the largest remaining library in Egypt, which was part of the temple complex. In the following year, Emperor Theodosius made such intolerance official policy, ruling that his own sect of Christianity was to be the only recognized religion in the empire, including Egypt.

In the midst of this cultural transformation, Alexandria continued for a time as the Mediterranean world's largest center for higher education and philosophy. The university at Alexandria still operated on a classical model, even if all "anti-Christian" philosophies were banned. And it continued as a major center for debate about Christian theology. The school had merely changed its spots and purpose—from a school of inquiry into nature, into a kind of debating chamber concerned with the emerging form of religious orthodoxy.

The "School of Alexandria" stood for an allegorical interpretation of the scriptures, and a constructive dialogue between classical culture and Christian faith. This philosophical project, however, seemed a contradiction in terms to many Middle Eastern Christians. The "School of Antioch" championed a much more literal and culturally exclusive interpretation of the Bible (*Encyclopedia Britannica* 1988). The Christian thinkers of Alexandria, such as Clement, Origen and Valentinus were classically educated men and were deemed insufficiently orthodox by the prevailing parties at the early church councils. Their ability to relate Christian and classical ideas made these philosophers seem less than entirely loyal to the faith. A.F. Shore writes, "the school [at Alexandria] received ecclesiastical approbation [and this] was of particular importance in the history of Christianity" (1971: 395).

Beyond the university, Egypt still held many classically educated Christians, many of whom saw Christianity as directly related to the older Greek and Egyptian mystery cults. Such "Gnostics" were also deemed unorthodox by the Middle Eastern majority at early church councils. As an example of the conflicting attitudes, Elaine Pagels (1989: 125) contrasts Gnostic and Orthodox views on suffering. The Gnostics generally saw suffering as being the result of insufficient insight into human nature. Orthodox Christians, on the other hand, viewed suffering as the wages of sin. Suffering was God's punishment for acts of deviation from an orthodox social code. It was beginning to seem that unquestioning loyalty to one communal creed was the wave of Egypt's future.

The determination to fix one religious orthodoxy eventually tore the Christian community apart. As the debate to establish one truth grew bitter, the university at Alexandria became a riot-torn verbal battleground. In this atmosphere, intellectual discussion itself was increasingly discredited. The desired result of dialogue was the silencing of opponents, and this was achieved. The Coptic Church of Egypt banned the Gnostics. Then the Roman Church excommunicated the Coptic Church. Or perhaps it was the other way around. Meanwhile, the university at Alexandria ceased functioning as an organized center for learning. Ancient Egyptian "wisdom" and Greek "pagan" learning were now firmly classed as heresy. The heretical centers of knowledge were closed, burned or turned into Christian churches and seminaries. The practice of traditional religions and philosophies, or the possession of books on those subjects, was considered both blasphemy and treason.

Of course, the priests and professors of classical knowledge did not meekly submit to the erasure of their disciplines. They spoke out in the face of official and semi-official persecution. But the cost of confronting the dogmatists was high. In Alexandria in 415 CE, Christian monks mobbed and murdered the female philosopher Hypatia. After that, it seems, the traditional pagans grew silent. Some Roman state and church authorities issued the death penalty for simple possession of heretical books. This is probably why the Gnostic Nag Hammadi manuscripts were hidden as they were, in buried jars, like secret time-capsules for another age. Greek works containing Egyptian lore went underground and resurfaced in Europe only nearly a thousand years later. Concerning the ancient literature and science which eventually did come to light, Praux (1971: 334) says, "It is certain that the manuscripts copied in Greece in the Middle Ages, and which reached the West during the Renaissance, owed their origin to Alexandrian texts."

According to popular legend, religious fanatics of one kind or another burned the library of Alexandria to eliminate unorthodox ideas. This is possible, but the same result could have been achieved without violence. Simple neglect would have sufficed. The old books of Alexandria, Memphis or [Egyptian] Thebes would have simply crumbled to dust if nobody was interested in re-copying them. And the original Egyptian texts were written in forgotten languages anyway. The last known piece of hieroglyphic writing dates from about 394 CE, which falls in the reign of Theodosius, the emperor who outlawed all religions but Christianity (Barb 1971: 138). Soon after that, the keys to understanding hieroglyphic script were lost somewhere in the tombs of Theban scribes. Finally, not a soul remained who could read any message from that lost cultural universe. And the spoken language of ancient Egypt also eventually faded away. By the 1400s or 1500s, even in remote villages, the "Coptic" language of ancient Egypt had been replaced by a tongue from West Asia called "Colloquial Arabic."

Egyptian Learning in the Islamic Empires

Under Islam, many pagan books passed into the privatized palace collections of governors or caliphs. The Muslims, after all, captured the existing intellectual property of Egypt, and this was a major reason for the Islamic world's intellectual superiority over Christendom in medieval times. Some Islamic governors, Fatimid caliphs or Mamluk generals, were avid collectors of old manuscripts. Some funded translations of old books into Arabic and encouraged scholars to build on pre-Islamic knowledge. Again we hear of libraries said to contain more than 100,000 volumes. Egypt would once more emerge as a center of scholarship drawing the intellectual cream of the Mediterranean world: scholars such as Ibn Al-Haytham, the pioneer in optics (d. 1039); Ibn Maymun (or, Maimonides, 1135–1204), the philosopher-physician; Ibn Al-Nafis (1210–88), who studied the lesser circulation of blood; physicians Ishaq Al-Israili (Isaac Judaeus, 855–955) and Ibn Ridwan (d. 1251); engineer Qaysar Ibn Al-Musafir (d. 1251); pharmacologist Al-Kohen bin Al-'Attar (c.1200) and historian Ibn Khaldun (1332–1406).

But these bright stars in the medieval sky brought little respite from the erosion of Egypt's intellectual heritage. The governors sent to rule Egypt (and deliver its tax revenues) depended for their authority on mercenary troops, and the loyalty of the troops depended on their pay. After paying the army, the second priority was payment of surplus revenue to the imperial caliph or sultan. After that, if anything remained, some revenue might go for establishing schools and universities, such as Al Azhar. These schools, however, would increasingly be Islamic religious schools devoted to the study of Islamic (basically Arabian) social tradition. They were centers of anthropocentric learning, with nowhere near the emphasis on learning about nature that had been found in the schools of ancient Egypt. The most famous and influential scholars of Islamic Egypt were not natural scientists and philosophers. They were men like al-Shafii (767–820 CE.), who pioneered in codifying the entire body of Islamic customs into one great set of social rules called the *Shari'a*.

During medieval times, Egypt was caught in wars between Islamic sects. In the 900s CE, the Shi'ite Fatimids rode in from northwest Africa to cast out the Sunni orthodox rulers. About two hundred years later, the Sunnis under Saladin returned in a *jihad* against both Christian and Fatimid heretics. Soon after this Sunni takeover, the great book collections of the Fatimid dynasty seem to have disappeared. They vanished underground, were confiscated by authorities or were burned by fanatics for social conformity. A rumor then spread that such an elimination of heretical books was authorized by Islamic tradition. According to a legend probably invented at this time, the first Islamic liberators of Egypt had struck a comparable blow for the faith by burning the library of Alexandria (Lane-Poole 1925: 77).

Demonization of Zoolotry

The Islamic teachers had certain religious reasons not just to neglect old Egyptian knowledge but to erase it. Many of Egypt's medieval rulers and clerics felt strongly that knowledge from pre-Islamic times was knowledge from the "time of ignorance." Ideas before Muhammad would have been expressions of pagan belief. And what use could knowledge from the time before truth serve, other than to lead the faithful astray?

The ancient Egyptians referred to their age of origins as "the time of God." That was the legendary time of Osiris and Isis, when the Egyptian state and religion had been formed. But in the view of most ancient Christians and medieval Muslims, this "time of God" was really a time of evil. In the Pyramid Texts, the prehistoric symbol of the tree with a python coiling in its branches referred to the axis of the earth and the ring of constellations rotating around it. The axis of the tree rose to a fixed point in the sky, to a constellation called "the thigh," meaning the thigh of the Great Mother of the universe. This still point in the sky was her womb, from which creation emerged. On earth, these symbols were represented as the central tree in the garden of life, the first woman and the circling snake (Finch 1989: 407). But the West Asian migrants who later predominated in Egypt altered the meaning of these creation symbols almost beyond recognition. The snake, which had been the image of time's cycles of eternal life, was made a symbol of the evil one who seduces the weak, corruptible female into an unholy knowledge of this world.

The ancient Egyptians deemed nature to be sacred. They honoured animals to the point of mummifying astounding numbers of deceased birds, cats, alligators and hippopotamuses. Perhaps no other culture displayed such lavish worship of Mother Nature's children, and today we see only a pale likeness of it in formal pet graveyards. But to the desert cultures of later Egyptian history, all this reverence for plants and animals seemed the vilest idolatry imaginable. By the standards of the Hebrew Bible, orthodox Christianity and Islam, ancient Egypt was guilty of "zoolatry," and it has been condemned down to recent times. As P. Renouf wrote in 1879 (in Karanga 1989: 353), "Many writers ... at present speak of Egyptian religion as one of the lowest and grossest forms of nature worship, as consisting in what is commonly called African fetishism."

The ancient Egyptians believed that the desert beyond the Nile Valley was a place of evil spirits. These spirits were supposedly embodied in the desert's wild animals, and in the various desert peoples who periodically invaded Egypt. Later, in Christian times, large numbers of monks went out to live in the desert, sensing that there they would wrestle with Satan. Many Christians in the Nile Valley felt the presence of monks protected their land from incursion by desert demons (Shore 1971: 403). All this would seem to accord with traditional Egyptian beliefs toward the desert, except for one thing: the Christians commonly believed that the aboriginal "animistic" religion of Egypt itself was demonic. Also, the Christian religion had come to the Nile from over the desert, borne partly by the many refugees from ethno-religious wars in West Asia. Whereas the old Egyptians basically demonized everything from the desert, the Christians demonized both the desert and the nature-deities of the Nile Valley. At least in theological terms, all of nature was now taken as most desert tribes took it—not as a friend and teacher but as a lawless threat to humanity.

Chapter 8

South Asia's Desert and Forest Heritage

TRADITIONALLY, MEN OF THE RAJPUT CASTE WERE WARRIORS, AND RULERS over those they claimed to protect. They controlled the borderlands of the Thar desert in northwest India for over a thousand years. Today, many Rajput males serve in the Indian army or police, as only somewhat befits their previous inborn role. Of course this is not the same as being both ruler and general of one's own kingdom (Harlan 1992: 36).

In Rajput custom, the eldest son inherited all his father's land. This left his younger brothers to go out searching for kingdoms of their own. When they had grown to be men, the younger brothers left their father's homes and ventured forth to win glory in the world, usually by offering their valiant arms in service to another Rajput maharaja. If they fought well and expanded their new lord's kingdom, they might receive a *thikana*, or grant of land. A *thikana* was a group of villages over which a Rajput noble could preside, drawing his income from the peasants' rents. In return, he would protect the villagers from exploitation by any other prince.

This practice of sending sons out to win their kingdoms came to an end with the rise of nation-states, as the military conquest of neighbouring villages became illegal. After 1947, the Indian government abolished the Rajput kingdoms as sovereign states. Yet some things stayed the same. "Little brother" Rajputs continued to leave their father's estates and seek some position their ancestors might respect. This, however, grew ever more difficult in the modern Indian job market. Lindsay Harlan (1992: 9) explains that many little brothers could only find jobs doing the work of lesser castes. Many became ordinary bureaucrats, shop owners, factory hands or even farmers. Some felt cheated of their inheritance. They still felt themselves born to the warrior caste, and superior in spirit to the lower-caste Indians their ancestors ruled (Harlan 1992: 27).

Rajput women traditionally practiced *pardah* (their word for *purdah*), or strict seclusion. Each home was divided into a women's quarters (*zanana)* and a men's area (*mardana*) (Harlan 1992: 37), almost exactly as in Arabia. This exclusion of non-family members from quarters of the opposite sex was very strict. To violate it was far more serious than entering a public washroom of the wrong sex in the West would be. Besides, a Rajput warrior's first duty was to protect his women's honour.

Rajput clans were divided into family patrilineages, much like those of the

Pathans in the Pakistani-Afghan border region. The Muslim Pathans called their family lineages *khel*, and the Hindu Rajputs called theirs *kul* (Spain 1963: 23). In both cases, the lineages were founded after a warrior ancestor conquered a kingdom for his line of eldest sons to rule (Harlan 1992: 27).

In recent centuries, the Rajputs often claimed to be quintessential Indian Hindus. Many of them served as a bulwark of Hindu resistance to various Muslim invasions. But recent research shows that the Rajputs themselves were originally invaders from the steppes of Inner Asia. "These invaders," Professor H.G. Rawlinson says, "carved out kingdoms for themselves and eventually settled down in the country, taking Hindu wives" (in Campbell 1976c: 199–200). In this they have much in common with other waves of Central or West Asian invaders who came before and after them.

In the Hindu festival of Navratri, Rajput princes presided over the ritual sacrifice of a water buffalo, which celebrated a mythological victory of good over evil. According to legend, the warrior goddess Durga destroyed the king of the demons (who was a great buffalo) in a cosmic battle at the dawn of creation. By killing a buffalo, the Rajput princes identified themselves with the victorious forces of good. The Vedas contain a similar story, in which the Aryan thunder god Indra kills the arch-demon Vritra. Both myths could reflect events near the dawn of Indian history, when migrating warriors from Inner Asia moved down the Indus Valley to conquer much of South Asia. The hosts of darkness defeated in these myths were probably aboriginal Indians. After all, the Aryan Vedas describe the aboriginals as *dasas* (devils) and "noseless, black-skinned demons" (Campbell 1976c: 159–60). The texts report Aryan attacks on the cities of these godless natives. Sure enough, archaeologists have found sword-cut skeletons of aboriginal people in the streets of ancient Harappa and Mohenjo-daro. It seems the legendary liberation of South Asia involved a transfer of land to the conquerors (Dimmitt 1986: 213).

Like other invaders, the Rajputs succeeded for centuries in ruling the aboriginal "lower caste" majority of their region. Recently though, they have seemed to be losing their grip. By the 1980s the Rajput caste could no longer control the state government of Rajasthan. Their status no longer protected them from the indignities of economic competition. Perhaps it is simply a law of life, like gravity, that Mother India will finally absorb all conquerors. But many Rajputs saw their "decline" as a major loss of Indian religious tradition. In defense of their old exclusivity, a new fundamentalism rose among them—as if the social superiority of higher castes from Inner Asia was the most essential element of Hinduism—as if the seclusion of high caste women from the surrounding population was the bulwark of Indian civilization. The traditionalist Rajputs developed a Hindu brand of fundamentalism. Yet it was like a mirror image of the Muslim fundamentalism of many Pathans in Pakistan.

Perhaps surprisingly, high-caste traditionalism proved widely popular in modern India. Just as high-caste groups seemed to be dissolving into the masses of democratic India, many lower-caste Indians, *Harijans* (untouchables) or

South Asia's Entryway from the Desert

tribal people, seemed increasingly bent on moving "up," to adopt the conventions of the higher castes. Accordingly, there was a popular rise in *pardah* for women. The general public picked up on dowry payment in marriages. Perhaps the high-caste cultures of former conquerors were actually losing little besides their exclusivity, and it was the diverse cultures of India's common people that were fading away.

Meanwhile the old environment of village India was definitely fading away. As high-caste landlords turned their lands into agro-business plantations, fields were often cleared of trees and peasant families alike. While lumber contractors cut down remaining pockets of jungle, pump irrigation for cash crops depleted the water table across the Ganges and Deccan plains. Thousands of villages found their wells going dry. As the land grew drier and more barren, the Thar desert crept deeper into Rajasthan and Gujarat.

Many villagers responded to these threats by organizing in new ways. "Anti-development" movements mushroomed in hundreds of locales across India. If these movements are successful, they might preserve the environments and values of aboriginal India. Otherwise the country might soon resemble the deserts of Inner Asia, from which so many of its invaders fled.

Timetable of Invasions of South Asia

c. 2500 BCE	Harappan towns rise over the ruins of Dravidian villages in the Indus Valley
c. 1500 BCE	Aryan invaders overrun Harappan civilization
c. 800 BCE	Aryans move into southern India
400s–305 BCE	Aryan states of northern India battle for supremacy till one Mauryan ruler prevails
327 BCE	Alexander's Macedonian Greeks invade the Indus Valley
78–102 CE	Kaniksha, emperor of the Kushans, conquers northern India
350–480	Hunnish invasions, ancestors of the Rajputs
610	Warlord Sri Harsha unites northern India
712	Arabs take Sind
1000–1600	Pathan tribes move into northwest Pakistan
1192	Ghurid Turks break past Rajputs to overrun northern India, forming the Delhi Sultanate
1296–1327	Mongol raids into India
1398	Tamerlane's horde sacks Delhi
1526	Babur invades, establishing the Mughul empire
1540–1555	Afghans briefly hold northern India
1555	Mughul empire re-established
1565	Mughul armies crush southern kingdom of Vijaynagar
1669–mid 1700s	Period of religious war between Muslim and Hindu princes
1700s–1800s	British move into chaotic regions in wake of the religious wars

The Conquerors' Women

On his wedding day, a typical north Indian groom rides through the streets in a parade of friends, relatives and, if affordable, a marching band. The groom is dazzlingly dressed, partly in the garb of ancient kings, and partly of heroes in the Hindi movies. With such pomp and display he comes to claim his bride. The groom traditionally rides a white horse, which recalls the ancient Aryan horse sacrifice in which a nomad king would set free a white horse, which was followed by a procession of princes. Wherever the horse roamed in the course of a year, the *raja* claimed as his own. This rite from the vast steppes of Inner Asia was imported into a land of small tropical villages, and its pre-Vedic imagery is still applied to the rites of marriage, whereby a man claims his lordship over a woman.

Rajput Women's Religion

Just as a Rajput man had certain exclusively male duties, so a Rajput wife had her female duties to perform. One of her responsibilities was to make the men of her family play their warrior roles unflinchingly. If the Rajputs were to rule, and not be made workers for others, then their sons had to be ready to sacrifice themselves in war. A Rajput mother's duty was to raise sons who would prefer death to a loss of power and position. To her daughters she taught female forms of self-sacrifice (Harlan 1992: 64). A woman's high-caste privilege required living in *pardah*. She was socially cut off and aloof from the common people over whom her men ruled. If she went out in the streets, she generally went veiled and accompanied by a male relative. Her honour required such a drastic display of exclusive loyalty.

According to Lindsay Harlan (1992: 52, 64), traditional religion for Rajput women concerned family deities much more than any greater deity of the universe. A woman's household shrine was normally dedicated to the protector of her husband's lineage, called the *kuldevi*, or goddess of the clan. This was a hybrid faith, combining the devotion of desert tribes to their patriarchs with the low-caste Dravidian worship of local village goddesses. But as in many desert cultures, the *kuldevi* was the patroness of a human clan, and not the spirit of an earthly place.

A *kuldevi*, according to Harlan (1992: 79), was like a divine Rajput mother and was normally addressed as *Mata ji* (respected Mother). Like a stern but loving matron, she nurtured her clan members, giving them the fluids of life: "The giving of fluids is to be understood as the quintessential mode of protection rendered by *kuldevis* and *pativratas* [good, dedicated wives] alike." The actual fluid of life was literally blood, which was replenished by water, or a mother's milk. These nurturing fluids were kept pure by the clan women's devotion and chastity. As in most desert religions, the water of life was seen to come through its human carriers, rather than simply flowing from the Earth. As Muhammad said in a different desert, the greatest act of merit is to give people water. But for most desert women, this giving of fluids was a sacrifice, rather than a boon for themselves. Harlan explains that women were to nurture rather than be nurtured: "Whereas the *kuldevi* serves as the dispenser of fluids to men and children, she does not serve in this capacity relative to women. Rather, she coerces women ... into playing an analogous role" (1992: 70).

Indo-Aryan Women

Over two thousand years before the Rajputs came to India, the great Indo-Aryan migrations moved into South Asia. The resulting conflict of Dravidian aboriginals and Aryans was among the earliest known instances of racism. The Aryan scriptures, the Vedas, display an explicit mythological link between skin color, vocation and inborn spiritual value (Panikkar 1963: 9). Wherever the Aryans prevailed, they tried to maintain themselves above and apart from the conquered races. The Aryan priests, or *Brahmans*, held that any social or sexual mixing of Aryan and Dravidian races was "pollution." To avoid such corruption, the law code of Manu set out elaborate taboos governing all relations between castes, especially concerning who could marry whom (Bumiller 1990: 17). As in most cases where conquerors tried to control their patrilineage, this had implications for the conquerors' women. Historian Romilia Thapar says, "To avoid pollution, you must control birth. But you lose control over birth if you lose control over women" (in Bumiller 1990: 17). This was a reason for *purdah* as it appeared in Hindu society long before the Muslim invasions. And probably every wave of invaders for over three thousand years regarded itself as a new racial elite, whose women must be protected from the lesser breeds.

The result of all this segregation is obvious in any modern north Indian village. As Elisabeth Bumiller reported, in the 1980s, the women of Khajuron village in Uttar Pradesh had basically two possible lots in life. Upper caste women stayed in their homes, almost as if in prison for life. Most were Hindus but were as firmly cloistered as any women in the Muslim world. The lower caste women on the other hand, "were free to leave the house, usually to work ... in the fields for less than fifty cents a day" (1990: 79). These women sweated in the sun, often for the same landlords whose wives were isolated in the biggest houses of the village.

Before migrating to India, the Aryans of Inner Asia had been nomadic, cattle-driving people. Their men were ancient cowboys, while the women made camp. The two roles were not easily interchangeable, nor were they equal. Still, nomadic women had considerable freedom. They worked in the open sun, looked after the goats and sheep, gathered plants and made butter. But when these women came as conqueror's wives to South Asia, all this changed. With the invasions, the Aryan women became a social elite who lived in a world of farming villages and conquered populations. To grow their own vegetables there, or tend their own goats, would mean associating with Vaisyas and Sudras (the farmer and laborer castes) by doing their work. Clearly, this was a problem for a conqueror's wife. The "status" of Aryan women now required aloofness from both common people and ordinary work. Their privileged position had made them economically useless. Other than housekeeping and bearing children, what was an Aryan woman to do?

The Evolution of Dowry

The low-caste aboriginals almost always regarded a woman as a valuable asset to her family. As in other parts of the tropical world, women were often the main food producers. This may explain the old customs of "bride price" among low caste people. When a Dravidian girl married, her parents were commonly paid to compensate them for the loss of a productive woman. This kind of "dowry" applied over much of tropical Africa also. A man had to save and pay handsomely for a wife, because she was valuable. But among the high-born of South Asia, "dowry" took on an opposite meaning. A high-caste girl was *supposed to be* aloof from labor and was therefore economically useless. Her male offspring might gather wealth, but she herself was an expense. With that expectation firmly in mind, dowry meant that the bride's family had to pay the groom's family—to take a non-productive female off their hands (Bumiller 1990: 49). Here again, "status" was in effect a fall. A high-caste woman may have been a carefully guarded sexual commodity, but on the marriage market she had a negative value.

The size of a high-caste woman's dowry was a rough measure of her expected worthlessness. The higher her family status, the more "costly" she would be. The size of the dowry was estimated to deflect the expense she would probably cost her new family. Her father paid because it was shameful if he could not unload a daughter, and because he would otherwise have to continue paying for her upkeep himself. A high-caste girl was an expense to raise, and a further expense to marry off, and none of this expense was compensated to the parents. To cut their costs, parents tended to marry their girls off as early as possible.

Sonless Women and Widows

In her delicate situation, a high-caste female was often left with only one useful role—mothering the landlord's children. If she ever lost this role, through her husband's or her own infertility, through widowhood or even through bearing only female children, she would lose the only status normally available to her. Then she would be deemed a failure and treated accordingly.

If a high-caste woman was widowed, at least two problems arose. First, she could no longer serve the purpose for which she had come to her husband's family. She would likely be seen, by her relatives and herself, as existing for no reason. Second, in her life to follow, there was always the danger that she might turn to a new man. K.M. Panikkar (1963: 113) notes that the remarriage of widows was normal among ordinary, low-caste people like the Jat farmers (1963: 113). But among those claiming descent from a conquering race, who lived as landlords over estates which must be inherited by the landlord's sons only, this was a problem. If a widow was free to go and marry again, she might beget half-brothers of her first husband's line, and a threat to the estate could then arise. Besides, if dowry had to be paid to dispose of a high-caste virgin, what dowry might the in-laws have to pay to unload a widow? Perhaps these are some reasons why the law code of Manu allowed no remarriage for widows or divorced women: "A wife sold or repudiated by her husband can never become the legitimate wife of another who may have bought or received her after she was repudiated" (in Singh 1981: 111). Where this was the prevailing attitude in centuries past, a widow had basically two options: she could live on in her father-in-law's household and be treated more or less as a hanger-on, or she could kill herself.

Panikkar (1963: 147) claims that in the age of Manu (perhaps 200 BCE), "the condition of widows was not miserable, but in the course of time more and more restrictions were imposed upon them." Then, the option of suicide grew more "popular." A widow sometimes made her grand exit by mounting her husband's funeral pyre, placing his head in her lap, and letting herself be burned alive as a *sati*. Thus she could solve all the problems of her existence at once and be gratefully remembered forever. The Muslim writer Al Biruni (in Singh 1981: 155) said of the practice, "As regards the wives of kings, they [the Hindus] are in the habit of burning them ... by which they desire to prevent any of them by chance committing something unworthy of the illustrious husband."

The early Dorian Greeks and Scythians, who were distantly related to the Indo-Aryans, also practiced widow burning (Panikkar 1963: 63). Bumiller (1990) says these cultures believed that a man could take his possessions with him to the next world by burning them on his funeral pyre. A man's wife was one of his main possessions, so he would want to keep her forever.

The Later Migrant Elites

It would be wrong to blame the Indo-Aryans for all the inequalities endured by females in India. Traditions like *pardah*, child marriage, *sati* (widow burning) and dowry developed slowly over many centuries. Likely these customs grew more established with later invasions. Many modern Hindus blame the greatest fall in women's status on the Turkic and Mughul Muslim invasions of the 900s to 1500s. Mahendra Pratap Singh claims there was no *pardah* before the Muslims. He feels the practice was adopted by Hindus in imitation of their new rulers, and as a measure of protection "from the atrocities of Muslims" (1981: 158, 198). He says (99) that some Hindus claim that child marriage was adopted to protect Hindu girls from covetous Muslim eyes. And *sati* is sometimes presented as an invention of the Hindu-Muslim wars, during which Hindu women burned themselves "to preserve their chastity and honour" if their men were killed in battle (Chauhan 1988: 148).

In contrast, Thapar (in Bumiller 1990: 17) shows clear evidence that upper-caste women's seclusion and *sati* were practiced in India long before the Muslim invasions. In the *Kathasaritsagar* (written about 400 CE) the heroine

Ratnaprabha protests against an existing practice: "I consider that the strict seclusion of women is a folly produced by jealousy. It is of no use whatsoever" (in Panikkar 1963: 60). The practice of burning women appears in the ancient Hindu *Ramayana*. Concerning child marriage, Singh (1981: 90) admits that several centuries before the Muslims arrived, a widespread desire to limit the expense of raising girls had pushed the age limit for marriage downward. The ancient Hindu *Dharma Shastras* say a girl can be married between ages seven and twelve, and the *Brahma Puranas* say she can be given from age four (in Singh 1981: 95).

Still, there is probably some truth to claims that the Muslim conquests brought a new decline in women's status. Every invasion of warlords from the arid lands brought a renewal of such inequalities. The Muslims were a fresh elite, pushing all who came before them downward. And whichever side blamed the other for its own actions, Singh shows that between the 800s and the 1200s CE there was a steady erosion of women's freedoms. Widow remarriage and divorce seem to have virtually ceased, and the custom of widow suicide became more generally favored, and not only in the ruling classes. Child marriage became the norm except for the lowest castes (Singh 1981: 81). *Pardah* became the standard for Hindu or Muslim families with the means to afford it. Panikkar (1963: 178) confirms that a "marked deterioration" in the condition of women was "a result of both invasions and of the rigidity of the system of purdah in Rajputana and the Gangetic valley."

The Strong Women of South India

Over much of tropical south India, a different tradition has remained influential to the present. Especially in the coastal regions, the south is rainy and lush rice country. Bumiller (1990: 107) felt the southerners of Kerala and Tamil Nadu "appeared physically smoother and somehow sweeter than the villagers struggling with a harsh existence in Uttar Pradesh and Bihar." In the southern economy a woman's farm work was often of greater value than a man's. Perhaps accordingly, the Nayar people of Kerala used to practice matrilocal marriage. The bride stayed in her mother's house, and the groom either moved in with her family or else visited as a night guest.

The southern regions used to be independent Dravidian states. Later, the south came under the influence of northern empires, especially after the 1500s when the northern Mughuls destroyed the southern kingdom of Vijayanagar. Cultural influence from the north then seemed to trickle down from the top of southern society to the bottom. Eventually it became normal for southern brides to leave their mother's families and marry into their husband's households. The forms of marriage grew more patriarchal, but something of the old style remained. In south India, marriages were alliances between families, and a bride came with the status of her original family. By contrast, most northern brides were seen as individual girls, starting new lives at the bottom rung of status in their husband's families (Hawley and Wulff 1986: xii). Vijay Lakshmi, a director of public health services for Kerala sate (in southwestern India), claimed:

> In northern India, women just do the housework and never learn to read and write. Here, women play more of a partnership role in society.... In Kerala a male cannot just brush his wife aside and make her a child-bearing machine. (in Bumiller 1990: 276–77)

In the 1980s Indian social scientists produced study after study on Kerala state. They hoped the deep south might yield some secrets of relative peace and well-being for the rest of India. Why, they asked, was Kerala's female literacy rate over 70 percent, compared to around 20 percent in the north? Why was Kerala the only state in India where females outnumbered males? Why was the average age of marriage for women 22 years, when in the north most women were probably married off (illegally) at below 18 years? Why was Kerala's birth rate about a third lower than India's average? Why was prostitution so little seen in the south? And why were there so many fewer dowry payments, and so few cases of killing brides over demands for dowry?

North India before the Fall of Women's Status

The status of southern women was once common over most of India. In the Maurya empire, around 200 BCE, large orders of Buddhist nuns wandered northern India, teaching in public (Panikkar 1963: 42). Perhaps of more interest, a female character in the *Mahabharata* called Drauppadi had five husbands. The commentator Nilakantha explains that having several husbands was an old custom among low-caste Dravidian women (Singh 1981: 115). In ancient Kashmir, according to the *Rajatarangini*, women commonly held their own land. The northern states were sometimes ruled by queens, such as Sugandha Devi, queen of Sankaravarman (early 900s CE), or Didda, queen of Kashmir (950–1001 CE). Kamala Chauhan (1988: 156) says that during periods of resurgence for aboriginal culture, "The women used to mix freely with the menfolk. They sat, played and drank wine with them. Queens accompanied kings on hunting excursions." They appeared openly in public, as native female nobles in a society of which they were not afraid.

All this, however, was before fresh waves of invaders re-conquered India, and re-emphasized the gap between ruling and subject people. This also was before the lush forests of the Ganges plain faded away, leaving a dusty, semi-denuded landscape. After that, the women of village India took a new kind of fall, bearing the brunt of an environmental collapse so pervasive that much of northern India deserves the term "emergency zone."

The Submersion of Dravidian Pakistan

The Great Indian Desert, also called the Thar desert, now stretches from the Indus Valley eastward deep into Rajasthan and Gujarat. It is not severely dry, but according to the Central Arid Zone Research Institute in Jodhpur, Rajasthan, it is "unique among the deserts of the world; not a single oasis or native cactus breaks the monotony of this vast ocean of sand" (in Walls 1980). Between about 2500 and 1500 BCE, towns and villages of the Harappan civilization dotted most of this region, usually along ancient watercourses such as the long-vanished Sarasvati River. In ancient times this region was a land of grassy savannas and thorn forests, yet soil samples suggest that the climate has changed little, if at all. Much of the desert gets enough rain for one thin crop of grain, but the countryside has been stripped of vegetation, mainly by hungry goats, sheep and cows. As Reid Bryson showed, "It is only necessary to exclude [grazing] animals for a year or two in order for a fine stand of native grass to spring up" (in Claiborne 1970: 295). For many centuries the grass has been eaten faster than it can grow. When the rains fall, the naked soil washes downhill towards the Arabian Sea.

Much of the Thar region's topsoil has accumulated on the ocean shoreline, from the Indus Delta to the Rann of Kutch. The Rann was an inlet of the Arabian Sea centuries ago but has since filled with runoff soil. Now it is a flat plain of seven thousand square kilometers, much of it crusted with efflorescent salt, gleaming white in the sun (Allchin, Goudie and Hedge 1978: 7, 16). Further inland, the site of ancient Mohenjo-daro now lies buried under ten meters of soil eroded from the north. As the Indus plain's sediments slowly covered these ancient towns, the groundwater also rose. The water level beneath both Harappa and Mohenjo-daro is now close to the ground level of those ancient cities. Since archaeological digs could not recover artifacts from below the water line, little evidence of earlier settlements has come to light. These major Harappan towns appear to spring fully formed upon the Pakistani plain, like "colonial emplacements" on a previously unmarked land (Campbell 1976c: 155).

More recently, other Harappan centers, such as Amri, Kot Diji and Kalibangan, were discovered on hills above the Indus floodplain. In these places, the archaeological record can be traced much further downward and reveals a farming civilization already ancient in 2500 BCE (Allchin 1982: 220).

Village Cultures before the Harappan Empire

On the western edge of the Indus plain, a neolithic site at Mehrgarh shows evidence of wheat, barley, date and even cotton farming. The ruins also hold unnaturally large numbers of water buffalo, cow, sheep and goat bones. These remains show dates from before 5000 and possibly to 6500 BCE (Allchin 1982: 108–9). The site is immediately recognizable as a typical ancient South Asian village. Already, it would seem, people had tamed the potentially deadly water buffaloes. We can picture the village boys and girls riding their buffaloes home from the fields each evening. As the sun set, a mist of smoke from the evening fires crept through the village. People sat outside their doorways in the cool of the day, grinding grain and talking.

If people in Mehrgarh worked like modern villagers do, their roles in gardening, herding animals or gathering wild plants would have been shared among members of the family. All of these nature-worker jobs went hand-in-hand where the land remained biologically rich. Only later, in places where the forests disappeared and the grass grew scarce, did these roles become separate vocations. After migrants arrived from arid Inner Asia, herding, farming and sylvaculture were viewed as separate functions for different ethnic castes.

Before the Harappan empire, the realm of Dravidian villages extended from the Afghan border to the Bay of Bengal. At Koldihwa in the mid–Ganges Valley, pottery containing cooked rice has been found, possibly dating to 5400 BCE. In the Nilgiri hills, people have practiced terracing for soil conservation since before recorded history. Large cattle pens have been found in the forests of Karnataka from near 3000 BCE, suggesting an interesting practice of forest sylvapasture (Allchin 1982: 118, 292, 123). Clearly, aboriginal South Asians lived in a realm of great biological diversity. In a forest world, they closely observed every visible species of plant and animal. Their technology was primarily biotechnology. Traditional Dravidian healers are recorded as using more than 2,500 medicinal plants (Grainger 1990: 272). Up to 400,000 varieties of rice were grown, as seed was selected for each place's seasonal micro-climates (Shiva 1988: 131). The aboriginals fit so closely in their bioniches that most latter-day city dwellers deemed them completely barbaric. The hill country tribespeople wore tiger skins. Bhila women wore peacock feathers, musk perfume and strings of fruit. They ate deer and roasted parrots (Chauhan 1988: 13).

The oldest religions of these Dravidians focused on the mysteries of wild nature. Their myths concerned birth and death, the Great Mother and her wild-man spouse, later known as Shiva (Campbell 1976c: 164). The Dravidians called their goddess by many names, such as Candimari, Candi and Kali. They worshipped her as the mother who gives birth to and destroys all things in the round of eternal life, and who lives in every woman. Their rites included sacred sex, in which the object was to honour and arouse the goddess, while avoiding insemination. These are just a few features of the Dravidian civilization which rose in the time of Old Europe and survived every invasion or oppression to follow.

The Harappan Empire and the Rise of Plantation Agriculture

About 2500 BCE, the old Dravidian villages at Amri (in Sind) and Kot Diji (in Punjab) were burned. When they were rebuilt, the characteristic square-grid layout of mature Harappan towns appeared from the ashes. Kalibangan was also transformed from a piecemeal collection of buildings into a town planned by central authority. The highest part of each town was a castle-like citadel, with granary and temples. The rest of the town lay at a lower level but was raised above the plain and enclosed within walls. The "downtown" areas were grids of straight streets, lanes and drains. Some blocks were divided in brick cubicles so small that they suggest slave-barracks (Allchin 1982: 142–43, 171–73, 179). These towns differ so strongly from Dravidian villages that Joseph Campbell is probably right in calling them "colonial emplacements" (1976c: 155). The skeletons and images of people found in Harappan towns indicate at least two racial groups—some Dravidian and some seemingly West Asian. It is likely that a new group of people with Middle Eastern racial and cultural traits had arrived.

Crafts and products of the Harappan civilization were as standardized as the town plans. Over an area larger than modern Pakistan, the uniformity of designs for flint tools, pottery, bullock carts and seal inscriptions is so striking that Bridget and Raymond Allchin (1982: 193) claim "it is possible to typify each craft with a single set of examples drawn from one site alone." Certain towns became centers for manufacturing certain products, in a kind of industrialization without machinery. Such standardized work probably increased efficiency, and this approach to production perhaps had its most serious effects when applied to farming.

In the old Dravidian culture and the ancient forested environment, families had worked for themselves, growing or gathering all the many plants they needed for food, medicine, fuel and housing. They nurtured some of these plants in gardens near their homes, and the rest they gathered from the woods. The Harappan rulers, by contrast, collected large quantities of grain into fortified storehouses, which suggests a major centralization of the land's wealth and a system of plantation farming. They claimed large tracts of riverside land and cleared them of trees. Then they dug ditches from the river to irrigate the newly denuded areas. Instead of growing scattered gardens of many useful plants, the supervisors probably reserved the best fields for crops of a single species and weeded out other plants. The workers who did this may have been slaves, hired hands, or villagers required to donate labor. They probably worked under command and not for themselves.

In this new system, farm workers who had once combined gardening, animal husbandry and forest sylvaculture became specialists. They tended to become either farm hands on the irrigated plantations or pasturalists in the dry areas. Irrigated fields were precious; to protect them, animals were made to graze outside the irrigated valley. A similar effect appeared recently with construction of the Rajasthan canal in India. As the channel arrived, local pastoralists and their animals were driven away. Communal grazing areas in villages all along the waterway were seized by the biggest landlords and used solely for cash crops (Agarwal 1989: 277). So the interlocking parts of the old village ecology were slowly pushed apart.

The wood-fired brick Harappan cities, the large granaries and the bones of contemporary wild animals (including bears, crocodiles, elephants, tigers and forest squirrels) all suggest that the environment of Pakistan was once far greener. Again, recent studies of soil and climate show no significant decline of rainfall over the course of history. Jacquetta Hawkes (1973: 30) says the old vegetation was destroyed "not by a loss of rainfall, but by tree-felling and the grazing of goats and sheep." Circles of deforestation probably expanded around the Harappan towns as firewood and lumber were cut but not replanted. As mono-cropped fields replaced riverside jungles, trees were systematically eliminated. Such deforestation tends to produce waterlogging on a flood plain. Trees constantly pump groundwater up to their leaves; removing the natural pumps halts this action.

The irrigation methods of the Harappans (and the Mesopotamians) channeled water onto the fields with little provision for drainage. When the soil grew waterlogged, mineral salts floated up to accumulate in the topsoil. We have evidence that the Harappans spread gypsum on their fields, which is an old treatment for salt contamination (Bryson and Murray 1977: 111). Given time, periodic floods from the Indus would naturally wash away saline deposits, but it seems Harappan farming methods polluted the soil faster than the river was able to cleanse it.

The slow decline of Harappan towns suggests a gradual slide toward environmental poverty. As the Harappans denuded the land around their rivers, the basins filled more rapidly with silt, choking the Indus with mud. Where accumulations filled the riverbed, the Indus broke its banks and roamed like a thrashing snake over central Pakistan. This is how rivers often behave in desert regions. The Harappan cities were built on massive earth embankments to escape the floods, yet even raised towns show signs of inundation. The shifting rivers left other settlements high and dry, such as the ruins by the banks of the old Sarasvati.

The decline of Harappan civilization was spotty. Some towns were destroyed quite early, around 2000 BCE. Others lingered about five centuries more. In those latter centuries, town residents increasingly cannibalized old buildings for brick and wood, and new construction no longer followed any master plan. Harappan writing fell from use. Finally, Mohenjo-daro and Harappa were sacked and some dead bodies were left in the streets. Probably most of the population had already left. Then nomadic invaders moved in, taking the land as better pasture than the fields they left behind.

The Aryan Waves

American Vedic scholar David Frawley feels sure that the authors of the *Rig Veda* were natives of the Indus Valley. The Vedic hymns repeatedly mention the Sarasvati River, and surely the chanters knew this river well before it disappeared around 2000 BCE. But composers of the Vedas were also pastoralists, whose word for war meant "desire for cows." Most likely, both herding and farming people lived in ancient Pakistan, and it was mainly the herdsmen's lore which was included in the Vedas (Frawley 1992: 26). The plantation economy could have slowly pushed local gardeners and herdsmen apart until they became separate cultures. Later environmental decline would have forced a greater reliance on pastoralism. But however indigenous the original Vedic herders were to Pakistan, other evidence shows that new pastoral tribes were arriving from Inner Asia before and after 1500 BCE. Perhaps at first they came in peace. But as if expecting attack from the northwest, both Harappa and Mohenjo-daro built their strongest defensive walls, with ramparts about fifteen meters high, facing that direction (Campbell 1976c: 171).

No signs of horses are found in the pre-Aryan ruins of Harappan towns—neither bones nor images. But in Baluchistan to the west, starting around 2000 BCE, the bones of horses and implements of horse riding appeared in the graveyards. The same graves contain Central Asian-style swords and metal arrowheads (Allchin 1982: 304, 311). Joseph Campbell writes, "The day of the peasant afoot and the nobleman ahorse had dawned" (1976c: 172–74). The people of the white horse sacrifice had arrived.

The Aryo-Dravidian Wars

At Morhana Pahar in central India, a series of rock paintings depict warriors in horse-drawn chariots fighting against spear-throwing, arrow-shooting men on foot (Allchin 1982: 82–83). These may be eyewitness depictions of the first Aryan incursions into the area, as seen by the forest tribals. The figures on foot seem determined and clearly unbowed in the face of superior force. The emotions of the charioteers are unreadable. Their minds could be filled with fear or with enthusiasm for blood sport. From the Aryan side we have the written record of the Vedas, which describe the conflict with India's aboriginals as a divinely ordained race war.

The region of Baluchistan (along today's Pakistan-Iran border) was once called "Gedrosia," meaning "country of the dark people" (Campbell 1976c: 156–57). Perhaps some Indo-Aryans also lived there, but the composers of the *Rig Veda* considered themselves utterly distinct from such people. "For the 'dark-skinned,'" Panikkar says, "they had only contempt" (1963: 6). The Vedas remained unwritten for many centuries, and tradition holds that the reason was not Aryan illiteracy but a desire to keep the sacred verses secret from non-Aryans. The holy words were supposedly revealed to the chosen race only, and not to unbelievers, untouchables (Dravidians) or women. Wendy O'Flaherty says that if such people learned the hymns, "it was believed the words would be polluted like milk contained in the skin of a dog" (1989: 80).

With a revulsion perhaps greater than the Islamic idol smashers, the priests of the *Rig Veda* condemned the ways of Dravidian India: "Let those whose deity is the phallus not penetrate our sanctuary" (*Rig Veda* VII:21–25, in Panikkar 1963: 5). The ruins of Harappan cities contain many phallic images and stone vulva rings, like those found at Hindu shrines today. But the early Aryans clearly did not view sex or fertility as holy. In fact, Campbell finds the Vedas hold "nothing either of the spirit or of the mythological world image of the later Hinduism—no reincarnation, no yoga, no vegetarianism or non-violence." The Vedic hymnists display ambition for conquest and wealth but "no mythology of salvation" (Campbell 1976c: 170, 183–84). Instead, the sacred hymns sing frankly of war and plunder, and the drunkenness conducive to both:

Like fire kindled by friction, do inflame me!
Illumine us! Make us rich!
For in the intoxication that you render, O Soma,
I feel rich. Now entering into us, make us really rich as well. (*Rig Veda* VIII:48, 6, in Campbell 1976c: 181)

The soma-drinking Vedic god Indra is described as lord of the thunderbolt and destroyer of cities. He was a great nomadic warlord in the sky, leading his chosen people to possess a richer land. In a tone resembling the Old Testament, the *Rig Veda* summons divine help to destroy whole towns of the enemy: "Strike down, O Maghavan (Indra), the host of the sorceresses in the ruined city of Vailasthanaka, in the ruined city of Mahavailastha" (I, 113, 3, in Allchin 1982: 308). The Vedic *Taittiriya Brahmana* then celebrates victory over the Indus civilization: "The people to whom these ruined sites belonged, … these many settlements, widely distributed, they, O Agni [lord of fire], having been expelled by thee, have migrated to another land" (II, 4, 6, 8, in Allchin 1982: 38). Sure enough, the Harappan towns commonly terminate in a thick top layer of ash. Panikkar (1963: 5–6) feels it is obvious that these were the towns Indra had been called to destroy.

In the Old Testament, a war of semi-nomadic tribes against aboriginal settlements is recorded as "sacred history." The Vedas, however, use a language abstracted into pure myth. The account of Indra's battle against the arch-demon Vritra seems to take place in the skies before time. But certain perspectives on geography, history and ecology can be detected. As Campbell (1976c: 182) transcribes the setting of battle, "Vritra, the archdemon without hands or feet, reposed amid distant fortresses, crouching on the mountains—having hoarded to himself the waters of the world, so that the universe, deprived for centuries of all fluid whatsoever, had become a waste land." This demon hoarded water rather than gold. To withhold water was perhaps the ultimate evil for people from the desert. And possibly this was what the controllers of irrigation seemed to do. According to the myth, Indra took a mighty quaff of soma, threw his thunderbolt and blasted the monster apart, liberating the waters of the Earth. Historically speaking, the Aryan patriarchs liberated the watered land from the enemy and took it for themselves.

In the Vedic liberation of the Earth, the land was claimed for the pastoral invaders, and the riverbanks were opened for grazing. Later, when the Aryans advanced further into heavily forested northern India, their aims were the same. The *Vedas* repeatedly mention fire as a holy force and a symbol of light fighting against darkness (Campbell 1976c: 180). Such lofty abstractions were often poetic expressions of practical things. Just as Agni, the god of fire, was praised for destroying the towns of the enemy, so ranchers use fire to make forests suitable for pasture in modern Brazil.

The *Mahabharata* contains a story in which Krishna and Arjuna share a picnic in the great Khandava forest which grew on the banks of the Yamuna River near present Delhi. While taking their food, these heroes are approached by a poor Brahmin priest who begs them for a boon, saying he has a great hunger. Krishna and Arjuna generously offer to give the Brahmin whatever he desires. The priest then reveals himself to be the Aryan god of fire, Agni, whose hunger is to devour the whole Khandava forest, with all its animals and tribal Naga people. Bound by their promise, Krishna and Arjuna mount their chariots and ride around the forest, letting flaming arrows fly. They oblige Agni's craving by driving back every creature who seeks to escape, till all are consumed in a great fire offering. Only one member of the Naga forest people survives, a man who happens to have been journeying beyond the forest that day. In relating this story, Madhav Gadgil (1989: 325–26) concludes, "[The] episode evidently illustrates the conflict between the indigenous ... forest people and the agricultural-pastoral invaders on the Gangetic plain."

Sure enough, the forests of the Ganges basin began to melt away following the Aryan invasions. The Himalayan foothills remained covered in forest till the twentieth century, and jungles remained over much of central and southern India. But the ancient Aryans managed to clear much of the forest down the Ganges to the Bay of Bengal. In the Buddha's day this deforestation was ongoing in eastern Bihar. The Buddha condemned Brahmanic *yajnas*, or fire sacrifices, in which whole forests and vast numbers of animals were "sacrificed" (Gadgil 1989: 325).

The Origins of Caste

Some Hindus claim that the Vedas never mention caste. But the Aryans' contempt for darker races was already the very soul of the caste system. As conquerors in Hindustan, the Aryan chiefs became hereditary rulers, warriors and self-appointed religious leaders. They assigned food production and physical labor to the aboriginals. After that, social roles in northern India were determined by birth and race, in other words, by caste.

If the Aryan tribes resembled other historic pastoral peoples, they probably had ranked their own members by caste even before they came to South Asia. The supposedly egalitarian Arabs and Berbers of old North Africa were divided into families with hereditary functions—vassal families, weapon-bearing families or those inheriting religious roles (Norris 1984: 313). The farm labor in their oases was commonly assigned to Black slaves—the parallel to Sudras and untouchables in India. The ancient Scythians from north of the Black Sea were similarly divided into Royal Scythians (the military and ruling caste), nomadic herdspeople (*Callipidae*) and, most lowly, farmers (*Alizones* or *Georgi*) (Melyukova 1990: 102–2). And to further suggest the caste-mentality of desert invaders of India, we have the writings of Zia Ud-din Barni, a leading philosopher in the Delhi Sultanate after 1300 CE, who seems to have regarded himself as a devout Muslim:

> Teachers of every kind are to be strictly enjoined not to thrust precious stones down the throats of dogs, or to put collars of gold round the necks of pigs and bears—that is the mean, the ignoble, the worthless and the low-born.... They [the low-born] are to be instructed in nothing more [than the labors for which they exist] lest it bring honour to their mean souls. (in de Bary 1958)

On conquering India, the Aryan warriors became the Ksatriya caste, and at first these hereditary fighters were the "highest" caste, above the Brahmins. Among the early Aryans, soldiers ranked higher than priests. According to the *Satapatha Brahmana*, "there is nothing higher than a Ksatriya, therefore the Brahmana sits down below the Ksatriya" (14.4, 1, 23, in Singh 1981: 53). The later rise in status for the Brahmin caste was probably influenced by Dravidian esteem for spiritual adepts; but the Aryan esteem for warriors also inspired the rise of professional soldiers among the Dravidians.

In ancient South Asia, battle was seldom a matter of total war. Some Dravidian tribes formed alliances with the Aryans; others were forced to recognize Aryans as their lords. Either way, the military elements in both communities tended to collaborate. The ruling Dravidian clans tended to become military elites, taking the caste status of Ksatriyas. So the Reddys of Andhra, the Maravas of Tamil Nadu and the Nayars of Kerala joined the great family of ruling-caste Hindus (Panikkar 1963: 13). Centuries later, the Chinese Buddhist pilgrim Xuan Zang (Hsuan Tsang) reported that the warrior clans in each Indian region considered themselves one race (Singh 1981: 54).

One of the chief tasks of the Ksatriya warriors was to enforce the social order, keeping each caste in its place. In the *Ramayana*, Lord Rama considers it his Ksatriya duty to kill a Sudra caste laborer who dares to perform spiritual exercises appropriate only for the high-born (Panikkar 1963: 16). The death penalty for such a "crime" suggests a determined effort to reserve religious power for certain castes. Of course, this effort met with only limited success. All the harshly stated restrictions on racial mixing in the *Dharma Sutras* and *Grihya Sutras* failed to stamp out intercaste love, or stop the absorption of Aryan tribes into the Dravidian population. As the populations of South Asia inevitably mixed together, Brahmin priests and Dravidian yogis began to resemble one another. In later centuries, members of the Brahmin and Ksatriya castes might be as dark-skinned as any Dravidian Sudras. However, as the Aryans were absorbed, Dravidian society absorbed the caste order of the dominators. Caste then became the constricting garment of Indian society. For the next three thousand years, vocational castes tried to exclude economic competition from outside their own hereditary groups (Panikkar 1963: 5). The outlets for each person's creativity were delimited, and talents other than caste-determined ones were supposedly ignored. The waste of human potential was probably enormous. But this was just a logical extension of the Aryans' desire to safeguard rulership by birth for their own people.

Popular Renunciation of This World

In reflecting on the period after the Aryan invasions, Panikkar (1963: 151) claims "healthy materialism is the normal outlook of even religious people when things go well"; where "other-worldliness" prevails, it is a sign that something is wrong. Religious renunciation of the world probably entered South Asia in the wake of the Aryan invasions and has continued as a recurring theme ever since. It is a religious theme which has grown or diminished according to the fortunes of ordinary people. During some invasions, the aspirations of village people were crushed till little hope remained for happiness in this world. The hopeless, however, could still hope for happiness in another world. Those who still professed joy in this life were seen as clinging to illusion.

In recent times, some observers have labeled almost all native Indian religions as forms of world-denial, or spiritual escapism. For example, the excavations of ancient Harappan cities turned up clay figures of people in yoga-like postures. The archaeologists who uncovered these items commonly presumed that yoga was even then a means of world renunciation and escape from rebirth. This presumption was probably false. Campbell (1976c: 212) points out that among most Dravidian villagers, yoga is still akin to traditional shamanism. The popular aims of yoga are bodily and spiritual health, acquisition of special powers and receptivity to spiritual insight. None of these aims are necessarily world-denying.

Before the Aryan invasions, there was probably nothing inherently otherworldly in Dravidian culture. Dravidian religion was concerned with nature, the harvests, seasons, human relationships and sex. The emergence of a new religious goal—of salvation *from* earthy life—can probably be explained by certain extremely painful events in history. As a later example, soon after northern India fell to the Turkic Muslims in 1192, India was swept by popular religious "devotional" movements. Panikkar (1963: 151) notes a difference in tone between the devotional movements of southern India (which was not yet conquered) and the north. The southern *bhakti* (devotion) movements were joyous and celebratory, but in the north, "mysticism" became "a psychology of national escapism."

After the Aryan invasions, new waves of Inner Asian migrants pushed into South Asia at least every several hundred years. In succession came the Kushanas, Ephthalite Huns, Shilikas, Rajputs, Turkic Moslems, Timurids, Mughuls and Afghans. The Kushana king Kaniksha, of around 100 CE, is depicted in a great stone statue at Mathura. Like hundreds of conquerors since Cyrus the Great, he bore the title "King of Kings." In the statue he stands gripping a sword in each hand. Campbell (1976c: 298–99) says his pose "suggest[s] dramatically the character of the Central Asians who had assumed leadership of India."

Perhaps it was under such rulers that Hinduism and Buddhism, which were often life-affirming faiths outside of north India, became identified with escape from *maya*, renunciation of sex, and the belief (contradicted by the Buddha) that women are generators of *maya* who can never attain release from this world unless reborn as men. Like most religions, Hinduism and Buddhism had both world-celebrating and life-negating sides, depending on the experiences of their various followers. But which spirit best accords with the seers who originally inspired these faiths? According to Campbell (1976c: 273), it was despair rather than life itself which the original Buddhists held to be the product of illusion: "By what principle of delusion, then, has it come to pass that so many beings ... are to such a degree self-concerned that they suppose their own and others' sufferings to constitute a cosmic problem, saying, 'Life is something that should not have been?'"

Resurgence of Aboriginal Religion

The most dramatic of all legendary encounters between Aryan and Dravidian religions is the story, included in the *Mahabharata*, of the yogic *rishi* Kapila, "the Red One." This mighty forest sage was sitting deep in meditation one day when, of all things, he was disturbed by several hundred Aryan princes on horseback. These impertinent men were engaged in a great horse sacrifice, following a white horse wherever it should lead, and claiming all the land they passed for their king. The large escort of princes and warriors was there to fight any locals who tried to block the horse's passage. Kapila turned upon these fools a gaze so scathing that, by the red glare of his eyes, the entire company was burned to ashes. So much for the impiety of interrupting a yogi's interior sacrifice with a "horse sacrifice," which was not really a sacrifice at all, but a ritualized exercise in stealing other people's land (Campbell 1976c: 257).

Until about half a century ago, it was widely believed that historic Hinduism was founded by the Aryans and based upon the Vedas. Now it seems obvious that the major source of Hinduism was Dravidian culture. As Paul Deussen, Joseph Campbell and K.M. Panikkar point out, the "forest books," or Upanishads, are utterly different in tone and content from the Vedas: "Between the Vedic and Upanishadic views the difference is so great that the former could not possibly have been developed out of the latter" (Campbell 1976c: 203). The Upanishads are concerned not with Vedic ritual or heroism in battle, but with Dravidian yoga.

Forest Wisdom and the Aryo-Dravidian Order

The Upanishads contain several stories of Aryan Brahmins who presume to teach Vedic ritual to the leaders of forest tribes. In the *Brhadaranyaka Upanishad*, such a Brahmin missionary comes to teach Ajatashatru, a Dravidian ruler of Benares. In the course of their discussion, Ajatashatru points out a man waking from sleep, and asks the Brahmin, "Where, when this man slept, was the person who consists of understanding; and whence did he come when he returned?" The Brahmin has no idea. Such a psychological question has never occurred to him. From his education as a Brahmin priest, he knows mainly rituals to petition divine support for his desires. Earlier in the text, Ajatashatru asks the Brahmin to summarize what he knows. The Aryan details rites concerning the evocation of lightning, wind, water and fire. Ajatashatru informs his visitor, "But that is not enough for the knowledge of Brahman [of God]." Perhaps most encounters between followers of Aryan and Dravidian religions ended at roughly this point, in mutual non-comprehension. But the *Brhadaranyaka Upanishad* tells a different story. With an almost astounding flexibility of mind, the Brahmin ponders Ajatashatru's strange questions, suspends his own missionary role, and answers, "I approach you as a pupil" (Campbell 1976c: 198–200).

What follows is a lesson concerning the *atman*, or the self. As translated into Western languages, atman means "soul," "consciousness" or "self." The Upanishads are then a guide for the evolution of self-awareness (Campbell 1976c: 198–99). Their ideas are not philosophy in the sense of classifying things into intellectual compartments, nor are they theology in the sense of naming the powers above and finding mutual accommodation between divine and human wills. The teachings in the first Upanishad more resemble a school of developmental psychology. When Sigmund Freud (1938: 164–65) was trying to separate the modern study of psychology from religion as he knew it (that is, Middle Eastern, Judeo-Christian religion), he claimed, "I believe that a large portion of the mythological conception of the world which reaches far into modern religions is nothing but psychology projected to the outer world." The Upanishads, however, are *consciously* about the internal world of the psyche. Campbell (1976c: 206) considers this body of Hindu lore "an essentially inward-turned, psychological system of thought ... in which much of what today is known of the unconscious was anticipated, and even, ... along a certain line, surpassed."

As Ken Wilber (1979) clarifies, the journey of the atman concerns the sense of identity. It deals with the psychological or cultural boundaries erected in the mind, which separate "me and mine" from the surrounding world. In the Upanishads, such boundaries of selfhood are taken as variable and potentially under conscious control. The "small self" may be transcended, and the sense of identity extended outward. Ultimately, the sense of separation between the small self (atman) and great self (Brahman) may dissolve, and the two may be realized as "one." In that case, the microcosm and macrocosm will be revealed as a seamless web. This "boundless" sense of identity is termed enlightenment, or *samadhi*. The language is distinctly Indian, but it resonates with the sense of other aboriginal religions around the world, from the Daoist Chinese, to the "animists" of the Congo and Amazon forests. As an anonymous Chinese text, the *Kuan Yin Tzu*, reads, "Knowing that I am of the same nature as all other natural things, I know there is really no separate self, no separate personality, no absolute death, and no absolute life" (in Needham 1956: 444).

In a barren environment like the Aryans' earlier homeland, a human community could easily feel itself the only significant actor on the world stage. It might then proceed, as Freud said, to project its communal values onto the surrounding universe, creating theology out of psychology and group culture.

"Man" (and a man from a particular tribe) might then presume himself the measure of all things. The value of other creatures would be no more than their usefulness for humans. The resulting attitude could be one of wandering over the earth, taking whatever desirable things are found and claiming this to be God's will.

In contrast, where plants and animals seemed to overshadow the village, as in India's monsoon country, the sense of identity would naturally include more than humanity. In a complex and interdependent environment, the boundaries between lives might appear blurred. Life could seem a web, with the one great life of the forest encompassing all. In ancient Dravidian villages, the farmers often viewed their land, trees and animals almost like family. All members of this extended family were needed for mutual survival. To keep a village sustainable for generation after generation required a balance among many living things. An inclusive sense of identity had survival value.

In the concept of reincarnation, all creatures were taken as fellow spirits who pass on a journey through all the myriad forms that life takes. The sense that one life was separate from another was presumed to be an illusion. In that case, the highest good was not defined as maximum enrichment of the human tribe and maximal destruction of its competitors. Instead, spirituality was a matter of overcoming blindness about relationships. So in the religions of Dravidian India the envelope of selfhood was to grow outward till it might encompass the universe of forests, villages, rivers and ponds. This was the kind of "self-realization" valued by the forest sages, not to mention the village mothers who raised them. The old Dravidians called such realization "forest wisdom."

After the Vedas and the basically Aryan *Ramayana*, the chief heroes of Hindu epics and holy books were no longer warriors, conquerors or chariot drivers. Instead, they were Dravidian-style yogis. Even the older Aryan myths were redone, editing out purely martial heroes. In the *Mahabharata*, the story of Indra slaying the demon Vritra is retold, with the meaning virtually upended. Instead of steeling himself for battle with a hearty swig of soma, Indra gains his power from the practice of yoga. When he slays the dragon, it is revealed that Vritra was a member of the Brahmin caste (Campbell 1976c: 248–49). Later yet, Indra was simply replaced, and the hero who slew the monster becomes the goddess Durga. Then, in the *Kena Upanishad*, the mother goddess resurfaces within orthodox Hinduism as the guru of the male gods. Campbell (1976c: 205) says, "She is represented as their mystagogue, their initiator into the most profound and elementary secret of the universe." The ancient practice of sacred sex became Tantrism, with yogins learning the mystery of eternal joy from their *devadasi* (holy woman) initiators, as in the *Kama Sutra*.

So, in the centuries after they had penetrated the forest country, the Aryans began forgetting their old gods. Indra the thunderbolt hurler, Mithra the keeper of covenants, Varuna the all-encompassing sky, and Agni the sacrificial fire were all neglected. Panikkar (1963: 6–7) says, "The same mantras were repeated, but to gods who were no longer worshipped." The mixed-race Indo-Aryans increasingly prayed to the Dravidian queen of nature and her spouse Shiva, the lord of the dance.

By the time of the later Upanishads and the Buddha, the Aryan invaders had been largely absorbed into Dravidian South Asia. Much was changed and would never be the same again. Where pre-Aryan society probably had its village shamans, yogis and *devadasis*, the Aryans imposed a new layer of spiritual leadership. The Brahmins claimed a birth-given (and not necessarily experience-given) authority over religion, including the "religious" segregation of castes. But yogis and saints continued to rise up among the lower-caste people. In ancient India, heterodox religious teachers were seldom persecuted, because villagers couldn't be stopped from choosing their own gurus. Campbell (1976c: 339–40) explains the paradox of a combined Aryan-Dravidian religion as "an essential affirmation of the cosmic order as divine, and an affirmation, equally, of the orthodox Indian social order as divine." That great catholic affirmation of nature, aboriginal culture and the Aryan social order imposed upon it, characterized all the great empires of India's classical age. A recognizably common civilization emerged across the semi-continent later called "Hindustan."

Hindustan's Shrinking Domain

When India and Pakistan broke into separate nations in 1947, a geo-social fact was officialized—"Hindustan" no longer reached westward to the Hindu Kush. Once, long ago, it had. But for over three thousand years Pakistan had borne the brunt of all invasions from Inner Asia and the Middle East. Sometime during the past millennium, the cultural boundary of tropical South Asia had moved eastwards, past the Indus, over the Thar desert, to where the land of Dravidian villages remains.

A Moving Ecological Boundary

The modern Indian desert, bare as it appears, supports a dense population. As of 1977 about twenty-three million head of cattle overgrazed India's portion of the desert, and some forty-six people per square kilometer scraped a living from its soil and sand. In Luni Block, near Jodhpur, a survey for the U.N. Conference on Desertification (in Walls 1980) found 88 percent of this denuded area was being plowed for rain-fed farming. The tilled fields yielded a steady stream of dust in the wind. Annual rainfall in Luni was about 350 mm, concentrated in the monsoon season, which could usually sustain one crop of grain (Allchin, Goudie and Hedge 1978: 1). It could also sustain vast fields of grass and scrub forest if the blades of grass and twigs were not consumed before they could flourish.

Excavation in this desert reveals alternating layers of desert sand and good soil—even rain-leached soil. The previous deserts here were caused by natural cycles of climatic change. This time, however, the shift is largely "manmade," because the desert is expanding in an age of relatively good rainfall (Allchin,

Goudie and Hedge 1978: 40–43). During the 1950s and '60s alone, areas of usable pasture in Rajasthan shrank by half, from about 60 percent of the state, to 30 percent (Grainger 1990: 84–85).

If West Asia is characterized by deserts and pastoralism, then the ecological boundary of West Asia is now the Aravelli hills, which roughly mark the edge of the arid zone from near the coast of Gujarat, northeast to around Delhi. Only southeast of these hills does the monsoon-watered, semi-forested country begin (Grainger 1990: 84–85).

Cultures of Desert and Forest in Collision

In the hothouse world of old Hindustan's forests, the myths and traditions of "Hinduism" rose in riotous diversity like a luxurious growth of jungle plants. In ancient towns, temples of diverse gods and goddesses crowded each other in the streets. Buddhist, Jain and Brahmanical images were commonly carved and painted together, like different trees in a sacred grove (Chauhan 1988: 184). Emaciated ascetics and practitioners of sacred sex were both depicted in religious art, each representing recognized spiritual paths. When Roman Emperor Theodosius I outlawed any religion but his own sect of Christianity, large numbers of artists and thinkers emigrated from Europe to Gupta India (Campbell 1976c: 326). Only in the 1100s CE would followers of one true faith impose their rule on northern India, but even then there was no question of expelling all the "non-believers."

In the lands from which the Muslim invaders came, there was usually no great diversity of life forms and human cultures. Different tribes had different customs, but these tribes were often enemies competing against each other for scarce resources. Communal loyalty tended to be the most often evoked principle of religion. The distinctions between tribes were commonly treated as marks for discerning friend from foe. From this, it was a short step to the idea of one holy community against all others.

It took various Arabian, Persian and Turkic Muslim armies from 712 CE to the 1190s to conquer what is now Pakistan. Only in 1192 did the Muslims break past the tenacious Rajput kingdoms, reach the Ganges plain and commence their seven-year drive to the Bay of Bengal. By that time, some desert areas of Sind province in southern Pakistan had been ruled by Islamic princes for five centuries.

During those centuries before 1192, the Islamic Pathan people had moved down from Afghanistan, through the Khyber Pass, to occupy much of the present North West Frontier Province. The Pathans are a Middle Eastern-style nation-tribe comparable to the Kurds or Armenians. The British colonists found them fascinating, partly because they seemed so "biblical." Many of the British speculated seriously that the Pathans must be a lost tribe of Israel, because everything about them reminded an Englishman of his Judeo-Christian roots. James Spain (1963: 28) lists the Pathan characteristics of suspected Old Testament origin: "their tight tribal structure, their stark code of behaviour, their striking Semitic features, their bearded patriarchal appearances, ... their ... biblical names (acquired from the Holy Quran)." Such Islamic and West Asian migrants did not utterly replace Dravidian culture, at least not in the Pakistani Punjab. As in all previous migrations, there was mutual survival and influence among ethnic groups.

Before the Muslim conquests, Hindus rarely thought of themselves as members of a single religion. The sense of a "Hindu civilization" was almost entirely a matter of hindsight. Panikkar (1963: 134) says the ancient Hindu dynasties such as the Guptas, Bharasivas, Pallavas and Cholas "never claimed to be the champions of a creed or the upholders of a society." But over centuries of warfare with Muslim kings, the Hindu Rajputs and kings of Vijayanagar became upholders of Hindu *dharma*, with their own versions of the holy law. Hindu caste taboos, with restrictions on socializing, eating, drinking and all other bodily functions, were increasingly taken not just as customs but as eternal laws imposed by a superhuman authority. Such guidelines were increasingly enforced, "with a rigidity which would have surprised Manu and Yajnavalkya [ancient Hindu legalists]" (Panikkar 1963: 110). Panikkar concludes, "There is a tendency of all societies on the defensive to be more rigid in their rules, to be more violent in their reactions towards non-conformity" (1963: 138). So, in a gradually rising reaction against the Muslim threat, certain Hindu groups came to more closely resemble their foes.

The Hindu-Islamic Spirituality of South India

In the south, a tropical and maritime India was in peaceful contact with the Islamic world for nearly a thousand years before the Muslim conquest of Vijayanagar in 1565 CE. Muslims came to the southern coasts as they came to Indonesia, as sea borne merchants carrying on a mutually beneficial trade with the local people. Most Muslims of Kerala and Tamil Nadu still retain that heritage of relationship with the local Hindu community. In the north, many Muslims speak their own language (Urdu) and go to separate schools. In the south, Muslims may learn Urdu and Arabic but normally do business in the language of their Hindu customers (Shahabuddin and Wright 1987: 162). In the south, dialogue between civilizations was not diverted by issues of military domination. The cross-fertilization of different social worlds had a chance to bear fruit there, and that fruit has been significant for the whole world.

In southern India, the Islamic ideals of brotherhood and equality among all believers were introduced as unsullied ideals, apart from the contrary examples set by West Asian rulers. Many Hindus were inspired by Muhammad's vision of social equality. Some converted to Islam, and many more applied Islamic ideals within Hinduism. As a result, popular Hinduism became more egalitarian and more theistic. There were still many Hindu gods and goddesses, but now each devotee was commonly dedicated to only one (Panikkar 1963: 135–36). The Islamic personal relationship with a sovereign Lord melded with the Hindu aim of psychological union with the divine. The result was a devotional religion of passion for a chosen deity. A new kind of sainthood emerged, in which the devotee could be described as God's lover.

Within Hinduism, this style of religion brought a series of *bhakti* movements, with their often sexual imagery for religious love. Especially with the saint Ramanuja (b. 1017 CE), the almost mechanical operation of karmic law was transformed (Panikkar 1963: 135–36). Salvation could be won by a personal appeal to grace, the way a suitor wins the favor of his beloved. Such religion distinguished two forms of devotion. The first was *vaidhi bhakti*, or formal and liturgical devotion, which Campbell (1976a: 63) says was called devotion only "by courtesy"; the second was *raganuga bhakti*, or passionate love, heedless of social decorum.

This second, or "true," devotionalism became perhaps most controversial when it reached the northwest of India and spread among women of the "purdah zone." In Rajasthan, for example, the female saint Mira Bai dropped all pretense of social conformity, abandoned her husband and *pardah*, left her semi-desert homeland and journeyed to the forests of Brindavan to be an ascetic in the woods. Allegedly she merged into an image of Lord Krishna, uniting eternally with her true lover, who was God (Harlan 1992: 206–7). The response of Rajput society to this story has been ambivalent. One Rajput woman interviewed by Lindsay Harlan (1992: 211) expressed deep respect for Mira Bai but added, "Mira ... had no faith in society. She didn't like Rajput society.... I don't think that it's a good thing; it's better to be a *pativrata* [a wife devoted to her husband]."

In the south Indian Islamic community, the spirit of submission to God, as before an arbitrary and omnipotent ruler, was modified. A new ideal of faith involved the *bhakti* sense of passionate union with the Lord. Of course, among Muslims such talk of "union" almost never meant a merger of *identity* with God. The union was best compared to the union of lovers. By implication, the most holy relationship, or the one best symbolizing the most important religious values, was not the relation of ruler to subject but the relation of lovers to each other. This kind of Islamic devotionalism spread through various Sufi orders, some of which originated in India. A state of grace called *fana* (rapture) could be attained through the loving instruction of a *pir*, or fairy princess. Such a guide was also called a *fravashi* (a spirit of the way) and the enlightenment reached through such a teacher was "the larger full surrender" (Walker 1996: 974).

Sufic devotionalism grew popular across the Islamic world, from India to Spain, Malaysia and West Africa. According to Campbell (1976a: 63–65), the Sufis' spiritualization of sexual love infected Christian Europe after the Crusades. The Sufi dervishes became models for the troubadours of Europe, the West's folksingers of a new morality. In that new faith, sexual desire could be an inspiration of greater spiritual value than any church sacrament. Love could become more important than any social convention, or any threat of eternal damnation.

But as such devotional movements spread to northern India, Panikkar feels their quality changed. There were great northern *bhakti* saints, such as Ramananda and Mira Bai. And some saints such as Kabir or Guru Nanak inspired Hindus and Muslims equally (Panikkar 1963: 138). Dara Shukoh, the son of Mughul emperor Shah Jahan, was an adept of both Sufism and yoga. He wrote the *Samudra Samagama*—the "Mingling of Oceans," otherwise known as the Persian Upanishad (Panikkar 1963: 174). But more generally in the Delhi sultanate and the later Mughul empire such dialogue and devotion spread as an "underground" movement. In the north, Islam was the faith of conquerors, and Hinduism the superstition of infidel subjects. In that context, unorthodox devotionalism and eroticism were often mere diversions, more a stimulus for private dreams than for social change. Commenting on this period, Norvin Hein (1986: 122–23) says the erotic became a refuge of satisfaction because it was the source of fulfillment hardest for the authorities to control. But in such a segregated society, the private erotic imagination ran to sickly sweet dreams of Radha and Krishna, or to harems of forty virgin *houris* in paradise. In the literature of Delhi court poets, Radha (Krishna's mythic lover) was no longer a lowly peasant saint. She became an upper-class princess, living perhaps in a Rajput or Mughul court harem where making love was her one and only vocation. Poetry had little to say of her, save to linger over her body parts. For example, her breasts were compared to the Himalayas (Schomer 1986: 90–99). The goddess here had no purpose save titillation.

The Muslim Conquerors of Hindustan

The first Muslim conquerors of northern India were more truly desert warlords than the Aryans. The Ghorid Turks were military adventurers with almost no understanding of agriculture, except how to tax it. Panikkar (1963: 131) claims "it was impossible for the Muslim kings, had they even desired it, to have the lands cultivated by the soldiers." Neither were the invaders skilled in trade. For them, "the elaborate system of *hundi* and credit on which Indian business was based was a mystery" (1963: 131). But, as in previous conquests, the warriors settled as tax collectors and landlords over large grants of land. Like the Aryan Ksatriyas, or the Rajputs before them, their vocation was one of fighting, ruling and collecting tribute.

When various Islamic invaders swept down from Samarkand or Kabul, they set their capital cities on the western edge of the Ganges plain near Delhi. Just as Beijing was the headquarters of China's occupation by the Mongols and Manchus, so was Delhi the capital for the Inner Asian Muslims. Since the eleventh century, the Delhi region has seen fifteen different capital cities rise, fall and rise again, as one empire succumbed to the next.

Tamerlane, who merely raided Delhi in 1398–99, was a descendent of Genghis Khan. Babur, the first Mughul emperor (from 1526), was a descendant of Tamerlane on his father's side, and of Genghis Khan on his mother's side (Panikkar 1963: 153). Babur was a king of Samarkand until the Uzbeks drove him south into Afghanistan. From there he rallied his subjects to conquer the greener fields of India. After Babur died, the next generation of Afghan nobles sought to steal northern India from Babur's son, Humayan. For fifteen years the Afghans held the Indus and Ganges regions, as Humayan's forces fled to Persia. Then, in 1555, Humayan returned to claim his father's empire. All this was business as usual in northern India. Emperor Akbar (r. 1556–1605) had only to include many high-caste Hindus in government, marry a Rajput princess, incorporate elements of native religion into his court and abolish the poll tax on Hindus. The normality of his position was then complete. Mother India might claim even these invaders.

Like all conquerors, migrants or refugees, the Mughuls felt a need to maintain some balance between assimilating with their new country and retaining their communal identity. Akbar experimented playfully with assimilation. Later, Emperor Aurangzeb (r. 1658–1707) recoiled against it. Aurangzeb believed that the subject masses must assimilate with the conqueror's culture, rather than the other way around. In his mind, the entire body of traditions from his Mongol and Muslim ancestors formed one homogeneous and absolute standard to which all the world must conform. So, upon the vast diversity of Hindustan, Aurangzeb tried to impose a desert-made social order. To a surprising extent he succeeded—though when masses of Hindus adopted his fanaticism, it brought the collapse of his kingdom.

The Rise of Hindu Fundamentalism

During the relatively peaceful reign of Akbar (late 1500s), there were several incidents of murder in the streets perpetrated by both Hindu and Muslim fanatics. Therefore, a Hindu holy man named Madhu Sudan Saraswati began organizing bands of Hindus and training them in martial arts. These groups assumed "holy vows" to fight and defend the Hindu faith (Panikkar 1963: 204).

Before the Islamic invasions there had been no "Hindu faith" as such to defend. There were instead hundreds of sects, and numerous feudal lords with men at arms. The various invaders of India had arrived and spread like fresh layers of social sediment over the land, pressing the lower castes further downward. But under the Mughuls, especially Aurangzeb, the strife among kings began to spread among the general population. The rivalry of Hindu rajas and Muslim sultans was being played out by their subjects. Whole communities of lower-caste Hindus were embracing Islam with its message of social equality. The lower levels of society were splitting into two camps, aligned with either the land's old or new masters. The new Muslim villagers of Dravidian ancestry were regarded as a new caste of traitors. If high-caste Hindus could not overpower the Muslim sultans, at least they could vent their rage on the poorest Muslims.

By the late 1600s, Aurangzeb felt the time had come to remake India in his own image. He therefore revived certain "Islamic" policies from ancient Arabia and Persia. In 1669 he ordered the destruction of all infidel temples and schools. Better ignorance among the people than unorthodox learning. A few years later he resurrected the poll tax on infidels and rammed it down the Hindus' throats.

By these means, Akbar's old high-caste Hindu allies were reminded of their place in God's order. With wrathful vigour, Aurangzeb's legions attacked the non-orthodox Shi'ite Muslim kingdoms on the Deccan plateau. In Aurangzeb's Islamic state, only Sunni Islam was recognized, and specifically a certain sect of the Hanafi school of jurisprudence (Faruki 1987: 53–54).

These policies managed to conjure up something quite new in India—Hindu nationalism. Before this, Muslim and Hindu kings had often formed alliances for practical reasons. But now, as revolts spread among the Hindu Marathas and Rajputs, or the Sikhs of the Punjab, Aurangzeb's armies faced a politicized countryside. The Mughul legions tried to crush not just rebel princes, but also hostile villagers. And their atrocities made matters worse. Shivaji, the rebel Maratha leader, wrote to Emperor Aurangzeb, "I wonder at the strange fidelity of your officers that they neglect to tell you the true state of things, but cover a blazing fire with straw" (Panikkar 1963: 163). The revenues of Aurangzeb's empire were mainly extracted from his Hindu subjects, and with these resources he tried to erase these taxpayers' culture. He pursued this "civilizing" mission until the expense in gold and lives broke the Mughul empire.

The rebellions against Aurangzeb escalated beyond any feudal disputes known in the past. These were people's wars of attrition and hate. "Religious war," as known in the Middle East and Christian Europe, had arrived in the heartland of India. Of course, this religious war was also an ethnic war. After Shivaji's death, the Maratha Hindu crusade became a *jihad* itself, with all non-Maratha people in its path liable to plunder, from Gujarat and Punjab to Bihar. The Islamic jihad and the theocratic state had conjured their Hindu match.

Likewise, in 1699, Guru Govind Singh formed the *Khalsa*, a religious-military organization of Sikhs against the Mughuls. Guru Nanak (1469–1539), the founder of the Sikh faith, had been so admired by Hindus and Muslims alike that when he died both communities tried to claim his body. But that was a time of relative peace, when the sharing of insight between cultures had been valued and possible. Now the Punjab was gripped in an ethnic holy war perhaps as brutal as the modern conflicts in Lebanon or Bosnia. Now only one culture was true, and only one would prevail militarily. The Sikhs embarked on their own religious war, fighting first the Mughuls and then the Muslim pastoral tribes of northwest Pakistan. Later they fought the Punjabi Hindus as well, with many of their best armed fanatics calling for a Sikh theocratic state.

By the mid-1700s, northern India was a world of plundering warlords, most of them claiming to fight for religion. The old, nature-oriented world of India had come to this. According to Panikkar (1963: 218), "In Bengal and the Gangetic valley there was anarchy of a kind comparable only to Germany during the Thirty-Years War." Into that chaos the British were able to move.

Public Memory in the Wake of the Religious Wars

On major hilltops across much of India, abandoned castles still stand, many dating from the Mughul-Hindu religious wars. Except for the Red Fort at Delhi, few of these brooding citadels are popular tourist attractions. Even the enormous fortress of Daulatabad in Maharashtra, once intended as a capital for all India, has few visitors. Many of the villagers around Daulatabad had been forced to build the fort, and today descendants of those conscripts still live nearby. When they gaze on the fortress, their feelings rarely include pride or nostalgia for past glory. Perhaps the most common attitude toward these forts and their rulers resembles Zahir Ahmad's remark about the old Nizams of Hyderabad: "What was their past, may I ask? Historically: Naked conquest and holding on to power, the purpose of which was nothing but idleness and a life of ease" (in Ahmed 1988: 166).

Those who tried to monopolize power and impose cultural conformity had a stultifying effect on the public mind. It was as if a forest of social diversity was being burned or clear-cut to make way for a human monocrop. The very memory of diversity suffered. Panikkar (1963: 217) writes that by the beginning of the 1800s the classic temples of "Ellora, Ajanta and Mahabalipuram meant nothing to Indians: nor did the magnificent sculptures of Elephanta and the temples of Orissa ... speak to us." Even the site of the Buddha's enlightenment under the bodhi tree had been forgotten and become overgrown. So successful were those who would erase other people's schools, traditions and minds. When millions of South Asians rallied to conserve their traditions, many seemed to envision simply a different cultural monocrop, made up of their own kinds of people and beliefs.

The Invention of Aryanized Hinduism

In the late 1800s, a movement arose to uplift Hinduism in the eyes of the world. For Swami Dayanand Saraswati, this meant highlighting elements of Hinduism which were respectable to India's upper classes and rejecting the superstitions of ignorant villagers. Saraswati proposed a return to a purely Aryan Hinduism—the pristine religion of the Vedas. To this aim, he established a movement known as the Arya Samaj. At a time when the European colonists were looking to Indo-European "Aryan" lineages for their original "races," the Arya Samaj stressed a pure Aryan identity for Hinduism.

According to Saraswati, the Aryan Vedas should be revered as Hinduism's sole canonical set of scriptures. The Vedas were reportedly "heard" from God, not composed by human beings as were the later "forest books," or Upanishads. With sure instinct, Saraswati preached the elimination of Dravidian culture from Hinduism. Only those elements which originally came from one set of desert invaders and corresponded to the values of later invaders would remain. Instead of many gods and goddesses, Hinduism would have one transcendent lord, to be named *Ishvara*. The faith of India would recover respectability before the monotheistic world religions. Veena Das (1989: 50, 48) presumes to call this reform a "Semitization" of Indian religion, influenced by both Islam and Anglican Christianity. In such Hinduism there was one god, one scripture, one true community of faith, and "contributions of the devotee as a means of maintaining religious bureaucracy rather than as offerings to the divine." For Saraswati and other reformers, the old diversity of myths and symbols was a sign of backwardness. The sense that different people have different spiritual goals—*dharma* (morality), *kama* (love), *artha* (achievement) or *moksha* (transcendence)—was replaced with the single goal of a unified Hindu social order (Das 1989: 52).

Perhaps Saraswati's version of Hinduism was only the latest step in a long journey toward "desertified" Indian tradition. Already the goddesses of India had been changed almost beyond recognition. David Kinsley (1988) points out that the goddess Lakshmi was once worshipped as the vitalizer of nature, standing on a cosmic lotus flanked by elephants spouting the waters of eternal life. But after the Muslim invasions, Lakshmi was ever less the upholder of nature and ever more the supporter of the patriarchal household. She was pictured kneeling before her lord and husband Vishnu, modeling the proper wifely role as prescribed in the Aryan *Dharma Shastras*. She was presented as a feminine sanction for the power of Hindu kings. As an ultimate *pativrata* (dutiful wife), Lakshmi seemed to answer the Muslim challenge, claiming that Hindu wives know their proper places at least as well as Muslim women do.

Scarcity and the Spiral of Communal Conflict

Natural scarcity drove waves of migrants out of Inner Asia, and environmental richness lured them to Hindustan. But under the invader's rule, the land of jungle villages turned increasingly desolate, with treeless fields and blowing dust. As in the legends of unrightful kings presiding over a wasteland, scarcity seemed to follow the refugee rulers to their new homes. Under the Mughuls, loyal soldiers received large grants of villages and lands. Whenever crops were poor, many of the peasants had to survive on loans from these landlords. If they fell behind in their payments, their lands were often confiscated. So the system generated "landless laborers," entrepreneurs with little or nothing in assets save their own bodies and wits. In rising numbers they drifted over the district lines into the towns and cities.

In a pure labor market, each landless laborer would compete against every other for the available work. In that case, the average adult's chances of employment in India would likely shrink as the population, and the number of displaced farmers, grew. But rather than accept such market conditions, many modern unemployed Indians tried to tip the scales in their own favor. As in situations of scarcity in other parts of the world, many tried to "look out for their own." Each ethnic or cultural group tended to help its own members compete, and to place other communities at a disadvantage. Such competition between groups often lead to a gradual escalation of intercommunal violence.

The constitution of India envisions a future based on legal and social equality for all ethnic groups. But after the democratic nationalism of the independence movement faded, the loyalties of many people ebbed away from the new state. An older view of government remained—a belief that politics is nothing but a device for the powerful to cheat the weak (Manor 1989). Within this view, loyalty to a local or ethnic group seemed an authentic moral stance, while support for national institutions (even in a democracy) smacked of collaboration with dominators. Within such realism, a moral reasoning like that of the Mafia took hold: if there is not enough to go around, and all power is corrupt, then people have to do what they must for their own group. Mahatma Gandhi had claimed, "There is enough for everyone's need, but not enough for everyone's greed." But by the 1950s this was no longer widely believed. The politics of scarcity was now indigenous to India, and no longer simply a style of rule imposed by foreign invaders.

After India's independence, in Maharashtra the memory of Shivaji's Hindu nationalist revolt inspired a political party called the Shiv Sena (the Army of Shivaji). This party argued in state and Mumbai (Bombay) elections that Maharashtra's "sons of the soil" should have legal preference for jobs and resources over "foreigners" from the other states in India. In Assam during the 1980s, vigilante sons of the soil eliminated some of their competition for jobs by staging several massacres of refugees from Bangladesh. In the 1980s and '90s, Gujarat was racked by communal rioting, partly over the issue of reserving jobs and university openings for former untouchables and tribal people.

With clan economic competition came clan religion. Ethnic self-defense begot the rise of reductionist fundamentalism. Hindu nationalists, such as those in the RSS, the Shiv Sena or the BJP, violated the spirit of Mahatma Gandhi by demanding a Hindu ethnic state. Sikh radicals also began disowning the role of Guru Nanak as a mediator between Hindus and Muslims. Das (1989: 57) says, "All these fundamentalist movements display antipathy to popular religious practices [that] evolved through centuries of living together." The most obvious result was the partition of India and Pakistan, which lead to one of the most horrific episodes of ethnic cleansing in the twentieth century. But that was just an act of realism of a certain kind. After all, much of the environment and culture of old Hindustan had already retreated east of the West Pakistani–India border. If these people of different cultural worlds moved towards a political divorce, no one could stop them.

South Asian Ecology in the Age of Development

The various waves of invaders from the northwest had come from lands where little would grow. In turn, they had little faith that resources could be renewable. Once in South Asia, they almost invariably made themselves into mere collectors of rent and protection money. They rewarded their loyal soldiers with estates, setting up the great *zamindar* class of landlords across northern India. For the zamindars, South Asia was a huge "resource" of watered lands with peasants to work them, endless forests, and exotic beasts to hunt. The subcontinent's economic product was measured by the amount of wealth extracted from the land and deposited in the royal treasuries.

The British inherited this legacy and built upon it. They also saw themselves as overseers of resource extraction. The British imposed their taxes on the zamindar landlords, and the zamindars passed their expenses on to the peasants in increased rents. To the colonists, South Asia was a subcontinent rich in the biomass of dead and living plants. To exploit that resource they converted villages into plantations and took the forests. First they promoted industrial-style farming of monocrops like cotton. Then they "opened" the forests of northeastern India for commercial operations in timber, tea and opium. They measured progress by the conversion of biological abundance into bank notes.

If not for their own national pride, most of Britain's own farmers might have identified with the position of India's peasants. After all, many English farmers were themselves descendants of peasants. They commonly displayed real sensitivity toward their land, and their farming craft often enriched the soil so that their fields literally rose on each side of an ancient roadbed. But the Englishmen who came to India were generally men from other backgrounds. And these uprooted adventurers proved themselves adaptable to the context of traditional rulers in South Asia.

Politics of Biomass Extraction

Before the industrial age, environmental poverty had been a local affair. But thereafter, depletion of natural resources in one region led to increased demand on almost all others. Thus, the old forces of desertification were amplified and transformed. Whereas before, the desert had seemed to advance locally from a certain direction, now environmental pressure on almost all lands was escalating. The implications for forests, farmlands and communities in every corner of the Indian subcontinent were dramatic.

The British administration claimed enormous tracts of jungle as government reserves. This did not mean the forests were being protected as nature parks. It meant that villagers were legally excluded from their forests, and the resources were awarded to British-backed industry. The colonial forest acts marked the first time that rulers of India had ever shown a strong interest in jungles. Before this, various kings and sultans had sometimes leveled forests for their construction projects, but they had normally viewed the jungle as unprofitable land. They had no way to tax the natural produce of forests. Only if the villagers cut the trees and farmed the land could the rulers measure the yield and take their cut. If left as jungle, the resource usually belonged to the villagers.

Before the British enclosures, the villagers had used the forests as communal foraging grounds for hundreds of useful plants. M.S. Randhawa (1980: 99, in Shiva 1988: 60) says the tribal people of Madhya Pradesh used thirty-one species of plants for roasted seeds, nineteen for baked roots, seventeen for fresh juice, twenty-five for leafy vegetables and ten for cooked flowers. They ate sixty-three kinds of forest fruit, and five kinds of fig. So in recent decades the Shantal forest people of Bengal often had a better diet than their high-caste neighbours (Parrack 1969: 40). Villagers generally prized the jungles in their natural state. With a forest nearby, gardening need be only a sideline activity. Forests provided the micro-climate, water, fodder and fertilizer to make a village viable for generation after generation.

Suddenly the British government wanted the forests. But its contractors did not value all the forest plants used by the local people. As Vandana Shiva (1988: 64) explains, of the three hundred tons or so of biomass per hectare, the contractors usually deemed only a certain amount of lumber as marketable. As a later forestry consultant put it, "By today's standards, most of the trees in these humid tropical forests are, from an industrial materials standpoint, clearly weeds." To extract the one resource they valued, the contractors tended to level the whole forest.

The British forestry department served partly as a police force to deny the villagers access to reserved woods. Supposedly, people should no longer trespass there to graze their cows, gather firewood or pick herbs. Of course, enforcing such regulations entailed some violence. Sometimes the right of contractors to cut down trees had to be protected by force. On some occasions, police and forestry officials felt compelled to open fire on protesting villagers, as in the May 30, 1930 massacre near Tehri, Uttar Pradesh. The villagers were now supposed to make do without renewable resources from the forest. They were to get all their food, fodder, fuel, medicine and building materials from the proceeds of their cultivated fields only. Worst of all, some tribal groups had lived entirely on resources from the reserved forests. When their woods were cut, these people lost everything. They wandered off to become field hands, urban laborers or "criminal castes." The old rulers had taxed what the peasants accumulated in their grain bins. Now British officials and contractors requisitioned the environment from under the villager's feet.

The new governments of India and Pakistan left the colonial forest acts in force. Apparently, trees still had to be reserved for industrial contractors and protected from poor village women gathering firewood. The new governments operated as sales agents for national resources. The villagers who had formerly used the woods for free were often willing to pay if they could regain access. Generally, however, the government preferred contracts with larger-scale interests. In Uttar Pradesh, the state forest development corporation rejected applications from traditional rope makers who needed *bhabhar* grass from the forests in favor of leases to pulp mills (Agarwal 1989: 275), perhaps because those larger contracts involved more money for the forestry officials.

In the south, the pulp mills chewed through virtually all remaining forests in Karnataka and Andhra Pradesh. The paper mills in those states then applied for cutting rights in northeastern India or the Andaman and Nicobar islands, or for freedom to import wood from elsewhere in Asia (Agarwal 1989: 273). Within 150 years of the forest reserve acts, the jungles of peninsular India were virtually leveled, save for certain pockets or tiny sacred groves around village temples.

When the forests were nearly gone, it suddenly became fashionable for governments to take a new role of *motivating* the people to conserve and plant trees. Of course, the villagers were to plant these trees on their own croplands or beside the roads. If they helped replant government forests, it would imply that these villagers had been given back the land. Here was a certain contradiction in the work of forestry officials. The resource police who had enforced the seizure of communal lands were now to motivate people toward social forestry. Surprisingly, they met with some success, but this success was mainly in stimulating something new—privatized tree plantations growing cash monocrops.

The Rise of the Mini-Zamindars

The old attitudes towards forests and farms changed most strongly in the 1950s after Nehru's Indian government abolished the zamindar landholding system. Under the new law, one man could no longer hold lands vastly beyond his own capacity to cultivate. "Excess" land was to be awarded to the actual tillers. Of course, the old landlords found many ways to evade the law. Often they simply re-registered land in the names of their close relatives. Still, there was some real redistribution of land. The landholding system inherited from the last wave of Central Asian conquerors was shaken up, and the class of farm owners considerably broadened. It was not that any vast amounts of land changed hands. The difference was that so many more "household heads" owned at least a small plot. The old distinctions between peasants and landlords were blurred. There were still old-style landlords and plenty of landless laborers, but between these extremes a broad middle band emerged, part peasant and part landlord. With land in these people's hands, farm production accelerated, contradicting the argument that larger landholdings must be more "economic."

Most of these new landowners were hands-on farmers, not rent collectors who stood above physical toil. But in other ways, many of these new landowners were just smaller zamindars who had fully absorbed the mentality of their former lords. Once the peasants got a chance, many of them tried to act the part of a farm boss. Whereas before, both husband and wife had labored together under the zamindar, a landholding man might now use his family as his workers, while he sat in the tea shop discussing markets for the crop.

All this showed a real change in the context of farming. For premodern peasants, a farm had been an end in itself. Land and family had fit together as both the means and goal of life. Most modern people would describe that life as one of aimless poverty. The animals slept in the courtyard like members of the family. Generations of people, beasts and plants came and went like a turning wheel. To the zamindars' view, this tapestry of organic life had no value save for the revenue generated. For them, the revenue from a piece of land *was* its value. Their view was the official one accepted by rulers, tax collectors and landlords alike. The new class of tiny landlords increasingly accepted it as well. Mahatma Gandhi may have dreamed of a popular return to traditional family farms and communal forests, but many new landowners wanted only to move "up" and make their fields miniature cash-crop plantations. The ancient cycles of crop rotation, which had been sometimes so elaborate as to require eighteen years to bring full circle, were widely replaced by continuous monocropping. The old goal of sustaining acceptable yields for generation after generation was often replaced by the pursuit of the highest possible revenue in each season. By that modern logic, the aim was to maximize farm output over input year after year.

Of course, revenue from a farm can be increased in many ways. Some actually improve the soil in the process. But one of the simplest ways is to increase the acreage planted, and this commonly involves eliminating the surrounding trees. In the new plantation economy, villagers increasingly viewed the forests as the old ruling class had seen them—as economic black holes. The remedy was obvious. By clearing the land, the logs could be sold, and then more crops planted. The clearing of small village forests accelerated.

By the mid-twentieth century, even the steep foothills of the Himalayas were being deforested. When it rained, silt from the hills began to wash down the slopes, filling the basins of new irrigation dams, and pouring directly into the Ganges. Downstream, the river repeatedly broke its banks, flooding the villages of Bangladesh. India, the land of jungles, was increasingly a land of dusty, sun-baked plains. Under the dry-season sun, goats scoured the countryside for tidbits of grass. Women and children wandered about searching for twigs or cow dung for fuel. To many foreign eyes, these villages seemed beset by problems inherited from the past—outdated ideas, noncompetitive technology or general ignorance. And many problems in India do have roots in the past. But viewed within the global projection of biomass exhaustion, these people's problems begin to look like quandaries of the future.

Anil Agarwal describes the difference between the ecological crises of industrialized nations and those of the tropical "developing" countries. The northern industrialized nations have been mainly concerned with proper disposal of industrial wastes, or the problem of pollution. Limiting pollution is generally seen as a non-essential boon which will further improve health and quality of life. In South Asia, however, it is the *biomass raw material* for industry and agro-business which is in danger of exhaustion. And since this "raw material" is the very stuff of life, its destruction is no mere inconvenience. In recent decades, around half of India's industrial output was directly "biomass based," as in the cotton textile, paper, sugar, soap, jute, rayon, timber and plywood industries. Each of these industries exerted enormous pressure on the shrinking resources of forest, soil and groundwater. In many areas of rural India, wood is now so scarce and expensive that the stock of houses, bullock carts and fishing boats is badly aging for lack of materials to replace them. Village women often work fourteen to sixteen hours a day just to find twigs or straw for fuel, fetch water and then do the cooking, field labor and childcare. The biomass famine is worst in the arid northwest, where many village women "may have literally reached their 'carrying capacities'" (Agarwal 1989: 273–75, 278–79).

The Market-Driven Ecology

In the late twentieth century, the idea that market forces could regulate the economy and environment better than any form of public trusteeship became globally prevalent. At the 1992 U.N. Conference on the Environment in Rio de Janiero, some heavily industrialized nations argued that the laws of supply and demand naturally balance the requirements of resource generation and extraction. Just as the global public grew convinced that exploitation of the environment must be somehow controlled, the very idea of social control over resources was virtually demolished. In India, Agarwal (1989: 278) protested that, "As the growing stock of biomass goes down, the demand for biomass from the cash economy goes up"; when demand begins to exceed supply, "pressure to exploit the remaining biomass increases dramatically." In such a market, every villager knows that the quickest way to get rich is take several truckloads of wood from a forest and sell them in a city. Perhaps a similar law of supply and demand was at work over the course of history, "balancing" the environments of Central Asia, North Africa and the Middle East.

In the Garhwal Himalaya in the 1970s, many local farmers tried to modernize by growing cash crops. In most cases they switched from their traditional rotations of grains to planting vegetables for regional markets. This increased their monetary income but also caused several side-effects. First, the old grain crops had produced a by-product of straw, and the village cows or goats got much of their food from that source. But the new vegetable crops produced no straw, and so the animals had to eat something else. Naturally they ate more of the grass on the surrounding hills. The straw from crops had also been a source of fuel for cooking. With that gone, the villagers also looked to the hillsides. The entire environmental culture began to unravel. With insufficient fodder, many of the animals had to be sold. This cut the supply of manure for the fields, which had been the main source of fertilizer. With grass stripped from the hill slopes, gullies snaked up the hillsides, sucking away the fields (Shiva 1988: 114). The villagers were eyewitnesses, not economists. In letting the market dictate what they would produce, they had abandoned their sensitivity to what that particular environment *could* produce. The famous Chipko movement grew in this region, not only as a movement to halt the clear-cutting of trees, but also to change destructive farming in the hills.

If the market always seemed to dictate the same acceleration of resource consumption, and neither big government nor big business could control this, perhaps only the local farmers could. Somehow, an economy was needed with biomass itself as the real measure of wealth. The lowly villagers who were previously seen as tax mules or the ignorant objects of development efforts began forming their own responses. They articulated the ecological implications of their religious beliefs. Starting in the 1970s, hundreds of local "anti-development movements," such as Chipko, the Kerala Sastra Parishad and the Mitti Bachao Abhiyan, formed and spread.

The Nature Worker's Alternatives

Some of the Havik people of the north Karnataka hills recently returned to their old ways of operating "spice gardens." These gardens are virtual forests of spice plants and trees, growing in layered canopies. On the forest floor are cardamom plants, plantain trees and climbing pepper vines. Above these rise coconuts, mangos and jackfruits. Most important are the huge betel nut trees. Betel nut is the main local cash crop, as chewing betel-pan is a small addiction for millions of people.

Over recent decades the Havik gardeners were drawn into the marketing and money-handling side of their operations. Many of them focused on traveling about to develop markets. Meanwhile, they began contracting out the actual betel and leaf picking to day laborers. The owners usually paid workers by the number of baskets of leaves or nuts they picked. On that basis, the faster the workers picked, the more money they made. To maximize their income, the hired hands often ripped down branches for rapid picking. By the early 1980s, large areas of the spice hill forests were, as Madhav Gadgil (1989: 328–29) described them, "a very sorry spectacle."

Fortunately, the problem was reversible in this case. Many of the Havik gardeners decided they were undercutting their own livelihood. The market demand on the trees was driven by sales orders, and the workers filling the orders were driven by economic incentive to speed. Neither of these values took account of the trees on which the whole operation depended. The gardeners' solution was, in many cases, to become gardeners again. They returned to paying themselves partly in pride about their gardens' beauty. But that was not the only benefit of renewed attention to the gardens. Some of the farmers in Hulgol village began experimenting with better trees for the most shady parts of their groves. They found that cocoa trees would thrive under the densest canopies of betel nut and mango, where plantains tended to die off. The cocoa trees filled this neglected bioniche very profitably. The gardeners also experimented successfully with a hedge plant called glyrecedia as a source of leaf-manure for the gardens (Gadgil 1989: 329). By such work, the Havik villagers returned to the creative interaction with plants from which their whole environmental culture had been created in the first place.

Vandana Shiva says the traditional gardeners in many regions of India remain unconverted to plantation-style, industrialized agriculture. Especially the women in these areas continue their old ways of mixed gardening, animal husbandry and forestry that can actually build the soil. Until the 1980s, outsiders almost invariably saw these traditional peasants as victims of backwardness. In development planning, these primitives had often been displaced in favor of more profitable schemes. But Shiva (1988: 108) claims, "It is from these remaining pockets of natural farming that the ecological struggles to protect nature are emerging." This argument might seem based on romanticism about the past. And everyone who knows rural India knows its past was hardly paradise. But growing numbers of Indian farmers have confirmed Shiva's view, returning to a more traditional gardening after trying the industrialized "green revolution" approach.

The Real Green Revolution

Plantation-style agriculture evolved where non-gardener overlords gained military or financial power over large tracts of land. The actual gardeners were then reduced to hired field hands, working basically for food and board. Particular plants were chosen for mass production, and the art of gardening was turned into a numbers game. So monotonous was the work on most plantations that the labor was easily amenable to mechanization. And where mechanization was later introduced, most of the former gardeners "became redundant." Still, in India and Pakistan, mechanization had made little headway by the 1990s. The holdings of "big" landlords were often tiny by Western standards. Farm machinery was still expensive, and labor dirt-cheap. But certain features of industrialized farming were widely adapted in South Asia. The bigger landlords increased their holdings, foreclosing on loans to their poorer neighbours. The old styles of farming with their complex inter-crops and systems of rotation were widely abandoned for a few cash crops. In service to this trend, the world's major agro-business research institutes engineered their new breeds of maize, rice, wheat or sorghum, which were the basis of "the green revolution."

The green revolution hybrids are gene-selected for maximum development of the seed-head. This growth comes partly at the expense of the leaves and stalks. Many kinds of "miracle rice" are therefore "dwarf" varieties, with smaller bodies and bigger heads. The underdeveloped stems sometimes fail to support the larger heads, so these plants are more easily blown down in the fields. The hybrids are also more fragile in other ways, requiring more water and fertilizer and needing more protection from insects and plant diseases than traditional breeds. Still, if the goal is to maximize production of grain for the mass market, the project makes sense. It is a case of increased risk for increased gain.

Probably a vast majority of Indian and Pakistani farmers tried growing the new miracle crops. In doing so they entered a mass market for standardized products, where efficiency in cost and scale of production were the main bases for any competitive advantage. Traditional Indian crops like ragi, jowar and mandua were shunted aside, despite their proven durability under local conditions. Even the nutritional value of the old crops was usually superior to hybrid grain, but now their lower yield per hectare made them non-competitive (Shiva 1988: 108).

Shiva reviews the case of HYV sorghum in southern India—a typical head-heavy, leaf and stalk-poor hybrid, introduced to replace the local jowar. In Kurugund village, Dharwar District of Karnataka, use of HYV took off as the wave of the future, going from 99 acres in 1970–71 to 835 acres in the early '80s, which was the entire area planted in sorghum. At first, the gains were impressive. With such production nationwide, India was able to stockpile vast reserves of grain. However, some critics claimed that the surplus came partly from lack of purchasing power, and average nutrition had actually fallen (V.K.R. V. Rao, in Shiva 1988: 129). But at least the food was there for a price and was usable for famine relief.

With irrigation, the productivity of hybrid crops was often spectacular. Those who could afford it sunk motorized pumps to irrigate their fields. With that dramatic rise in groundwater consumption, the villagers watched the water levels in their wells fall, often by several feet a year. By the 1980s, thousands of additional villages each year faced the strangely terrifying prospect of water famine in the dry season.

Larger heads of hybrid grain meant reduced straw. And, as in the Garhwal Himalaya, this meant less fodder, less natural fertilizer and greater demand on the local grass and trees. The loss of trees as natural groundwater pumps allowed further decline in the water table. If the groundwater fell too far, almost all the trees would die.

Since hybrid crops were vulnerable to local pests, the farmers often tried to protect their investment with pesticides and herbicides, some of which also eliminated "weeds" in the fields that had formerly been picked and eaten by field workers. Now, use of chemical poisons tended to eliminate both the weeding jobs and the weeds as food. The village women, who had earlier been deprived of food from the forests, were now deprived of natural food from the fields as well (Shiva 1988: 130).

The gains of the green revolution were well publicized. In Kurugund village, at first the HYV sorghum yielded seven to eight quintals per acre. But due to a complex environmental decline, this yield fell to an average of less than four quintals by the early 1980s, so low that the traditional breeds were again competitive in strictly quantitative terms. Villages across the country had similar results. Shiva (1988: 107) warns that, "It is precisely because ... essential links in the food chain have been ignored and destroyed by 'developed' and 'scientific' agriculture that the croplands ... are rapidly being turned into deserts." But the "green revolution" was still officially seen as a cure for, rather than a cause of, India's problems. It was mainly the supposedly ignorant villagers who started to think otherwise. By the 1980s, the tide in the countryside began turning back toward traditional kinds of gardening. Kurugund villagers in 1986 planted only 460 acres in HYV sorghum, which was down by nearly half from several years before (Shiva 1988: 124). The village culture was showing its own sense of balance, trying new things but moving away from all that was unsustainable.

The traditional village ecology was a resilient partnership of local plants, animals and humans. Perhaps some measure of respect for that trans-species culture is now returning, so it is no longer seen as something to be brushed aside from the path of progress. Perhaps science can now be used, not to dismantle traditional ecosystems, but to fine-tune them for synergy and sustainability. To bring back green cover to the country, Agarwal says, would be "the real green revolution" (1989: 282–83).

There seems to be an ancient Hindu myth relevant to every situation. In the *Vishnu Purana* and the *Padma Purana*, the Aryan God Indra abuses Lakshmi, the goddess of the sap of life. She therefore withdraws herself from the three worlds, taking with her all color, beauty, warmth and vitality, leaving the universe a sterile shell. The world is rendered an empty shadow of what it was, until she returns (in Kinsley 1988: 27).

Chapter 9

The Impact of Desiccation on China

IN TIMES OF DROUGHT, THE ANCIENT NOMADS OF INNER ASIA DROVE THEIR animals to the borders of China, looking for pastures near arable land. During the worst years they came in droves, crowding into the frontier markets, offering their skinny beasts for grain. In a buyer's market they bid to sell, inadvertently driving down the price of meat. Meanwhile the prices of grain and vegetables soared to meet the clamoring demand. At times like these, the Chinese placed their armies on alert. Everyone knew that if the "barbarians" could not get what they needed by trade, they were ready to take it by force.

By around 300 BCE the northwest Chinese states of Zhao (Chao), Yan (Yen) and Qin (Ch'in) all raised border walls against the desert people. And against those walls the Xiongnu (Hsiung-nu) nomads came riding in units of tens, hundreds and thousands. Fixed walls were always inadequate against the nomads' lightning raids. So the king of Zhao learned from the enemy. He ordered his troops to dress like the Xiongnu and practice their ways of war on horseback. By fighting in the barbarian way, Zhao general Li Mu was able to drive the raiders back in several campaigns (Yu 1990: 118–19). Then the borderland kingdoms found they could use the nomadic arts of war against their Chinese neighbours to the southeast (Pulleyblank 1983: 450). It was one of many lessons China learned from the so-called barbarians in subjects such as war, politics and religion.

If we divide China from northeast to southwest, the northwest half is a land of scanty grasslands, barren mountains and deserts, obviously another world from the farming villages of Han China. But the people of this desert world were far more than nomads and outcastes on China's doorstep. Many times they invaded and ruled vast fertile regions. They formed several major imperial dynasties and from the emperor's throne made their values "official" within Chinese civilization.

China's Heritage of Warlord Rulers

In the villages of China's southeast half, countless popular legends tell of an ancient "golden age" before the days of armies, warlords or dynasties. The legends may reflect simple romanticism about the past or real folk memories, however distorted. Many archaeologists describe the cultures of neolithic China (c. 6000–2500 BCE) as "matriarchal societies." The most ancient settled villages were clusters or rows of houses, each of roughly equal size. "Goddess figurines" sometimes presided over the fireplaces. As in Old Europe, there was a near absence of weapons of war or defensive walls. In predynastic writings, women commonly appear as clan heads. The prehistoric "emperors of perfect virtue" bore their mothers' family names. The ancient word for "family name" (*xing*) is a compound of symbols for "woman" and "bear," suggesting a typical matrilineal totem-clan. Perhaps such fragments of the past represent the legendary golden age, so fondly recalled as "the time of unspoiled nature and uncorrupted human virtue" (Min 1995: 553–55).

Like many countries, China's heritage seems split into various levels. Along with the bedrock of village traditions ascribed to the golden age stands another legacy of ruling warlords and emperors, who also claimed to represent their subjects' values. As Min Jiayin (1995: 555) explains, sometime around 2000 BCE, China slowly "turned into a patriarchal society, which can be symbolized as blades in the hands of male warriors." Some historians claim this shift happened naturally, without major invasions of warlike nomads such as those which overran Old Europe. But some series of events must have turned China's reigning ideals almost upside down between the "matrilineal" Yangshao culture of 5000–3500 BCE and the rise of the first imperial warlords.

The Shang

Around 1760 BCE, the Shang people established an empire over much of northern China. Their soldiers rode horses and chariots, and they took patrilineal names. They buried their leaders in large pits full of gifts and human sacrifices. According to Min Jiayin (1995: 559–60), "They were a semi-agricultural and semi-nomadic tribe who migrated eight times before they … established the Shang dynasty." The Shang apparently valued militarily capable men so much more than women that they constantly consulted oracle bones about the sex of unborn children. They practiced, according to the *Hanfei Zi*, "extending congratulations on the birth of a boy and killing a girl when she comes" (Du 1995: 158). Graveyards dated to Shang times show an adult population ratio of nearly two men for every woman. Du Jinpeng (1995: 159) concludes, "The high adult imbalanced sex ratio for the Shang dynasty population must … be due to some manmade reason."

Accentuating the imbalance of men and women, the strongest Shang men took many wives for themselves. King Wuding (c. 1300 BCE) apparently kept sixty-four women, at least thirty of them bearing the title *Fu*, meaning "king's concubine." Some of these women represented marriage alliances, and some were taken in war (Du 1995: 133, 162). We know that such polygamy, plus the general population imbalance, left many men without wives. The unmarried men often served as soldiers in the emperor's bands of tribute collectors.

Many Shang oracle bones list numbers of human sacrifices offered on specific occasions. The approximately two thousand known sacrifice lists (as discovered through the early 1990s) record a total of nearly thirteen thousand people killed. Apparently a Shang slaveowner could use his slaves for labor, sex or sacrificial offerings. Modern Chinese historians describe this phase of history, along with much of the following Zhou era, as "the slave age" (Du 1995: 163–64).

Mozi (Mo Tzu) (470–391 BCE) conveyed an old story about the establishment of the Shang dynasty. And although Mo opposed military aggression in his own time, he made the Shang war seem strangely justified. In this tale, the older Xia (Hsia) leaders (of about 2200–1766 BCE) were overthrown by a Shang hero named Lord Tang. The last Xia ruler, Zhieh, was so morally corrupt that both heaven and earth moved against him. Drought scorched the land, wilting the five grains. Winds and demons howled their protests, and people growled that there was not enough to eat. Yet more ominously, the cycles of the sun and moon grew irregular. Obviously a new pivot of heaven and earth was required. Finally, the nine ceremonial cauldrons in the Xia court, which were tokens of divine favor, vanished in a puff of smoke and reappeared in the court of Lord Tang. Tang then knew himself divinely chosen for power. He and his loyal men mounted their chariots and rode forth to execute the will of heaven. Like Lord Rama attacking the evil Rakshashas in ancient India, Tang smote his virtually subhuman foes, driving them in terror from the field. Then, as a true paragon of virtue, he offered the emperor's throne to any who dared take it. None dared. Tang was therefore obliged to make himself the pivot of the universe (Karlgren 1946: 329–33, in Campbell 1976c: 393–96).

Joseph Campbell (1976c: 395, 407) suggests this legend echoes the victory of mounted and chariot-driving warriors from the northern steppes over the early village cultures. At about the same time the Harappan cities of aboriginal Pakistan were being sacked by Inner Asian Aryans, the Shang warlords overran much of northeastern China. Civilization as we know it had arrived in the Far East.

The Zhou

Around 1100 BCE, another set of migrants from China's northwest borders displaced the Shang. The new Zhou (Chou) rulers are recorded as descendants of the semi-nomadic Huaxia, Rong and Di tribes. The Zhou were partly farmers and partly pastoralists, as befit the border zone with Inner Asia. Their nobles sought to control and tax productive villages, which led them to expand their conquests into the main zone of arable land.

Like many aspiring rulers, the Zhou princes claimed to bring a higher civilization to the benighted villagers. Subjection to a Zhou lord supposedly uplifted the peasants to a meaningful existence. Through paying taxes to a militarized state, the villagers become useful for something more important than themselves. The peasants, however, had their own views about the march of civilization. For example, a song from the *Shi Jing* (recorded sometime before 600 BCE) says,

> Great rats, great rats,
> Keep away from our wheat!

> These three years we have worked for you
> but you despised us:
> Now we are going to leave you
> and go to a happier country,
> Happy land, happy land,
> Where we shall find all that we need (Needham 1956: 106).

Historical evidence confirms that waves of northern farmers did migrate southward in Zhou times. At the Huai River they entered wetter, warmer country where rice could be grown. The harsh northern plains with their grasping warlords lay behind. But those who went south could not escape the warlords so easily. When large numbers of their taxpayers moved, the rulers simply sent soldiers and tax collectors after them. In hindsight, the Zhou princes even portrayed their fleeing subjects as obedient colonists going forth to settle rich new lands for their rulers.

So the subjects and then the rulers of the north arrived in the tropical lands near the South Pacific. There, Han migrants learned the arts of rice and pond farming from the southern aboriginals. The political and military power of northwestern warlords met the ecological sophistication of southeastern villagers, and these elements meshed to form a distinct Chinese civilization. On the one hand, China was a vast realm of nature-worker villages still steeped in the culture of a former age. On the other hand, the Zhou princes divided the country into a quiltwork of states competing for power and prestige in a slowly escalating arms race. All this lasted till 256 BCE, when the Qin state began to overshadow all others.

The Qin

The state of Qin (Ch'in) lay on the northwestern perimeter of arable land and perhaps bore the brunt of the Xiongnu nomad raids. The Qin ancestors were "compared to nomads … and rejected by the tribes in China" (Min 1995: 571, 564). In the Qin homeland, fighting was more necessary, and rain-fed farming less fruitful, than in "softer" states to the southeast (Claiborne 1970: 343). Qin talents were warped accordingly, and their rulers evolved the most effective military machine of the Warring States period. Around 300 BCE, they began swallowing state after state, advancing south and east. By 221 BCE, they had eliminated all rivals, "unified" the semi-continent for the first time and bestowed their kingdom's name on all of China.

The Warring States period (403–221 BCE), which the Qin victory brought to a close, had involved clashing ideas as well as armies. It had been an age of "one hundred schools of thought." The ideology championed by the Qin empire is often called "Legalism." And during the relatively few years of Qin rule (221–207 BCE), it seemed possible that all other schools of thought might be erased forever. The first emperor, Qin Shihuang, tried to wipe the Chinese mind clean and write only his favored doctrine upon it.

The content of Legalist philosophy was simple and straightforward. It was an almost flawlessly consistent expression of what Riane Eisler (1987) calls "dominator values." Basically, the Qin victory had seemed to prove that might makes right. Now all of society was to be reorganized according to this principle. The followers of old China's other "hundred philosophers," including the disciples of Confucius, Mozi (Mo Tzu) or Laozi (Lao Tzu), were either silenced or killed. The Qin prime minister Li Si (Li Ssu) advised the emperor this way:

> Your servant suggests that all books in the imperial archives, save the memoirs of Ch'in [Qin], be burned.... Those who dare talk to each other about the [Confucian] *Book of Odes* and *The Book of History* should be executed and their bodies exposed in the market place. Anyone referring to the past to criticize the present should, together with all members of his family, be put to death. (in de Bary et al. 1960: 140)

Back around 400 BCE, Mozi had asked, "If a small crime is considered crime, but a big crime—such as attacking another country—is applauded as a righteous act, can this be said to be knowing the difference between righteous and unrighteous?" (in Griffith 1963: 22) However, this question now fell under the Qin ban against criticizing the present by standards from the past. Surviving copies of Mozi's works were there fore hidden from the police.

Where force was king, the king counted warriors as the most productive members of society. The Qin rulers valued farming villages mainly as a source of supplies and manpower for the army. The philosophers and artists, previously employed in most princely courts, were now counted as a waste of resources. According to the Legalist *Book of the Lord Shang*,

> If a country is strong and does not make war, there will be villainy within and the Six Maggots, which are, to wit: rites and music, poetry and history, the cultivation of goodness, filial piety and respect for elders, sincerity and truth, purity and integrity, kindness and morality, detraction of warfare and shame at participating in it. In a country which has these twelve things, the ruler will not be able to make the people farm and fight, with the result that he will become impoverished and his territory diminished. (in Waley 1939: 236–37)

The Qin prime minister Li Si maintained (as the Mongols later would) that the reason the Chinese people existed at all was to serve their rulers:

> Now, if a ruler will not ... apply the system of censure in order to utilize the empire for his own pleasure, but on the contrary purposelessly tortures his body and wastes his mind in devotion to the people, then he becomes the slave of the people instead of the domesticator of the

> empire. And what honour is there in that? When I make others devote themselves to me, then I am honourable and they are humble: when I have to devote myself to others, then I am humble and they are honourable. (de Bary et al. 1960: 142)

How could such beliefs be resisted? Certainly not by pointing out logical inconsistencies in the Legalist philosophy. Besides, argument with the Legalists was out of the question. The Qin rulers were so contemptuous of others that they commonly killed anyone who dared contradict them. If the dominator ideology could not be defeated in debate, it would have to fail in practice. And fail it did, because when Legalist ideas were followed to their logical conclusions, the emperor had almost no enemies left to make. According to the *Han Shu (Treatise on the History of the Former Han Dynasty)*, Qin taxes on farmers quickly rose to half or even two-thirds of the crop. These warlords set up massive irrigation works but displayed a most exploitative attitude toward farm workers. The *Han Shu* says the Qin drafted "half" the able-bodied men in China for military expansion and construction of fortifications such as the great northern wall. The combined tax on food and manpower brought on a general famine in which "half the population perished" (de Bary et al. 1960: 212). With little or nothing left to lose, groups of conscripts and peasants across China rose in revolt. Their armed bands gathered strength like typhoons and converged on the second Qin emperor from several directions. The emperor's aides tried to avoid telling him the bad news.

After the fall of the Qin in 206 BCE, Legalist values fell from favor for a time. The new Han dynasty rulers again recognized intellectuals, sages and farmers as valuable in themselves. Most farmers conscripted into the Qin army went home to focus on production rather than conquest or extortion, and the warlords were temporarily banished to the frontiers of China.

Warlord Rule in the Dynastic Age

Out in the deserts to the north and west, the nomads needed China as much as ever. When the rains failed, they came to it like a tide, repeatedly breaking the military dikes raised against them. The Chinese armies posted on the frontier depended on the farmlands as well and were often tempted to ensure their supplies and pay by force. For the next two thousand years, through dynasty after dynasty, hired guards on the walls repeatedly joined hands with the nomads and turned into warlords themselves. Even in the early twentieth century the desert borderlands produced a stream of "northern warlords," such as Yuan Shikai, Wu P'ei-fu and Chang Tsung-ch'ang, the "governor" of Shandong in the 1920s, commonly known as "the dog-meat general." So the Qin "first emperor" of China was an example for many northern warlords to follow. The following brief summary suggests the extent of warlord rule over the past two thousand years:

When the Han dynasty dissolved in 220 CE, the country split into three main states. The strongest militarily was the northern frontier state of Wei, whose rulers came from a Turkic community called the Toba. Around 316, "the five barbarian groups," namely the Xiongnu, Jie, Di, Qiang and Xianbei, moved in to carve out kingdoms in north China. As the *Times Atlas of World History* (Barraclough and Stone 1989: 95) explains, "The result was the emergence of a distinctive Sino-nomad aristocracy, many of whom spoke both Chinese and Turkish, ... [and] it was from this ruling aristocratic group that emerged the ruling houses of the Sui (581–617) and Tang (618–907) [dynasties]."

Among several "barbarian" kingdoms in north China, the Sui state slowly eliminated its competitors. In 589 it conquered southern China as well, allowing an unprecedented concentration of wealth and power. But in 618, certain military aristocrats on the northwestern frontier rebelled and established themselves as the Tang dynasty. The first Tang emperor, Tang Taizong, called himself "the Heavenly Khan" (Gao 1995: 273).

In the 800s and 900s, Tang China faced damaging wars with the Turkic and Yugur (Uighur) desert people. At one point the Yugurs managed to sack the Tang capital. Finally, as the Tang administration fell apart with new military rebellions in 907, the Qidan (Kitan) tribes from the Gobi moved in, conquered northern China and ruled as the Liao dynasty.

The Liao were overthrown in 1125, not by the Chinese but by the next wave of pastoral invaders coming behind them. This time the Nuzhen (Jurchen) tribes of Manchuria advanced into northern China, establishing the Jin dynasty (1125–1234) up to the Yangze River.

Next came the greatest eruption of pastoral warriors in world history, as the Mongol horde swallowed the Jin and then rode south to conquer the entire Middle Kingdom. For over a hundred years (1260–1368), China lived under the Mongol yoke. A similar experience marked Russian society down to the present time, and this conquest left enduring marks on China as well.

With the Mongols' decline, the Ming Chinese resurgence lasted 276 years. Then semi-pastoral warlords from Manchuria overran all of China again. This Manchu (Qing, or Ch'ing) attack came in the midst of "the Little Ice Age." A series of unusually hard winters and droughts had reduced the growing season by around two weeks, and northern China was racked by famine and tax revolt. And just when the farmers were weakest, the Manchurians had the greatest incentive to move south (Spence 1991: 21). They formed an alliance with Mongolian and northern Chinese military units and were able to dominate the semi-continent from 1644 to 1911. After that, from the 1911 revolution till 1927, control over much of China fell into the hands of various "northern warlord" armies which were often as predatory as any known in the past.

Timetable of Warlord Invasions in Chinese History

c. 5500–2200 BCE	Legendary golden age
c. 2200–1766 BCE	Xia (Hsia) dynasty, rise of patrilineal inheritance of power
c. 1766–1100 BCE	Shang dynasty, possibly established by horse and chariot-riding warriors from the northeast steppes
c. 1100–256 BCE	Zhou (Chou) dynasty established by semi-pastoralists from the northwestern frontier
403–221 BCE	Warring States period, competition of the "hundred philosophies"
221–206 BCE	Qin (Ch'in) dynasty, all of China conquered by one warlord from the northwest borderlands
206 BCE–220 CE	Han dynasty
316 CE	Northern China overrun by "five barbarian groups"
589–618	China reunited under the Sui dynasty
618–906	Tang dynasty arises from coalition of northern generals
907–60	Period of division and competition among warlords
907–1125	Qidan (Kitan) tribes from Gobi establish Liao dynasty over much of northern China
1125–1234	Conquest of northern China down to the Yangze by Nuzhen (Jurchin) pastoralists, Jin dynasty
1234	Jin dynasty falls to the Mongols
1260–1368	Mongols conquer all of China, creating Yuan dynasty
1368–1644	Ming dynasty
1644–1911	China conquered by Manchurian armies, Qing (Ch'ing, or Manchu) dynasty
1916–27	"Warlord Period" of chaos over northern China
1931–49	Civil War and World War II
1949	People's Republic, modern China's assault on the desert

In sum, from the time of the Shang dynasty (1700s BCE) until the twentieth century, "northern warlords" ruled all of China more than half the time and the northern regions almost all the time. These various conquerors left their influence like layered deposits in China's composite heritage. The "top" of Chinese society was colored more than the "bottom." The north was affected more than the south. Matrilineal societies survived down to the present, but only in the southern mountainous areas, or on islands in the South China Sea. "The conclusion may be drawn," Min (1995: 555, 571) says, "that the semi-farming, semi-nomadic races, among which the patriarchal model was more advanced, imposed their culture … on the farming races." In 1898, during the last years of Manchu rule, T'an Ssu-t'ung reached a similar conclusion. But his expression of it was a raging, nationalist, frankly racist outburst:

> For the past two-thousand years the ruler-minister relationship has been especially dark and inhuman, and it has become worse in recent times.... How could we have allowed the unworthy tribes of Mongolia and Manchuria, who knew nothing of China or of Confucianism ... to steal China by means of their barbarism and brutality! After stealing China, they controlled the Chinese by means of the system they had stolen.... Instead of burning the books to keep the people ignorant, they more cleverly used the books to keep the people under control. Compared with them, the tyrannical emperor of the Ch'in [Qin] dynasty was but a fool! (in de Bary et al. 1960: 89)

The Desert at the Barbarians' Backs

In the most recent Ice Age, up to 10,000 BCE, glaciers scoured the regions northwest of China and scraped the topsoil into giant dunes, which blew southward in the prevailing winds. The soil fell like yellow snow drifts over much of northern China. So the rich soil of the Huangho basin came at the expense of the Gobi (Walls 1980: 53–54).

As the glaciers withdrew, Manchuria, Mongolia and Xinjiang were frigid sub-Siberian lands with thin soil and scant plant cover. But the Gobi was not yet a desert. For thousands of years more, glacial streams and snowmelt watered the land. And as long as the environment sustained it, hunting, plant gathering and gardening were more important to the Mongolians than herding. Farm settlements dotted Mongolia before 2000 BCE. The residents dug their houses halfway into the sod for warmth, and ground their grain on millstones. Later, however, as the land grew dryer, people abandoned these dwellings. For the next several thousand years the Mongolians were mainly herdspeople (Walls 1980: 70–71) who followed the fleeting rain and good grass over the steppes north and south of the expanding Gobi Desert.

Between the Desert and the Walls of China

By the second millennium BCE, the pastoralists were raiding farm communities near the edges of the steppe. Urbanized sites like Altyn Tepe, fortified villages on the Amur River such as Kharinskaya mound, and Kirovsk (near Vladivostok) were abandoned suddenly. In some huts, pots of millet have been found, burned to carbon on the fire, because the cooks never returned (Walls 1980: 72–74).

Behind the raiders came the desert itself, and this too appeared as an implacable enemy of the farmers. The old Silk Road from Hami to Kashgar is dotted with abandoned towns. During the 600s CE the kingdom of Kucha reportedly stretched nearly 650 kilometers down this northern arm of the Silk Road, skirting the growing Taklamakan Desert. The oases there grew rice, grapes, pomegranates and plums. The people mined various metals and consid-

The Eastern End of the Afro-Asian Arid Belt, and Greater China

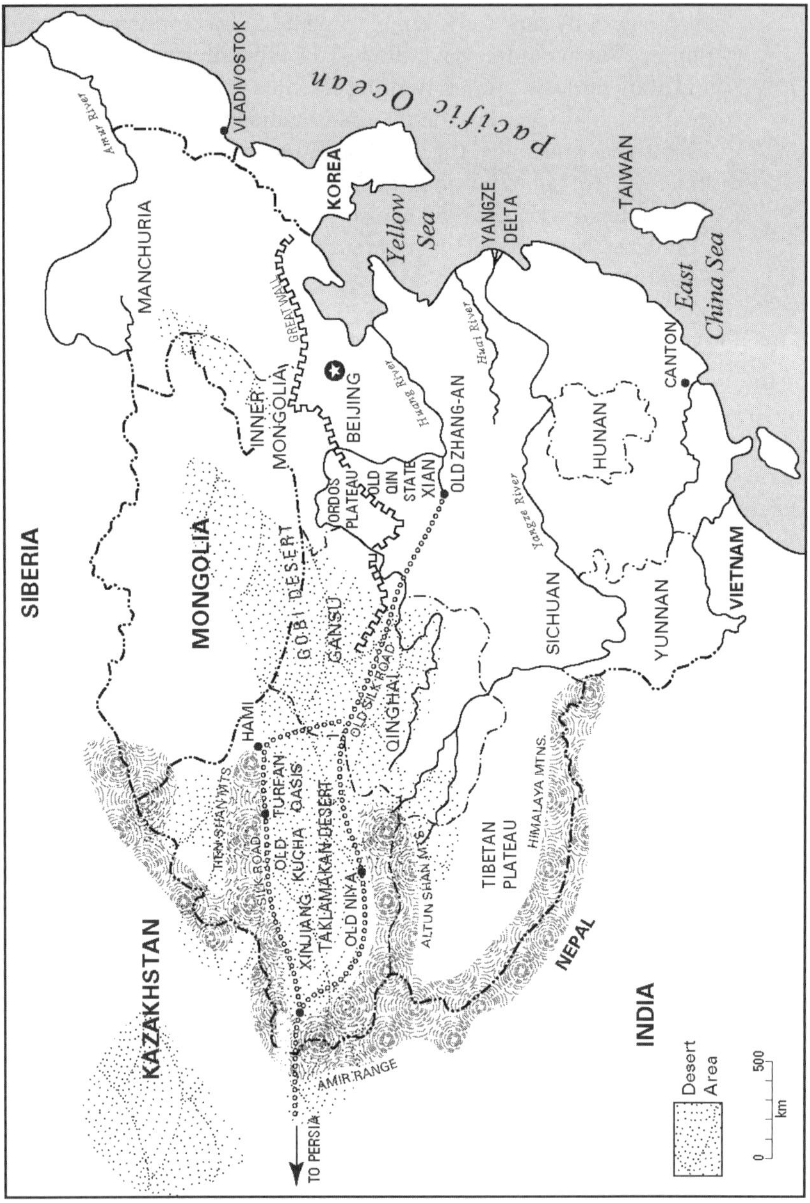

ered themselves rich (Hopkirk 1984: 139–41). Then the water supply slowly and "naturally" diminished. On mountains ringing the Xinjiang basin, the Ice Age glaciers gradually melted. Each century the meltwater flowing into the basin decreased. The Taklamakan Desert began in the center of the basin and slowly grew outward. The farmlands of Xinjiang shrank to narrow belts near the mountain ranges and then shriveled into islandlike oases. Some villages channeled mountain streams to themselves through hand-dug tunnels and survived as if attached by umbilical cords to the hills.

On the southern arm of the Silk Road, the residents of old Niya abandoned their town to the desert in the third century CE. They traveled southward and built a new Niya closer to the Tibetan plateau, where waters from the highlands still flowed (Hopkirk 1984: 30). Around 1900, Aurel Stein led a team of explorers back to old Niya. There, as the team removed sand from several ruined houses, Stein realized he was standing in an ancient garden. The trunks of poplar trees stood arranged in avenues or squares. With an overwhelming sense of obliterated time, Stein walked down the lanes between rush fences. Beside the houses he recognized remains of apple, peach and apricot trees (Hopkirk 1984: 91). According to Yugur tradition, the name of the Taklamakan Desert means "homeland of the past" (Allen 1996: 43).

By around 350 CE, the old Konche River dried up before reaching Lou-lan (near Lop Nur), and the Lop lake dried into a vast plain of salt. The nearby town of Katak was likely deserted near that time. But the sixteenth-century Moslem historian Mirza Haiden set a tale of Katak's destruction in the Islamic era. In this story, Allah passed judgment on the town dwellers' sins. The mullah and muezzin were aloft in the mosque's tower calling the people to prayer when a fine rain of sand began to fall. As the holy men watched, it piled up in the streets, rising over windows and doorways. Soon the very housetops disappeared. When the level of sand rose near the top of the tower, the mullah and muezzin jumped down. Finding no other survivors, they fled the godforsaken place (Hopkirk 1984: 31, 33).

With such treatment from the powers above, most people in Xinjiang gave up gardening to become nomads. But the pastures slowly retreated towards the basin's mountain walls and east towards China. So down the old Silk Road to Chinese markets the pastoralists brought their horses and sheep. Their trade routes skirted the edges of the desert. The passing herds commonly overgrazed the route, "laying a welcome mat of barren earth for the desert to claim" (Walls 1980: 29).

As the Gobi Desert slowly widened to nearly one thousand kilometers across, the forests of central China faded and thinned. The watersheds of the upper Yangze and Huangho rivers then no longer supported sufficient trees to absorb the rain. Like most arid-country watercourses, these rivers grew increasingly subject to flash floods, which periodically tore through the heart of China. The rivers became conduits of massive soil erosion, their waters the color of Inner Asian topsoil. With this gradual collapse of the environment, the people of Xinjiang, Gansu and Inner Mongolia were slowly squeezed between the advancing deserts and the walls of China.

The Barbarians' Answers to Poverty

China's farming peasants were often so productive that they could meet almost all their own needs while paying heavy taxes. Most were poor but more than self-sufficient, and over time their numbers mounted. But the pastoral economy of Inner Asia was usually incomplete. To supplement their diet, the nomads normally needed to import grain and vegetables. The pastoralists of Xinjiang and Mongolia were basically single-resource sellers, and the Chinese appetite for their animals was limited. Horses did not thrive and had little use in confined semi-tropical villages. So, while the nomads needed to maximize their sales of horses, the Chinese tended to offer very low prices. Apparently, the law of supply and demand meant that the pastoralists could get little for their products. But they were commonly unwilling to abide by that law. Rather than go hungry in deference to Chinese tastes, they would rather take economic law into their own hands. They had other means of making the Chinese pay.

Extortion for Peace

In 209 BCE, the great warlord Motun became Khan of the Xiongnu (Hsiung-nu) nomads. He was the son of a Xiongnu king, but his father chose another brother to inherit the mantle of leadership. To win power for himself, Motun had to strategize. First he trained his own companies of archers to obey him without question. He rewarded those who obeyed when he told them to shoot his favorite horse. Later he upped the ante. At a shooting display, he ordered his men to turn toward the reviewing stand and shoot one of his wives. With that test passed, Motun knew his men were ready to obey him in a rebellion against his father and brother (Sinor 1990b: 120–21). In such contests of "succession," the winner could always say the best man had won.

The Xiongnu had recently watched their neighbour, the king of Qin, rise up to seize control of all China. Motun may have concluded that if the Qin could learn military tactics from the Xiongnu, then he could learn political tactics from the Qin. It was a common prejudice among the Chinese that the barbarians could never unite themselves politically. As the Sui dynasty annals record,

> The Turks prefer to destroy each other rather than to live side by side. They have a thousand, nay ten thousand clans who are hostile to and kill one another. They mourn their dead and swear vengeance. (Yu 1990: 307)

Motun would prove the Chinese wrong. First he crushed the Dong-hu tribes, winning their submission to a stronger general. Then he drove the Yueh-chih so far to the west that they ended up in India. A few more campaigns, and Motun became a kind of first emperor over all the arid northwestern half of China. Next his horde encroached on the borders of China proper. Most Chinese settlers there had already fled as the Qin empire collapsed. But as Motun advanced, he came up against the new Han dynasty.

With his centralized authority over the nomads, Motun was able to offer the Chinese a new bargaining chip. He could offer to turn his people's raids on or off. He could offer the Chinese not just horses but also, if he willed it, peace. Starting in 198 BCE, Motun gained a series of "peace agreements" with the Chinese. He dealt with them not as a begging merchant before an emperor, or as an outlaw before the police, but as Khan to Khan. The Chinese would offer grain, vegetables and luxuries. Motun gave them horses and peace. Peace was a new commodity. It could be bought for a price, or else it would be taken away. In a moment Motun could unleash his raiders. If the Motun could not prosper by selling the Chinese things they wanted, then the great Khan could make them pay to escape pain.

Extortion rather than ordinary trade became the nomads' main business. The price of peace for the Chinese was measured in food, wine, silk and Chinese wives for the Xiongnu chiefs. Between 198 and 135 BCE the Xiongnu made ten peace agreements. Each time they raised the price. In 50 CE, Emperor Guangwa sent the southern Xiongnu 10,000 pieces of silk cloth, 10,000 catties of silk thread, 25,000 bushels of dry rice, 36,000 head of cattle and sheep, and various other "gifts." Like Rome's payments to the Huns, the treaties became a constant and increasing drain on the empire. And still the Xiongnu did not completely abstain from raiding border areas to supplement their take (Yu 1990: 143, 125).

The protection racket, of course, had its limits, because it tended to drive away its own customers. For example, it was the protection racket that killed the old Silk Road. As Inner Asian chiefs along the route imposed ever higher fees for safe passage, the cost of transport often rose higher than the worth of the goods. Trade then drained away till the Silk Road was virtually abandoned. In the future, trade crossed in boats through the South Pacific and Indian Oceans.

Protection money was worth paying only so long as peace cost less than war. By about 50 CE, most ministers in the Han court felt that this point had been passed, and further "appeasement" of the barbarians fell into disfavor. The ministers then redirected peace payments to the Chinese army. Mass conscriptions of farmers swelled the ranks. The mobilization effort took two decades, as millions of people were diverted from production into readiness for war. Then from the 70s to 102 CE, General Ban Zhao (Pan Chao) made his career flinging the armed might of China against the Xiongnu. His strategy was to redivide the pastoralists. He treated each tribe in the confederacy as a separate nation and attacked them one at a time. After each victory, he conscripted the conquered tribe's men to help him fight their neighbours, paying them to settle grievances with their traditional enemies.

With sufficient force, nomads could be scattered into the wastelands from which they had come. They could be driven deep into Inner Asia, causing shock waves of migration all the way to Rome. They could be made to go away for several years, and when they returned to China's borders, they could be treated with naked contempt. But they could not be made to disappear. With incredible tenacity, the people of the drylands survived the fiercest droughts and wars of extermination. They often managed to increase their numbers in places where most other life forms came close to extinction. Inevitably, the nomads would return to the borders of China, starting over, trying to find a place in the greater Chinese economy. They had not only animals to offer, but also themselves as workers. And their work experience was generally in war.

The Mercenary Business

During the first Han dynasty (202 BCE–9 CE), Emperor Wu sent an expedition seeking military alliances with "far barbarians" (the Yueh-chih) against the "near barbarians" (the Xiongnu). Later, General Ban Zhao redivided the Xiongnu, turning barbarians against barbarians. This was possible because the pastoralists saw military service as employment, and Ban Zhao offered them jobs. After the war, many nomad army veterans clung to their positions as frontier guards, and in this role they were more willing than most Chinese conscripts. Most Chinese farmers viewed service on the frontier as like a prison sentence, but young men from the pastoral tribes often saw regular soldiering as a step up in life.

Like Rome, China accumulated a frontier army full of mercenaries from the border tribes. The tendency now was to use near barbarians to guard against far barbarians. Of course, this arrangement could backfire at any time, because hired mercenaries could always turn on their masters. Therefore the Chinese emperor was often compared to a man riding a tiger.

The guards on the Great Wall were supposed to protect their Chinese paymasters from their own unemployed kinfolk. They were expected to lay down their lives for an empire which did not recognize them as real citizens. All of this could be galling to proud men. Still, the mercenaries usually served their masters well. Only occasionally did they put their own ambitions first. In 691 CE, for example, the Turkut tribes demanded vastly increased tribute for their role in guarding the western frontier. The Chinese government treated this as treason and moved to suppress the Turkuts by force. However, the rebels under Khagan Mo-tsh'o received inside information from other Turkuts within the loyalist forces, allowing the rebels to trap and massacre a Chinese army sent against them in 706 CE.

In 907 CE, bandits under Zhou Wen overthrew the Tang dynasty, and China entered a fifty-year free-for-all contest between former-mercenary armies. The Cha-to soldiers, who had served as guards on the borders of Tibet, temporarily seized the Chinese capital. Next, in 942, Qidan (Kitan) tribe mercenaries on the northern frontier rebelled over lack of pay. Their leader, To-kuang, took the north of China around modern Beijing and proclaimed himself emperor.

Conquest of the Middle Kingdom

Beyond trade, extortion or military employment, the ultimate means for arid-country people to get what they wanted was to conquer the entire Middle Kingdom. After the collapse of the Tang, northern nomads mounted a series of escalating incursions into China. The Qidans of the Liao dynasties took most of China north of the Huai River. The Nuzhens (Jurchens) swept in behind them and pushed their Jin dynasty down to the Yangze. Next, in the 1200s, the Mongols took it all.

When the Mongols overran China, at first they practiced a traditional policy of killing males and children, and seizing women and craftsmen. As Genghis Khan reportedly said, "It's the greatest pleasure for a man to defeat his enemies and stamp them out ... take their pretty wives and concubines, and let them attend upon him in bed" (Liu 1995: 327–28). Basically, this is what the Mongol generals did in northern China. Their early victories there featured massacres of Chinese men and mass suicides of Chinese women. Some Mongol ministers recommended exterminating the whole population, taking all the people's money and using the land for horse pasture. But as the Mongols moved south, wiser heads prevailed. Obviously more could be squeezed from living peasants than dead ones. So the conquered Chinese subjects gained an officially accepted right to exist as producers of revenue. In 1239, a merchant named Abd Al-Rahman approached Ogedei Khan and said that if he were made chief tax-collector for northern China, he could double the take. He won the contract (Morgan 1987: 100–103). Under Mongol taxes, many farmers could not survive without overexploiting their land. All told, due to war, famine and plague, the population of China seems to have dropped by about 40 percent during the period of Mongol rule, from roughly 100 million to 60 million.

The Mongols brought in about 300,000 of their countrymen to oversee China. Under such an administration, the traditional Confucian examination system for selecting government officials was suspended (until 1315). Like the Qin Legalists before them, the Mongols did not recognize any source of authority other than military might (Morgan 1987: 111).

The Mongol invasion was probably the nadir of relations between the Han Chinese and pastoralist "outsiders." After that, Sino-barbarian relations had nowhere to go but up. However, the Liao, Jin and Mongol conquests forced the farmers and pastoralists to live together. By the reign of Qublai Khan, many Mongol officials had "declined" into adopting the sensibilities, tastes and scruples of the Chinese. So the invaders faced the constant danger of being absorbed into the culture of village China.

The Warlords' Women

The "historical defeat of the female sex" happened in China largely through an escalating race for military control. Of course many women participated in that race. According to Du Fangqin (1995: 217), the Warring States period (403–221 BCE) was like a storm of "buffeting waves whipped up by the quest for power." In that uncertain situation, many women "wanted to be patronized and supported, and exchanged their moral qualities, dignity, and freedom for patronage and support." A paradox arose which sometimes eludes discussions of women's status. On the one hand, the women of conquerors could gain great "status" as partners in domination. On the other hand, a culture and economy based on coercion recognized no true partnership, but only varying degrees of dominance or submission. As the *Yi Jing* oracle text advised,

> If Yin [the female quality] is the dominator in a family, then a man becomes an outsider.... Therefore Yang will become weak and Yin will be strong.... If you meet a woman who is physically stronger than you, do not accept her. (in Sun and Pan 1995: 252)

Such attitudes could serve as a counter-natural force of evolution, promoting the survival of submissive women. But even if the warlords' women had to be weaker than their husbands, they could still be far stronger than most men in the land.

With very few exceptions, women in the ruling circles served as the warlords' wives or concubines. The crucial distinction between these two evolved during the Zhou period (c. 1100–256 BCE). Basically, the children of concubines were not allowed the same rights of inheritance as the children of wives (Du 1995: 176). Such concern for both legal polygamy and patriarchal control over inheritance seems to reflect nomadic rather than village customs. Liu Ruzhen (1995: 327–28) lists various reasons for polygamy among the ancient northwest nomads. First, the nomads' constant wars over scarce resources made for high losses of men and a "surplus" of women. Second, in traditional pastoral war, the maximum self-interest of each party was to eliminate rival males and absorb rival females. By contrast, Zhang Zhijing (1995: 419) claims that polygamy was never widely accepted in village China. Villagers were seldom inspired by their rulers to go raiding for concubines.

Powers of Nomadic Women

It is simplistic and dangerous to blame the problems of Chinese women on bad influences from the pastoral tribes. No ethnic group should be made a scapegoat for the suffering of the past, and the relative status of village and nomadic women was uneven. From the southern rice fields to the northern plains and northwest deserts, all people tended to value each other according to their economic roles. In the open Mongolian plains, a woman's economic role was obviously crucial. Her tasks of herding smaller animals, shearing sheep for weaving, or fetching water and fuel commonly required expert horseback riding. This presented an image of female power and freedom which was shocking to many Han Chinese. Child raising itself was highly valued, and as the men were often away with the herds, pastoral women usually did their work without male supervision. These women were more accountable for meeting the needs of nomadic households than for conforming to strict social rules. Some nomad women were prepared to defend their camps with proficient archery, and Mongolian widows often succeeded their husbands as family heads. All these rights and powers were impressive to the often-conquered villagers of northern China (Jagchid and Hyer 1979: 95).

But so long as the steppe nomads lived chiefly by herding horses (which is dangerous for mothers with infants), and so long as the Mongols solved the problems of drought by riding those horses to war, men remained the central actors in the nomadic economy. Like the Arab Bedouin, the old Mongols practiced polygamy for men only. When they conquered areas of northern China, it was their customary privilege to seize local women as war booty. Before the establishment of Buddhism in Mongolia, Ming dynasty historian Xiao Daheng (in Jagchid and Hyer 1979: 103) noted, leading Mongol warriors were buried in "simple" chieftain graves. These commonly contained a wooden coffin, a suit of armor and weapons, some clothing, a favorite horse, plus the dead man's concubines and secondary wives.

Status of Dominator Women

When the pastoral nations invaded China, many formerly nomadic women became members of the Middle Kingdom's ruling elite. This, of course, was a vast rise in "status." But what was the *quality* of that status?

The Liao dynasty (907–1125 CE) emperors came from the Yel clan of the Qidan (Kitan) people, and their empresses came from three "empress clans," namely the Bali, Yishiyi and Shul. According to Liu (1995: 330), "People outside the clans of the emperors and empresses had no opportunity at all to hold the power of the state." Even within the Qidan tribes, the men and women at the top were a dominator elite, holding power through military or marriage alliances. This tribal elite then placed itself atop the whole society of northern China, also by military means. The women involved in this enterprise were partners in domination. Some historians cite this as evidence of high status for women among the nomadic nations. But was this a case of "status" for women or a problem of the culture of domination involving women?

The official histories of "nomadic dynasties" record incidents of court women exercising power. The Liao dynasty empresses presided over the court whenever the emperors were unable to attend. Reportedly, they were usually competent and well respected in this role. But Emperor Xingzong felt that Empress Fatian grew too influential. He therefore put her under house arrest and dethroned her (Liu 1995: 322). In the Jin dynasty (1125–1234) which followed, Emperor Shizong decreed that "the empresses are forbidden to interfere in politics even when they have high ranks" (in Liu 1995: 327–28). But this was difficult to enforce. Empresses were part of the ruling family household. Later in the same dynasty, Emperor Xizong's wife, Daoping, displayed ambition and took over the functions of prime minister. Xizong grew angry and finally ordered her executed.

In the late Jin dynasty, court women grew increasingly assimilated to Han Chinese culture. Empress Xiaoyi studied Confucian classics, and especially loved the Daoist works of Laozi (Lao Tzu) and Zhuangzi (Chuang Tzu). Like many court women, Xiaoyi "adopted the easy-going attitudes of the Taoists [Daoists] and acted in accordance with the rites" (Liu 1995: 327). Many men of the Jin aristocracy felt that these women were "going soft" and abandoning traditions which made their people superior.

The Jin were said to be going soft, and the Mongols felt hard enough to replace them. Mongol empresses then displayed their own drive to "interfere" in politics. The second empress of Zhizheng persuaded her crown prince to invade Korea. Then she entered into a conspiracy with the eunuch Piao Buhua and the warlord Koketema to overthrow the Zhizheng emperor and place her crown prince on the throne. This intrigue exploded into civil war. Both the empress and emperor fielded personal armies, "which greatly exhausted the country" (Lin 1995: 326). Again the rulers tried to limit women's interference in politics, since it compromised the principle of unified command under the strongest man.

North China's Women under Warlord Rule

Patriarchy and militarism reached their most extreme forms, not among the nomads in their steppes and deserts, but in the towns and villages of northern China—the farming regions most often ruled by nomadic invaders. Whenever nomads invaded northern China, they changed from wandering pastoral communities into military elites ruling over the most populous regions on Earth. There they gained power and wealth beyond their ancestors' dreams and were subject to all the temptations such power brings. In that situation, their overriding concern was the preservation of their dominance over others, and this was the leading principle they displayed to their subjects. Subordinates were to obey superiors, as if all society was a chain of command. The implications for both subject and ruling-class women were quite clear.

As the invaders usually lived in towns and cities, their brand of ruling-class culture was most influential in the urban areas. In the crowded towns, the ruling families lived as if constantly on display. As in the Middle East, urban women were often subject to patriarchal conventions far more than their rural sisters (Jagchid and Hyer 1979: 95).

One glimpse at the conditions of life for women in the north comes from the village of Daoyi, in southern Manchuria, where village records have survived from the late 1700s. In those days, roughly half of all Daoyi children died before age twenty. But this death rate was unequal for boys and girls. Substantially more men than women reached marriageable age. This imbalance was further aggravated, because the most powerful local men often took several wives and concubines. So, while almost every woman was either married or widowed by age 30, some 20 percent of the men never had a wife. We can imagine the ferocious jealousy which would surround women in this situation

(Spence 1991: 95–96). We also know how gangs of "surplus" men could become loose cannons in society. Such men might be channeled into the army to vent their aggression on outsiders, they might make themselves a force for patriarchal-style religion, or, if such semi-honourable options failed, they might fall into banditry and prey on their own communities. The source of Manchu army manpower was obvious. Manchuria at this time ruled all of China with its superior resources in unattached men. North China offered one of the world's best examples of a society suffering from an imbalance between yin and yang.

China's laws concerning women grew more complex and restrictive dynasty by dynasty, and the records show which rulers did the most to limit and control females. In predynastic times, women could initiate divorce; no authority beyond the family held a woman subject to an abusive husband, and no law prevented women from marrying again. It was the Qin dynasty (221–207 BCE) which first criminalized remarriage—for women only (Zhang 1995: 421). Later, the originally nomadic Sui and Yuan (Mongol) rulers decreed that women above a certain rank could never remarry if their husbands died, thus protecting the patrilineal inheritance of great estates (Zhang 1995: 421).

During the Han dynasty (202 BCE–220 CE), Chinese women had a legal right to take their dowry property with them if they divorced, and Mongolian women had similar traditional rights in their homeland. But strangely enough, after the Mongols conquered China, they imposed a new law for divorced women: Her "dowry and other properties are to be left at the disposal of her former husband. She is not allowed to take them with her" (*Yuan Dianzhang*, in Zhang 1995: 430). Such was the difference in mindset between nomads and formerly nomadic emperors.

The Yuan dynasty Mongols were the first rulers to allow men to punish adulterous wives by selling them into slavery. The Yuan rulers also permitted husbands to kill "cheating" wives for "resisting arrest." The Yuan law said that male and female adulterers would receive the same punishments, but any children born of the love affair were given to the male offender, and the woman could then be sold by her husband—to anyone except her lover (Zhang 1995: 436). Yuan law also removed all legal limits on the numbers of wives and concubines a man could keep (Liu 1995: 337). Zhang (1995: 449) concludes that for over two thousand years, "women's subordination to men became ever more entrenched," and that "the legal codes of the Yuan and Qing (Manchu) dynasties … contributed most to this development."

Under such rulers and laws, many higher-class women lived in a kind of *purdah*. Strict rules, armed guards, raised walls and "talk" protected women's "purity" from outsiders. Especially in northern China, many women's public and economic roles grew so delimited that they were commonly foot-bound to enhance their sex appeal even at the expense of every other human function. During the last years of the Qing (Manchu) dynasty, the fiery young martyr Qiu Jin (1875–1907) protested:

> When men said we were useless, we became useless: when they said we were incapable, we stopped questioning them even when our entire female sex had reached slave status. At the same time we were insecure ... so we did everything to please men. When we heard that they liked small feet, we immediately bound them just to please them. (in Ebrey 1981: 248)

It seems her words found an echo among the majority of Chinese women.

Village China's Age of the Goddess

As in the West, China's warlords altered history to suit themselves. They made it seem that civilization began with armies and autocrats. The First Qin emperor tried to make it impossible for people to criticize the present with standards from the past and destroyed books and records from previous times. So the predynastic age before warlords was largely purged from the official record. The supposedly ignorant villagers, however, lived by their own records, and in their vast oral heritage, the memory of predynastic cultures lived on. The peasants commonly idealized their pre-imperial past and onto the slate of antiquity projected their fondest dreams. No one could prove these dreams wrong. So, in the chaotic decades after the fall of the Han dynasty (220 CE), Pao Chingyan claimed:

> In ancient times there were no Lords or officials. Men (spontaneously) dug wells for water and plowed fields for food. Man in the morning went forth to his labor (without being ordered to do so) and rested in the evening. People were free and uninhibited and at peace: they did not compete with one another, and knew neither shame nor honours. There were no paths through the mountains, and no bridges over waters, nor boats upon them.... Thus invasions and annexations were not possible, nor did soldiers gather together in large companies in order to attack one another in organized war. (in Needham 1956: 435)

Where memory and vision combined, the legendary golden age emerged, reflecting the values of village nature workers. As Zhuangzi (Chuang Tzu) remembered it,

> In the days of Shen Nong ... they knew their own mothers but not their fathers. They lived with deer, they fed on the food they grew and wore the cloth they wove, and harboured no ill will against one another. This was indeed a supremely ethical society. (in Min 1995: 592)

Where did such tales come from, if not bedtime stories told by grandparents to grandchildren?

Persistence of Paradise Myths

The West has its own myths of an "age of the garden." And both the biblical and quranic stories of an original earthly paradise are still widely believed to be an historical fact. But these "western" scriptures also claim that this paradise has been lost forever. It can never be recovered in this world, but can only be attained in another life. In China, however, perhaps most people continued to believe it possible that the time of unspoiled nature and uncorrupted human virtue could return.

The primitive Chinese reverence for unspoiled nature was akin to that of North American natives. Early village shrines celebrated nature's beauty, like the original Shinto temples of Japan. The old Daoist shrines were called "kuan," meaning "to look and see," and the thing looked upon was nature. The Daoist temple of Heilongdun, for example, was a place for viewing the surrounding countryside. Inside all was bare save one tablet bearing the words, "Nature—the Mother of All Things" (Needham 1956: 56–57, 164). The ancient villagers evidently felt their land full of the promise of paradise. By nurturing the plants, trees and animals, they hoped to make their land bloom ever more beautifully. And perhaps their approach to development did just that, because these nature workers evolved perhaps the most sustainably productive, biodiverse system of agriculture in the world.

Emperors of Perfect Virtue

The later rulers of China presumed that the values of the golden age had come from enlightened despots. The Yellow Emperor and other mythical emperors of perfect virtue were portrayed as sitting upon their thrones, looking for all the world like latter-day supreme patriarchs. These were conventional-looking vessels for social mores, but inside were stored values from pre-patriarchal times. The folk stories about how these emperors *behaved* reveal values so different from those of the military age that Westerners might term them "otherworldly." Here was no *realpolitik* of force-backed ranking. The emperors of the golden age were portrayed as leaders in a meritocracy of village-style spirituality.

According to the *Shu Jing*, the Confucian history classic, the legendary Emperor Diku was a leader due to his tireless service for others. In his old age he sought a successor of comparable spirit. The one he selected was not from his family or any noble clan, but a man from "among the poor and mean," namely Shun (Campbell 1976c: 386). Shun's style of leadership is described as follows:

> The farmers of the Li-shan encroached upon each other's boundaries. Shun went there and farmed; and after a year the boundaries were correct. The fishermen on the Ho bank quarreled about the shallows. Shun went there and fished and after a year they gave way to their elders. The potters of the eastern barbarians made vessels that were coarse and bad. Shun went there and made pottery. After a year their vessels were solid. (in Campbell 1976c: 388)

Likewise, Yu, a minister for Shun and later an "emperor," is said to have labored, "in accord with the natural conditions. When he entered the land of the naked he stripped himself to accord with native custom" (Campbell 1976c: 389). Yu reportedly spent his years like some hero of the People's Republic, digging channels for floods, draining swamps and demonstrating the arts of farming. He led not by command but by example. He was not the sort of role model most future landlords or rulers would wish to emulate. According the *Book of Master Zuang*,

> Yu wielded a wooden spade himself, and kept dredging all the rivers and waterways. His long exposure to the wind, the rain, and the parching sun roughened his body. Finally he helped thousands of tribes settle down. (Min 1995: 558–59)

Incidentally, recent geological evidence seems to confirm that coastal China was subject to major incursions by the sea, most recently around 3000 BCE. There may indeed have been leaders who organized the Yangze and Huangho valley people to build dikes and drainage channels against the floods (Min 1995: 558–59).

According to official histories, these legendary good emperors were teacher-kings, revealing their wisdom to an ignorant population. The tales have a Confucian slant, casting the prince as the author of his world. Therefore all efforts to recover the golden age depend on reforming the prince. However, in the days before empires, there were probably no emperors at all, but only farmers and village elders. According to Min (1995: 556), the perfect emperors Huangdi, Zhuanxu, Diku, Shun and Yao all bore their mother's family names. They were not heads of royal patrilineages, but men who had married into important matrilineal clans. The good works ascribed to these "emperors" were the kinds of work commonly done by village elders. And village elders were probably the ones who built the culture, economy and religion of the golden age. They were the ones who passed on village wisdom about the ways of fairness, good work and partnership with nature, which was later called the Dao.

Roots of Daoism

Joseph Needham (1956) says that Daoism arose from two main sources. First, it evolved from the ancient shamanic and nature-worshipping traditions of aboriginal villagers. When the warlord age arrived, this village Daoism became the pagan culture of rustic people, as opposed to the official culture of rulers in their cities. Wang Bo (in Min 1995: 590) claims that the Daoist heritage is a survival of culture from before the Shang dynasty. Min (1995: 593) concurs: "The philosophy of Daoism stems from the matriarchal society of the southern Xia nationality living in the Yangze basin."

The second source of Daoism was the educated philosophers who dropped out of the empire-building rat race. Even before the Warring States period, a stream of educated idealists, or losers, abandoned the courts of princes and went back to the land. Some of them wrote books on traditional shamanic lore (Needham 1956: 33–34). The resulting Daoist classics became a scholarly source of village wisdom. As Laozi (Lao Tzu) wrote,

> The Valley spirit never dies
> It is named the Mysterious Female
> And the doorway to the Mysterious Female
> Is the base from which Heaven and Earth sprang. (Campbell 1976c: 149)

And,

> It will not do to use military power to rule over the world....
> Weapons are not auspicious tools, not gentlemen's tools,
> and must not be used except under extraordinary conditions.
> It is far better to lead with benevolence. (Min 1995: 591)

If this sounds simplistic or naive to our modern ears, what does that say about ourselves?

From the perspective of most feudal-style rulers, Daoists were just rustics without power. The villagers, however, believed that their values were those of the golden age and that they, not their rulers, were the ones living correctly. As Needham (1956: 72) expresses their logic, "Finally, go to Nature and not to Authority, make your own fire and dig your own well."

Such aboriginal traditions can still be found in the most isolated regions of southern China, in the tropical mountains of Yunnan or Sichuan, or on the islands of Taiwan and Hainan. For example, the Naxi people of the Yunnan-Sichuan border lived in matrilineal clans down to the twentieth century. The Naxi did not practice marriage as established in northern China, but took lovers as *ah xiao*, or "special friends." At thirteen years of age, girls were given rooms of their own. Lovers could then visit each other freely but did not set up separate households. They usually offered each other gifts but did not depend on one another economically. Their children and property were held in common by each matrilineal clan. All children in a clan were called "brothers" and "sisters," regardless of who their fathers were. This culture clearly valued female and male children alike. It knew no distinction by which some children were more "legitimate" than others. "Special friends" did not "own" each other and had no authority to claim exclusive rights over each other. "Marriage" in this case could be based on mutual attraction, without additional economic or political requirements (Cai 1995: 61).

When the Mongol armies came in the 1200s CE, they were surprised by the ways of southerners like the Naxi. The Yuan (Mongol) dynasty writer Zhou Zhihong said of the Loulou people (of north Yunnan), "Their leader was a woman, and mothers were in charge of everything. A noblewoman had as many as one hundred husbands" (in Cai 1995: 61). Despite this judgmental tone, the Mongols were somewhat tolerant, or at least indifferent. They allowed the locals to live according to their own traditions, provided that "they accept the command of the central authorities, hand in tribute and taxes and allow their headmen to be appointed by the higher authorities." So long as these conditions were met, "all the rest would remain unchanged" (Cai 1995: 63). But after the appointment of male administrators who were accountable to foreign conquerors there was not a great deal left unchanged.

In such ways the dominators of China tended to suppress ancient village cultures. The villagers, however, were resilient, comparing themselves to bamboo. Their memories of a pre-military and pre-patriarchal world lived on like dreams imprinted inside their eyelids. For them, the Daoist golden age was an historical memory, a living ideal and a religious metaphor. It was the goal of psychological unity with the source of life. It was the lost tenderness between mother and child around which the world should turn. It was a real time in the not-so-distant past, with known traditions and moral standards by which the present age would be judged. So, according to popular Daoist legend, the Queen Mother of the West came to the court of Han Emperor Wu in 110 BCE, to deliver her judgment against him. This emperor had launched victorious wars against the barbarians, building the might of the Chinese empire to rival Rome. He had adopted an official version of Confucianism as the state religion, in which people's main moral obligations were to serve their social superiors. In his political and spiritual roles, Emperor Wu would be the rough equivalent of Roman emperors Augustus and Constantine combined. And to this great figure, the goddess reportedly said, "You were born licentious, extravagant, and violent; and you live in the midst of blood and force—no matter how many Taoists [Daoists] you invite here in hopes of immortality, you will only wear yourself out" (Cleary 1989: 3–4).

Partnership and Dominator Forms of Chinese Religion

China has suffered not one but several "dark ages." Whenever a major dynasty collapsed, it was comparable to a second, third or fourth fall of the Roman empire. At such times the state imploded, as frontier armies rushed toward the center battling for supremacy. For decades afterwards, the chaos could be more fearsome than any organized tyranny. As in most regions touching the Afro-Asian wastelands, desert invaders repeatedly imposed their authority and values on the indigenous people. In retrospect we can untangle the competing strands of partnership and dominator traditions that have often divided China's people against themselves.

Confucian Protest in the Empire-building Age

The semi-mythical Confucius lived in a time of slowly escalating war (reportedly 551–478 BCE). He spoke as if the princes of his time had long departed from the ways of the golden age. According to the *Book of Rites*,

> By now the Great Dao is disused and eclipsed. The world has become a family inheritance. Men love only their own parents and their own children. Valuable things and labor are used only for private advantage. Powerful men, imagining that inheritance of estates has always been the rule, fortify the walls of towns and villages and strengthen them by ditches and moats. (Needham 1956: 168)

Northern China was then divided between some 770 princely warlords who milked the peasants to equip their armies and expand their kingdoms. Confucius wandered from court to court, trying to persuade rulers to return to the values of the past. The words and sentiments ascribed to him were perhaps those of many people. But if Confucius was only a symbol for many like him, perhaps that would make his legend an even more authentic expression of ancient values.

In appealing to the Zhou princes of his day, Confucius claimed that the founders of the Zhou dynasty had been men of most austere virtue. In turn, these early rulers (who lived around 1100 BCE) reportedly drew their inspiration from the still more ancient "emperors of perfect virtue," who had lived before the Xia dynasty (which likely began around 2200 BCE). In that past time to which Confucius ultimately pointed, the "princes" were merely aboriginal chiefs who held no military power over other villagers. Needham (1956: 104–5, 61–62) claims those primitive chiefs, "half-apologetically exercised leadership from within, and vied with one another in ... potlatch [gift giving] ceremonies." According to the potlatch ethic of aboriginal communities around the Pacific Rim, the greatest person was the one who *gave* the most, not the one who possessed the most. For these people the most respected kind of power was not military might, but the "feminine" power to give and nurture life. According to Mengzi (Mencius), the villagers of the 400s BCE expected this kind of leadership and were still amazed by the new breed of force-backed rulers:

> The ruler of Theng is indeed a worthy prince, but nevertheless he has not heard of the Dao. Real leaders cultivate the ground in common with the people, and so eat. They prepare their own morning and evening meals, carrying on government at the same time. But now the ruler of Theng has his granaries, treasuries and arsenals, which is oppressing the people to nourish himself. How can he be deemed a real leader? (in Needham 1956: 120–21)

The roughly contemporary *Tao Te Ching* similarly observed,

> So long as the court is in order,
> (rulers are content to) let the fields run to weeds

and the (village) granaries stand empty.
They wear patterns and embroideries, carry sharp swords,
glut themselves with drink and food,
have more possessions than they can use....
These are the riotous ways of brigandage;
they are not the Dao. (chapter 53)

When Confucius tried to revive the spirit of the pre-military age, he referred constantly to a web of mutual obligations among all creatures. The spirits of nature and the ancestors were omnipresent and had to be honoured. Likewise, mothers, fathers and children needed to honour their obligations to one another. As *The Doctrine of the Mean* said, "The beginning of the way of a gentleman can be traced to the alliance between a wife and her husband" (in Zhang 1995: 418–19). Without such mutual support, the web of life would come undone. Calamity would befall first the family, and then the whole village. For village elders this was common sense.

But when Confucius urged such common sense on the warring princes, they found it irrelevant to their ambitions. It interfered with their dreams of making themselves lords over all. As mentioned before, when one of these princes, Qin Shihuang, actually achieved emperorship over all of China, he declared Confucian values an anathema. To make himself the center of all obligations, he brushed aside the cobwebs of old-fashioned ethics, killed Confucian teachers and burned their books. If Confucian ethics had come from the *literati* alone, this might have worked.

Confucius was a consultant, or would-be consultant, somewhat like those who wish to advise business and government today. The princes, however, rejected his advice. Rather than proposing new ways of maximizing the personal benefits of their rule, Confucius offered a holistic view on the good of the realm. He urged rulers to "serve" the villagers, which would mean reinvesting tax wealth in the country. But the successful rulers had already invested the country's wealth in military strength, by which they hoped to control the land and its fruits for themselves. With such people, Confucius argued that true power grew from the observation of traditional virtue.

Centuries later the disciples of Confucius were still arguing with Han dynasty emperors: "If you foster high standards in the temple and courtroom, you need only make a bold show and bring home your troops, for the king who practices benevolent government has no enemies anywhere. What need can he then have for expense funds?" (Ebrey 1981: 24). These moralists, however, were arguing with an heir to the imperial throne as established by Qin Shihuang. And this fact alone showed that power in the warlord age grew from swords and armies. Accordingly, almost all dynastic rulers located their courts near the northern or western frontiers. From that position, they could control the frontier armies and still collect revenue from the farming regions. The balancing of military and economic power required a fulcrum near the border zone. So the Shang capitols rose in the northeast. The "relatively barbaric" kingdom of Qin rose on the northwest frontier (Campbell 1976c: 408). During Han times (202 BCE to 220 CE) the capital was Zhangan (Chang-an), at the end of the Silk Road. In the Sui dynasty (589 to 618 CE), the emperors sought to feed their courts and armies by building a Great Canal, nearly 1,900 kilometers long, from their northern capital to the rich southern farmlands. Up this canal they brought hundreds of thousands of tons of rice every year, supporting over a million non-farming government and military men near the northern frontiers (Ponting 1991: 93). Later, the Jin and Mongol invaders established China's capitol at Beijing and supported their ruling classes in the same way. Such were the geopolitical facts which the early Confucianists were unable to change.

Later Emperors' Lip Service to Confucius

In the Han and certain later dynasties, some semi-idealistic emperors claimed to rule according to Confucian values. They taxed the villagers in moderation, leaving the farmers with reserves of grain against a drought. But the rulers who invaded China from the wastelands seldom took such a long-range view. They usually saw China as a cornucopia of wealth to be seized while the opportunity remained. These emperors often taxed the peasants to a subsistence level, forcing them to walk a fine line between starvation if they paid in full, or military retaliation if they didn't. The Qin Legalists were so vehemently self-serving that they simply killed farmers who could not pay, and murdered any Confucianists who urged mercy. Then, with all voices of restraint stilled, the Qin destroyed themselves through sheer greed. In the future, most warlords would pay at least outward respect to their taxpayers' values.

Over the past two thousand years, most rulers found that paying lip service to Confucius was an effective way to control ethical judgment. If they hired Confucian advisors, they could claim their decisions had been taken with due consideration for popular tradition. In exchange for such employment, many Confucian consultants willingly offered whatever kinds of advice their rulers wanted. With a little encouragement, they displayed a tendency "to flock around the worst robber barons, vying with one another to become their counselors" (Needham 1956: 102). So, in the process of offering advice that would sell, the Confucianists slowly altered their own values almost beyond recognition. In the hands of these advisors, the old mutual obligations among all people were changed—into the unilateral requirement that inferiors must serve superiors.

It was the Legalist Han Fei (whose philosophy was official in the Qin dynasty) who articulated a morality based on "the three bonds"—"the minister serving the ruler, the son serving the father, and the wife serving the husband" (Du 1995: 224). Later, the Han dynasty scholar and consultant Dong Zhongshu proposed a similar doctrine to Emperor Wu (140–87 BCE). Dong offered a doctrine of the "Three Cardinal Guides"—"a ruler is a cardinal guide to a minister, a father is a cardinal guide to a son, and a husband a cardinal guide to a wife." This, Dong insisted, was the real point of Confucius' teachings. Emperor Wu was pleased with this interpretation. And just as the first emperor Qin Shihuang had banned all the hundred schools of philosophy save Legalism, so Emperor Wu tried to ban all other doctrines save this Legalistic Confucianism (Min 1995: 566–67).

In Dong Zhongshu's doctrine, dictatorship gained a cosmological justification. Dong claimed that the hierarchical social order was modeled on hierarchical relationships between heaven and earth, yin and yang. As Sun Xiao and Pan Shaoping (1995: 236) explain, this gave Dong's philosophy "a kind of theological touch.... For this reason, it is not an exaggeration to say that Dong Zhongshu ... was the founder of the teleological theory supporting the autocracy." So, after the writings of Confucius had been banned and burned in the Qin dynasty, the Han emperors oversaw a reconstruction of Confucianism. Each part was separated from its original context and, as Campbell (1976c: 380) says, "remounted carefully in a late, highly sophisticated setting, like an old Egyptian scarab mounted as a ring for some fine lady's hand."

Dominator Confucianism was concerned with enforcing a chain of command between each class of human beings. In this legal system, it was a far more serious crime for a son to abuse his father than for a father to abuse his son. The same uneven standard applied in crimes between husbands and wives. Confucius was thus proclaimed a prophet of patriarchy and the divine-right of kings. He was lifted entirely out of the context of the golden age and mounted like a gem on the emperor's ring.

The Official Cult of Confucianism

Despite the state's efforts, the conversion of Confucianism into a dominator ideology was slow and uneven. During some periods, the Confucian establishment recovered something of its old spirit. During the indigenous Song (Sung) dynasty of 960 to 1279, the emperors tried to avoid military revolts by appointing Confucian scholars to oversee the provinces and troops. In that case, the administration valued scholars over generals, and farmers over soldiers. The results were weak armies, a regime of semi-gentile bureaucrats, and economic prosperity. Farming, trade and letters were non-martial arts which flourished under the seemingly anti-military Song government.

Gao Shiyu (1995: 278–81) claims that widespread abandonment of military careers for literary ones during the Song dynasty "tended to weaken males." It seems Confucian culture produced a tradition of sensitive, idealistic, "soft men." In the eyes of the Mongol horde, these were not real men at all. But Chinese women commonly favored such men. Gao (1995: 282) notes that in popular romantic literature of the Song, Ming and Qing dynasties, the "heroes and ideal figures for female love are almost all gentle and frail-looking scholars ... mirroring the social mood." So Confucian society became famous for soft men, and these were blamed for China's military weakness. The softest, however, were not the men of literary ambition, or the millions of hard-working farmers, but the hereditary despots and hangers-on in the imperial court.

During the Song period, "Neo-Confucianism" emerged as a medieval synthesis of Buddhist, Daoist and relatively authentic Confucian traditions. All these strands were woven into one comprehensive standard for Chinese culture. For would-be dominators, this was a project of cultural imperialism to unite the empire. For others, Neo-Confucianism was a path of self-cultivation and benevolence. The most famous Neo-Confucian teacher, Zhu Xi (Chu Hsi) (1130–1200 CE), interpreted his tradition as an expression of golden age ethics. Universal harmony, he said, was not a matter of all creatures conforming to one superior will. It was the spontaneous cooperation which all would share if they followed their own instincts and honoured one another. The moral order, according to Zhu Xi, was not the political order of the emperor, or that of a deity above the emperor. It was the order of nature which arose, as Needham (1956: 562, 453) explains it, "because the universe has the property of bringing to birth moral values and moral behaviour when that level of organization has been reached."

For Zhu Xi, human fulfillment included the fulfillment of nature. As with anti-dominator Buddhism and Daoism, enlightenment was a realization of inclusive unity with the natural universe:

> If there is a single thing not yet entered [by an outreaching spirit], the reaching is not yet complete, and there are things not yet embraced. This shows that the mind still excludes something—for selfishness separates and obstructs, and consequently the external world and the self stand in opposition. This being the case, even those dearest to us may be excluded—therefore, the mind that excludes is not qualified to be one with the mind of heaven. (de Bary et al. 1960: 497–98)

So morality flowed from identifying with all creatures, and then acting for the good of them all as if they were part of the self. Here again was the attitude of village elders enshrined in high-sounding philosophy. In case this philosophy sounds hopelessly mystical, Needham (1956: 465) asks his fellow modernists, "What, in terms comprehensible to us, were these Sung [Song dynasty] philosophers [such as Zhu Xi] affirming? Surely the conception of the entire universe as a single organism."

The backlash against such Confucianism came soon enough. The Mongol and, later, Manchu rulers firmly redefined Confucian orthodoxy to mean obeying the ruler's will. In the classic *Zhong Yong* (*Chung Yung The Doctrine of the Mean*), Confucius had reportedly said:

> There are five relationships which concern all men, and three virtues by which they are fulfilled. The relationships of ruler and subject, father and son, husband and wife, older and younger … and of intercourse between friends—these five are the relationships which

> pertain to all men. Knowledge, humanity and courage—these are the virtues which apply to all men, and that by which they are practiced is one. (de Bary et al. 1960: 120)

The sage here seems to recommend one set of virtues for all. No separate list appears for each class of humans, such as "obedience" for inferiors only, or "strength" for superiors only.

But by the Qing (Manchu) dynasty (1644–1911), such subtle points were habitually swept aside. So in 1898, Chang Chih-tung defended the only kind of Confucianism he had ever known, against the "Western" ideas of liberty and equality:

> If we recognize the bond of subject to sovereign, the theory of the people's rights cannot stand. If we recognize the bond of son to father, then the theory that the father and son are amenable to the same punishments cannot stand.... If we recognize the bond of wife to husband, then the theory of equal rights for men and women cannot stand. (de Bary et al. 1960: 84)

Chang held that this almost military-style ranking was the very essence of Chinese civilization: "In order to protect the race, we must first preserve the doctrine.... How is the doctrine to be maintained? It is to be maintained by strength, and strength lies in armies" (de Bary et al. 1960: 83). But the army Chang spoke of was the Manchu-led army, whose main function was to control the Chinese themselves. Chang's "Confucian orthodoxy" was identical with the ideology of the occupying army.

Such cultural imperialism convinced most modern Chinese that their own religious heritage had been a cult of oppression since the age of slavery. But if they commonly rebelled against Confucianism as they knew it, they did so while affirming their own morality, which was often more authentically Confucian than they knew. So, more than two thousand years after the rise of the first emperor, the Communists came to power, partly by proclaiming themselves champions of China's communal village values. And no matter how many waves of warlord armies washed over China to seize the emperor's throne, no matter how often bureaucrats put their own interests first, each generation of Chinese villagers still felt it possible to recover the legendary time of harmony and peace, described in legend as "The Great Togetherness."

Popular and Official Forms of Buddhism and Daoism

Strangely enough the Buddhist "doctrine of peace" came to China from over the desert. Among the nomads of Turkestan, Buddhism spread as a kind of counterculture, by which ordinary people adopted a different set of values from their rulers. When it reached China's farming zone, this religion was able to meld into the local cultures of village elders, somewhat as it melded with village

Shinto in Japan. For the rulers, however, the Silk Road also brought a "dominator form" of Buddhism.

One striking convergence of Buddhist and native Chinese spirituality was Chan Buddhism, the source of Japanese Zen. According to this school, enlightenment was an awakening to the sacredness of this world. All things were to be realized as "Buddha things," just the way they are. This was not a Buddhism of arcane metaphysics or renunciation of life. It was a practice of living simply and intimately in the world. In retrospect, much of Chan Buddhist teaching seems indistinguishable from the Daoist teachings which had come before in China. For example, in the following story, try substituting the words "Buddha nature" for "Dao":

> Tungkuo Shun-tzu said to Chuang Tzu [Zhuangzi], "Where is this so called Tao [Dao]?"
> Chuang Tzu answered, "Everywhere."
> The other said, "You must specify an instance of it."
> Chuang Tzu said, "It is here in these ants."
> Tungkuo replied, "That must be its lowest manifestation surely."
> Chuang Tzu said, "No, it is in these weeds."
> The other said, "What about a lower example?"
> Chuang Tzu said, "It is here in this earthenware tile."
> "Surely that must be its lowest place?"
> "No, it is here in this dung also."
> To this Tungkuo gave no reply. (Needham 1956: 66)

Such earthy mysticism which had flowed from the wisdom of Chinese village elders was melded with a similar village wisdom from India. It was "otherworldly" only in its peasant-style rejection of warlords and "worldly ambition." For such Daoist and Buddhist sages, enlightenment and harmony with nature were close to the same thing. Harmony and enlightenment were practical things. Perhaps the writing of Yeh Meng-te (c. 1150 CE) illustrates the views of practical villagers, who were part Buddhist, part Confucianist and possibly entirely Daoist:

> If you ask those who live in cities, they know nothing.... I often arose very early in the morning and with an empty mind concentrated on the clouds, mountains, rivers, fields and trees in all their beauty, and found I could predict the weather aright seven or eight times out of ten.... Thus I realized that it is only in quietness that the cosmos can be observed, the body's moods felt, and real knowledge obtained.(in Needham 1956: 85–86)

Even Buddhist monastics tended to make their religion accord with local village wisdom. As Daoism had many popular goddesses and female adepts, so

Buddhism offered the goddess Guan Yin (Kuan Yin), the bodhisattva of universal compassion. So Buddhism connected with the values of village women and entered the heart of Chinese culture. For vast numbers of Chinese down to the present, Guan Yin has been better loved than the Buddha himself.

Other forms of Buddhism, however, were less convergent with the values of village China, and more influenced by those found in the homelands of pastoral warlords. In "dominator Buddhism," kings and court clerics chose to explain that the poor and downtrodden deserved their fates because of bad deeds in previous lives. The powerful and wealthy, on the other hand, obviously owed their positions to past spiritual merit. This was the lesson which the warlords of the Silk Road drew from the Buddha who gave up his kingdom.

According to "dominator Buddhism," women were born to low status as a punishment for evil deeds. This spiritual wisdom was taught to little girls, especially girls of powerful families. So, around 550 CE, a female nun called Tao-jung paid a stone worker to carve her edifying words of penance on a temple colophon:

> Results are not born of thin air: pay heed to causes and results will follow. This explains how the Buddhist disciple and nun Tao-jung—because her conduct in her previous life was not correct—came to be born in her present form, a woman, vile and unclean. (Ebrey 1981: 53–54)

There was also a Buddhism of the *dominated*. The conquered villagers, who had been turned into virtual serfs after being free farmers, often saw their lives as hopeless. They commonly took their oppression as the nature of this world and saw life itself as something to be escaped. The Buddha had said that suffering comes from attachment, and life may be lived free from attachment. But the conquered peasants of both China and India tended to hear a different message: they heard that this world is vile, and existence is unavoidably painful. Needham (1956: 430) says, "it was natural that they should unite in calling the visible world ugly because they could not make it happy."

For those who took Buddhism as a confirmation that this world must be escaped, it seemed logical to live in celibacy as a monk or nun. And during the dominator age, this response grew popular even in family-oriented China. By the Tang dynasty (618–906 CE), so many people had embraced celibacy and "world denial" that the emperors grew alarmed. They felt it necessary to unleash the army to slaughter Buddhist monks and nuns, as if only naked force could prevent China's taxpayers from renouncing the world en masse.

In the post-Han age of chaos (220 to the 500s CE), dominator culture influenced the Chinese whichever way they turned. When educated Daoist gentlemen retreated to form communes in the forest glades, they were fleeing a militarized world. Their lifestyle, like that of many Indian forest yogis, took on an air of escapism. Traditional village Daoism also began changing into a religion of personal salvation from a cruel world. Religious professionals soon staffed the gateways to salvation. Previously there had been no "Daoist Church," any more than there had been a "Hindu Church." But now Daoism started to compete with "dominator Buddhism" as an organized religion. Wing-tsit Chan (de Bary et al. 1960: 298) writes that Daoism "imitated Buddhism in a wholesale manner in such things as temples and images, a hierarchy of priests, monasticism, and heavens and hells." Daoist theologians developed "a vast system of celestial bureaucracy," reflecting the hierarchical world they saw around them. Then, instead of upholding ancient village traditions against the cult of the state, Daoist teachers began preaching obedience and civic morals to the peasants (Needham 1956: 161, 159). Some Daoists imitated the Confucianists in trying to serve warlords as consultants. So the Daoist way of harmony with nature came to involve concocting potions to ensure the eternal life of the emperor.

Legions of Chinese Goddesses

Not all Daoists were co-opted into dominator values. In remote and mountainous villages, many people clung to a Daoism from the time of matrilineal clans. Min (1995: 593–94) points out, "Daoism has always enjoyed the favor of women in Chinese history." And southern China especially remained densely populated with quasi-Daoist goddess religions. Among these goddesses of the south are Xin Qi Niang, Lady Lin Shui, Lady Jin Hua, Ma Xian Gu, Long Mu, Qi Xing Nai and Ma Zu. Ma Zu alone now has perhaps 100 million worshippers worldwide (Zhao 1995: 406–15). Many millions more worship the Buddhist goddess Guan Yin. All told, southern China is perhaps the world's greatest province of religion by and for women.

The Confucian bond of wife to husband was more a matter of patriarchal superiority in the north than in the south. In 1925, young Mao Zedong reported that among the poorer peasants of his home province of Hunan in south China the authority of husbands over wives was weak. These women worked in the fields like men, had a roughly equal say in family matters and even "considerable sexual freedom" (in Spence 1991: 376). Mao saw this as evidence that working-class women were stronger than those of the landlord class. But environmental factors also figured in the status of southern peasant women.

The villagers of Mao's home province lived in the southeastern Asian rice-growing world. Women there had always excelled on their small tropical household farms. Zhao Zhewei (1995: 414) writes, "The social status of women in the south used to be higher than that of women in northern China, as they … often replaced men in all kinds of economic activities." In the monsoon wetlands, farm work involved long hours of planting and weeding, but seldom the sheer muscle power needed to break dry ground on the northern plains. A southern woman's endurance and patience often produced more than a man's strength. Also, farmers in the south could produce year round. They were not so subject to the northern seasons of famine, which were usually hardest in the summer while crops were ripening. The southerners had a larger green cushion to fall back on, protecting them from scarcity and the ruthless competition scarcity can unleash. As the south was less often dominated by warlord rule, it retained a heritage of relative equality between the sexes, as is often found in Burma, Thailand and Bali.

Concerning southern women's muscle in business, an old directory for the Southern Song capital (in the 1100s and 1200s) lists numerous businesses managed by women. Among them were Five Sister's Fish Custard shop, Granny Cao's Meat Pie, Ma Wang's Tea Shop, Aunt Weng's Wine Shop, Granny Li's Custard with Mixed Vegetables, Ma Chen's Herbal and Medicine Shop, and Ugly Granny's Herbal Medicine Store (Gao 1995: 301). Southern women have a long-standing reputation for competence in management, which still stands out when comparing Canton or Hong Kong with Beijing.

Along with economic importance went religious importance. The old village religions involved both male and female shamans, who served as channels of the powers of nature. The term for such spiritual women was "wu," which Christian missionaries sometimes translated as "witch" (Needham 1956: 134, 152). From the ranks of such witches, Wang Jianghang listed 145 "female immortals" in his "Stories of Immortals through the Ages" (Min 1995: 594). Many people believe that all of China's thousands of goddesses were real women who lived lives of leadership and compassion, and were deified after they died.

Modern China's Assault on the Desert

By 1850, the Han Chinese population had approached 400 million, vastly outnumbering the formerly powerful pastoral people. This was the real victory of the farming culture—the people of the bio-rich environment had sustained themselves better. Rather than outfighting the desert people, the villagers outnurtured them. The victors then began encroaching on the desert.

The last big victory won by the pastoralists was the Manchu conquest, in 1644, but its results were contradictory. On seizing power, the Manchus incorporated their own land into greater China. Then they sent mounted banner units on campaigns beyond the Gobi and Taklamakan deserts. The Manchus annexed all conquered areas outside the Great Wall between Russia and Tibet into the Middle Kingdom, making the Great Wall irrelevant.

Reversal of the Migration's Winds

Perhaps until the early 1800s, the pastoralists of Inner Asia flocked to China, trying to gain access to a land of relative plenty. As late as the 1500s, village China may have boasted the highest general standard of living in the world (Sowell 1996: 179). After that, the quality of both life and land declined. In the arable heartland, which comprised about 14 percent of China's area, subdivision of land and expansion of estates steadily reduced the size of family farms. By 1850, many millions of families could no longer feed themselves.

Population density in the lower Yangze provinces pushed past 385 people per square kilometer. Then from 1919 to 1921, a series of droughts and famines in that region killed at least half a million people (Spence 1991: 309). The villagers had kept an organic balance in almost every aspect of their environment, except the human population. Now, almost every village produced a stream of "surplus people," flowing towards the cities or the remaining 86 percent of China's "non-arable" land. Whereas the winds of migration had previously blown toward the heartland, now they turned and blew the other way. New waves of economic refugees flowed outwards to Xinjiang, Inner Mongolia, Yunnan or Taiwan. The migrants cleared hills above the upper Yangze. They plowed sod near the Gobi. On virgin soils their first yields were encouraging, but the settlers' compost and manure was often insufficient to replace eroded soil (Spence 1991: 95). In a land famous for quality farming, the farmers' "surplus" children would either reinvent a sustainable economy under harsher conditions or slowly turn the countryside to dust. If they degraded the pasturelands to desert, then their farming civilization might still lose to the nomads in the end.

The Chinese frontier was perhaps more peaceful than the American West, with more mutual benefit for migrants and natives, but it could be violent too. In 1911, as soon as the Qing dynasty fell apart, Mongolia declared independence from China. Migrant farmers had taken the best grazing fields, forcing Mongolians onto inferior pastures (Jagchid and Hyer 1979: 314). A similar encroachment of farmers helped spark Muslim revolts in Gansu and Xinjiang. In dealing with such controversies, China's rulers showed almost no flexibility. The supposedly Confucian obligation of subject to sovereign tended to preclude legal protest. The authorities labeled local dissent as treason, especially in the border regions. Demands for local management of local resources were commonly answered by unleashing the army. After all, this was the age of colonial empires, and the race of the Great Powers was on.

Warfare's Crop

For some three thousand years the desert people of Inner Asia had been China's main enemies. To subdue them had been a national dream. And finally, in the late 1800s, as a series Muslim revolts were suppressed in the west, the fulfillment of this dream seemed at hand. Imperial China stretched from the Pacific to the Pamir Mountains. But just in this hour of triumph, a new set of enemies appeared from the opposite direction, from off the eastern coast, where no walls existed to hold back barbarians. China found itself defeated on coastal battlefields. For the next century, from the 1840s to the 1940s, China stood on the defensive in the global race for colonial empires. The country never quite fell to any aggressor, but it remained under siege, with the attackers' violence slowly escalating to the murderous crescendo of World War II.

According to the Communists, these decades of war caused a general decline of agriculture in the arid areas. As in past centuries, war interrupted the careful husbandry of the oases and allowed the deserts to move in. The English missionaries Mildred Cable and Francesca French witnessed such a process while traveling through western China in the 1920s and '30s. In Gansu a defecting army general commandeered the Muslim revolt he was sent to repress. As the missionaries passed back and forth between rebel and loyalist areas, the Chinese army terrorized the local population with almost indiscriminate killing. Cable and French noted the rise of criminal gangs, as impoverished young men turned to banditry. Meanwhile, rumors spread of worse to come as the Chinese civil war and the Japanese invasion began. But perhaps these women's most disturbing reports concerned the state of the land:

> Moreover, this aridity is not stationary but increases steadily, and in the course of only a few years the decrease in water supply is perceptible, and the inhabitants of a very small oasis may see the water level in their wells decline, their ponds drain away, and a whole line of arable land become useless for cultivation. As years go by, the effects of evaporation, the increasing depth of saline deposit in the marshes, and the slow choking of lakes and riverbeds with drifting sand . . . cause a ceaseless encroachment of desert conditions. (Cable and French 1984: 95)

According to the Lanzhou Institute (1982: 4), the desert was advancing largely as a result of human neglect. It was encroaching where war and banditry made careful nurture of plants and animals next to impossible. In the decades before 1949, the farmers of Turfan oasis (in Xinjiang) had tried to hide their grain from pillaging soldiers. They tried to look like they had nothing to steal in a land where green fields stand out for miles. Local men hid in the hills to avoid conscription by one army or the next. Meanwhile the irrigation tunnels fell into disrepair, and the tree belts died. Only a few scattered trees remained around the houses, and sand dunes blew through the oasis perimeters (Sinkiang 1977: 3). This was warfare's crop in one oasis.

Many Westerners might expect that neglect of the land would allow regeneration of plant and animal life. But the Chinese version of events reflects a traditional view in which humanity is a positive force in the environment. This positive thinking has some basis in local history. In the past, most Chinese farmers assumed they must make their land support their families forever. For them, the game of life was to work the soil so that both the tillers and the land would grow richer over time. The contrasting Western assumption that humans are inherently harmful to nature also comes from historical experience. In lands such as Spain or California, naturally forested regions were reduced to near-desert by human actions. But most of China's land is naturally drier than Spain, and some is drier than Death Valley. Climatically, much of China fits the definition of "true desert," as land with a water-loss factor of ten or more. In other words, this terrain naturally loses at least ten times more water through evaporation than it gains from precipitation.

In Turfan, the average precipitation of rain or snow is 16.6 mm per year. But the heat and dry wind will evaporate 3,003 mm per year from an exposed dish. If a real desert has a "dryness ratio" of ten units potential evaporation for each unit of actual precipitation, then Turfan has a ratio of 180 to 1. Without "artificial" human nurture, almost nothing would grow (Sinkiang 1977: 47–48). The land would naturally revert to a moonscape like the nearby Flame Hills. Yet local farmers have made Turfan support crops, not just for a few years but for century after century.

Just as the Dutch claimed land from the sea, so the oasis farmers had to make and sustain their own soil. To create a farm plot, they poured water over the surface to wash off the noxious salts. Next they gathered cartloads of grass, and spread it in layers to compost in the sand. They added buckets of humus-rich "guest soil" from their established gardens to impregnate the new soil like yeast. For the first several seasons they planted leguminous grasses such as sweet clover or alfalfa, which built up the nitrogen content (even though these farmers had never heard of nitrogen). Meanwhile they planted rows of bushes and trees around the farms as dikes against the scorching wind. Beside their houses they built arbors for grapes, gourd vines and creeping peppers. Such gardens in the desert required constant care because they could wither and die after one season of neglect.

Harnessing the Land and the Farmers

The most important thing the Communists brought in 1949 was peace. Masses of soldiers and refugees returned to their farms to focus again on nurturing plants and animals. With this first and most important step, the vast postwar reconstruction began. Next, the Communists broke up large landholdings from Manchu times, giving millions of peasants a new lease on life. Then, in a banker's nightmare, the government simply cleared the books of debts owed by peasants to landlords and moneylenders. With those three strokes, the Communists basically achieved their mandate.

For the first several years after the revolution, almost all gains in agriculture came through peasant labor and applied traditional knowledge. The peasants were given peace and allowed to work their magic. But the Party was not content with healthier traditional villages. The visionary modernizers felt it obvious that all this grubbing in the soil was a backward stage in history—Marx's "Asiatic mode of production." The Party was now to lead the peasants "beyond" this. Surely the future lay with ever more industrialized farming, as seen in the Soviet Union. Obviously the peasants would be more efficient if their farms were consolidated into plantations larger than those of old-fashioned landlords.

The new rural production brigades of two hundred to three hundred households gained certain efficiencies of scale. But where the peasants had lavished care on their own farm plots, commune officers now supervised workers paid by labor points. The farmers' intimate relations with micro-environments were rendered unimportant. Instead of gradually improving yields through soil enrichment, the new system required rapid growth to meet imposed production quotas. The caretakers of plants and animals began behaving like assembly-line workers. The countryside began to resemble a factory floor, and the modernizers called it progress. On the arid plains of Inner Mongolia, tractors began plowing wide areas for a great leap forward in grain production. The Russians had just set the pace by plowing the steppes of Kazakhstan. Within a few seasons, the new plowed earth of both regions began disappearing in storms of dust.

The government set optimistic production quotas. At the same time, it reduced the prices it paid for farm produce. In the Great Leap Forward, agriculture was to accomplish "more, faster, better, cheaper." The money saved was to go for a dramatic expansion of industry. To allow full thrust to this effort, even the remaining family gardens were discouraged. All China, it seemed, was united in a campaign of self-sacrifice to build a modern nation. Unfortunately, the newly-formed communes suffered a series of poor harvests, and the production brigades were paid for their reduced yields at the new lower prices. The return, in many areas, was too low for the farmers to escape malnutrition. Between 1959 and 1962, people were starving to death in the countryside. The Communist Party found itself operating as the biggest landlord in history, and doing so in a time of hunger. It was a corrupting experience. After those hard years, the Party was never the same. Too many of its members had changed from patriotic idealists into power-brokers in a land of scarcity.

What the Peasants Had to Teach

Of all lessons drawn from the Great Leap Forward, the most loudly proclaimed was that the bureaucrats should learn from the peasants. Mao started out emphasizing this and periodically recalled it. "You know," he told André Malraux, "I've proclaimed for a long time that we must teach the masses clearly what we have received from them confusedly" (in Walls 1980: 89). The peasants were not more "red" than the bureaucrats; they were more "expert"—in traditional rather than industrialized farming. They did not naturally think in terms of meeting production quotas or maximizing outputs over inputs. Their way was to feel out a middle path, where both the soil's fertility and the harvest could be slowly increased. Their dream was of endless growth, but their kind of growth was based on building an ever richer environment. After perhaps seven thousand years of hands-on learning, their art had evolved into the most consistently productive system in the world.

The art of southern Chinese pond agriculture was described in 1149 CE, in a treatise by Ch'en P'u. Ch'en advised that between a fifth and a third of each farm should be dug out for ponds:

> On the embankments plant mulberry and pomegranate trees on which cows can be tethered. The cows will be comfortable under the shade of the trees: embankments will be strengthened because the cows constantly tread on them: and the mulberry trees will grow beautifully because of the nourishing water. (Ebrey 1981: 110)

The ponds were stocked with fish and ducks, which also enriched the water. In the pools, lotuses flowered as a water crop. In the summer, pond weeds placed in the pig sty protected the piglets from heat stroke. In the monsoon season, the ponds rose to flood the rice fields. The fish in the fields produced extra food without harming the crop, and grass carp in the irrigation ditches ate the weeds (Gabor et al. 1981: 199). We now know that algae grows in the shallow water of a flooded field, and this algae fixes nitrogen in the soil. The pond water was a fertilizer soup and the pond a liquid compost pit to recycle organic "wastes" almost immediately (Ponting 1991: 93–94). Such a bio-intensive system could sustain two harvests of rice a year, plus other crops, including a variety of vegetables, fruits and sugar cane. If beans and rice were intercropped, the yield was much improved. Ch'en P'u continued,

> Anyone who knows the right timing and follows the order can cultivate one thing after another, and use one to assist the others. Then there will not be a day without planting, nor a month without harvest, and money will be coming in throughout the year. (Ebrey 1981: 110–11)

According to Zhao Songqiao of the Chinese Academy of Sciences, such farming in the Yangze delta is "probably the best ecological system in the world. With the constant creation of new paddy (rice) soil, it is actually improving" (in Kohl 1989: 295). The output in energy terms is roughly fifty times the input (Ponting 1991: 291–92). And this is no temporary peak of productivity, but something sustained over centuries. After due consideration of the peasant's methods, in 1940 Sir Alfred Howard was moved to praise: "The agricultural practices of the Orient have passed the supreme test—they are almost as permanent as those of the primeval forest, of the prairie, or of the ocean" (in Shiva 1988: 105). This ecological sophistication was the real basis of Chinese civilization, and these were the kind of ultra-conservative peasants whom Mao was at least talking about learning from.

Within two decades of the Chinese revolution, the moves toward Soviet-style agriculture had proved disappointing, both economically and ecologically. The peasants gravitated back to their old ways of intensive gardening on family plots, and government policy bent to their will. But if industrialized farming was unsustainable, traditional farming had also reached certain limits. The government imposed its one child per couple policy, yet it still hoped for unlimited economic growth. The two main escape routes to that growth were industrialization and cultivating the arid zones.

The National Reclamation Movement

Like Israel, modern China mounted a state-backed assault on the desert. This was a multi-sided offensive involving economic development and political domination. In 1945, local leaders in Xinjiang had proclaimed an "Eastern Turkestan Republic." The Communists crushed this republic after 1949, and sent in hundreds of thousands of troops to occupy border regions. With the military occupation came an influx of Chinese settlers, on a scale which dwarfed Israel's efforts. Within decades the migrant Chinese were the largest ethnic group in Xinjiang and Tibet.

But the assault on the desert was also a movement for environmental change. With fine enthusiasm the Party exhorted frontier villagers to "Tame the wind, harness the sand, and transform the Gobi." The government provided the rhetoric, and self-interested local people supplied the know-how and hard work. Government research and extension workers documented local efforts and publicized good results in "holding" sand dunes, sustaining tree belts or generating pastures.

The Oasis Dwellers' Progress

In the Turfan basin, the oasis farmers faced not only punishing heat but also desert winds of up to fifty, or even eighty km per hour. These "black hurricanes" could hurl gravel like buckshot and drive drifts of sand to bury trees, crops and homes. "The conditions," as the Lanzhou Institute staff (1982: 48) understate them, "can be destructive to agriculture." To live in this place, the Turfan people had built a system of tunnels bringing snowmelt from the Tien Shan Mountains. After 1950, the Turfanese repaired and expanded the tunnel networks, gradually increasing the water supply from thirteen cubic meters per second to over thirty (Walls 1980: 54–55). This, however, would make little difference without improved protection from the searing heat, wind and creeping sand. The Turfanese also had to re-establish their organic walls against the desert.

The main tree belts facing the desert were made thirty meters thick, with five rows of trees and five concentric rings of irrigation ditches. By the outermost two ditches, the villagers usually planted sand date palms. These had the necessary tolerance for hot wind and alkali salts. Behind the palms came two rows of Xinjiang poplar, or elm. The innermost ditch commonly supported mulberry bushes, partly for silkworm food. Behind such belts, wind velocity was cut by more than half, greatly reducing evaporation by hot wind. In front of the tree belts, the villagers planted desert grasses or shrubs such as camel thorne, deer horn grass, or salt ear tree. They periodically watered the grass zone with any remaining "tailwater" from the irrigation ditches. The grass areas grew into green belts three hundred to five hundred meters across, which stopped most blowing sand from reaching the trees (Walls 1980: 55).

Even these grass and tree belts were not enough. In Turfan, the heat and wind combines to create such a rate of water transpiration that one external belt of trees and grass could do little for central fields. There had to be additional interior belts of trees. Not only did the soil have to be built up from sand by gradual composting, but a bearable outdoor microclimate had to be created as well. As the wind direction was variable with the seasons, the internal tree belts could not be placed in parallel rows but had to form a cross-hatch of wind breaks. The Turfanese therefore turned their land into a quiltwork of fields about fifteen hectares square, each surrounded by trees on all sides. Later, many of the local farmers concluded that even more protection was needed and put in additional tree belts, reducing the size of open fields to about ten hectares. By the 1970s, Turfan boasted some 1,400 kilometers of tree belts. The oasis had re-established its defenses. Now the black hurricanes of blowing sand could rage for days, and the green dikes would hold.

Outside the oasis, the local government set limits on the size of goat and sheep flocks so the grass could grow faster than the animals ate it. In favorable parts of the terrain, recovering grass slowly built the sand into something resembling soil. Along the tailwater channels in the grass zone, the villagers planted grapes. They carried "guest soil" with high humus content from the interior of the oasis and potted it around the vine roots. Finally, when the areas planted in grapes and fodder grass were sufficiently fertile, the farmers moved in to claim new fields. Around the new gardens they made fresh belts of trees. The old trees were left behind as internal forest-crop areas (Walls 1980: 56). By 1980, the oasis was double its 1950 size (Thompson 1981: 113). Turfan had become a humus-generating organism, making a clear profit in its real currency—living soil.

By their example, the people of Turfan and thousands of other communities taught the government's staff "confusedly." The researchers then proclaimed these methods across China "clearly." The government sometimes tried to claim credit for motivating the farmers, but the farmers were already motivated. Even the national forestation program was mainly a matter of villagers helping themselves (Thompson 1981: 115).

The central authorities were prone to seek World Bank financing for capital-intensive, environmentally disruptive techno-fixes like the Yangze dam. But most rural people still looked for bio-intensive solutions. The pastoral communes of Inner Mongolia increasingly managed their pastures like farms, with fodder as a crop. This grass farming also tended to enrich rather than deplete the soil. In the Kurban-Tangut desert, three years after the pastoralists planted alfalfa in sandy areas, soil analysis found that "each mu [1/15th of a hectare] can accumulate 57.6 catties of nitrogen compounds, 10.9 catties of phosphorous, and 16.5 catties [one cattie equals 604.6 grams] of potassium—the equivalent of 20,000 catties of animal manure in terms of nutrition" (Lanzhou Institute 1982: 36). If these herdspeople could manage their animals and fodder plants in a soil-enriching way, this would be the biggest advance in pastoralism since the horse.

To "hold" sand dunes, the frontier farmers planted desert grass upwind from the dunes, and lines of trees downwind. This was called "blocking in front, and pulling from behind." Along the entire northwest frontier with the desert, such efforts by thousands of villages slowly linked together, till a roughly continuous "green wall" stretched over 6,000 kilometers.

The Greatest Environmental Challenge on Earth

Perhaps no country has done more to revive its dying lands than China. Yet all these efforts have been insufficient to turn the tide of desiccation. Losses of trees from insects, fire, acid rain, drought, and the voracious demand for wood has almost certainly reduced total tree cover over the last several decades. By the 1990s, only a few African nations had fewer trees per person. Despite all the cases of local progress, China still lost near one million hectares of grassland per year in the 1980s, largely to human-caused desertification (Smil 1993: 65, 61, 10).

In the big picture, all of China is an oasis. If, for example, a map of China is imposed over a map of the United States, the Great Lakes and most of east Canada would be replaced by desert. Almost all land west of the Mississippi would be too dry for rain-fed crops, with farming mainly confined to small oases. The remaining green area would support a billion plus population; and even within its ephemeral "green wall," the relentlessly growing economy would be swallowing, polluting and building over ever more of the farming heartland. The number of surface hectares in China has remained absolutely finite. By 1990 the amount of arable land per person was below that of Bangladesh (Smil 1993: 190). And most of the remaining farmland was being stripped of its former fertility by new farming methods.

The grain crops on the northern plains consume about fifty kilograms of nitrogen from each hectare per season. Traditional farmers generally replaced this by heavy manuring, recycling crop residue, rotations of legume crops, and some resting of fields. The farmers of the 1980s and '90s may have re-emphasized family plots and intensive gardening, but they also increasingly applied the methods of industrialized plantations to tiny farms. As the ever more profit-minded farmers tried to maximize grain harvests year after year, the old rotations of legume crops were often squeezed out. Almost all crop residues were burned for fuel, and input of natural compost and human waste was falling fast. By the late 1980s, the organic content of soil over much of the northeast was down to two percent, from nine percent before the onslaught of modern farming (Smil 1993: 58, 101–2, 164, 179). The main remedy was a massive rise in the use of chemical fertilizers, along with a massive drain on groundwater for irrigation. In the near future, pressure could mount to bring the arid northwest under the plow.

By the early 1990s, the Chinese Academy of Science estimated that 110 million peasants had recently fled villages to escape environmental poverty. These masses had become an army of unemployed, operating outside the system. Conditions for those remaining in the villages were often so harsh that kidnapping and selling women reappeared. When Jan Wong (1996: 327) asked a village farmer if he knew of anyone buying a wife, he said, "Loads. Otherwise, nobody could find a wife. The women in our village all want to marry someone in the city."

The cities were providing an outlet where new resources could support more people, but even this had certain limits. For example, from the 1950s to the early 1990s, the water table beneath Beijing fell from near five meters to over fifty meters below the surface (Smil 1993: 167–69). The whole of northern China drew closer to exhausting its water supplies. The most populous region on Earth faces the possibility of strict water rationing, or political conflict over water among people, farms and development projects.

Dominator and Partnership Legacies in the Age of Ecology

By the time of Mao's death, the era of militant anticolonial nationalism was fading from memory. In the future loomed the issues of the ecological age. With rising unemployment, loss of good farmland and a gently falling per capita output of food, "national security" was threatened more by ecological ruin than by any "enemies of the state." Unfortunately, the pressure of scarcity tends to heat up competition for control. So in the early 1970s, Mao and his close associates launched their last public education crusade, attempting to capture the public's rising frustration and turn it against the critics of centralized power.

In the "Anti-Lin Biao, Anti-Confucius" campaign, Mao presented a carefully crafted version of Chinese history. He presented Confucius and the recently discredited army head Lin Biao as symbols of the reactionary forces in every age. Confucius was labeled a spokesman for the old "slaveowning class," who had stood against the rising feudal princes. It was the first emperor, Qin Shihuang, who had swept aside these old slaveowners and imposed a more advanced feudal mode of production that had laid the basis for a strong and united China.

In the classic Marxist theory of history, humanity must evolve through five stages of evolution, namely an original stage of primitive communism, followed by slavery, feudalism, capitalism and socialism. If we compare this to Mao's last vision of Chinese history we note that Confucius was made to represent the slave economy rather than the "primitive communism" of the golden age. Or perhaps Mao just dropped the golden age and "primitive communism" out of the picture entirely. Then, as in conventional Western history, the story of civilization would start with the first warlords, and there would be no mention of any age of the Goddess before that.

Confucius, Mao said, was a slaveowner who had tried to stand against the process of nation-building. So, Mao argued, the future in every age belongs to those who grasp the newest modes of management and use them to build an ever stronger state. The whole twentieth-century Chinese revolution was fought for this cause. Before the revolution, China lay prostrate before the colonizing powers. Now, the Chinese government had the power to crush enemies from without or within. This was the meaning of China's liberation. In the future, the people would only grow more united behind their leaders, loyally serving an ever more powerful state.

In his youth, Mao had been a fiery idealist from southern China, opposed to every form of domination. Then he rose to a position of power rivaling that of the old emperors. He became "first among equals," responsible to oversee and enforce the division of scarce resources. If in his last years he came so close to identifying with the tyrannical first emperor, it was no coincidence that young protesters began calling for the "overthrow of Qin Shihuang." So the high-pressure politics of environmental scarcity had proved powerful enough to reduce even this heroic soul to a self-protecting autocrat.

In 1995 a number of Chinese scholars produced a book with a different vision, called *The Chalice and the Blade in Chinese History*. This book traces the interaction of dominator and partnership cultures through the ages, building off Riane Eisler's analysis of Western history. It shows the evolution of relations between men and women, emphasizes the values of the pre-military age, highlights the goddess religions of China and points to a possible future return to values from the age of the Goddess. This book was prepared and translated into English in time for the 1995 United Nations Conference on the Status of Women in Beijing, but some bureaucrat withheld authorization for the book to be sold at that event. Thus, editor-in-chief Min Jiayin and his colleagues in the Chinese Partnership Studies Group were stuck with the stock of the English edition, and the printing bills.

The members of the Partnership Studies Group are obviously concerned to write history fairly, without providing justification for any prejudices among the Han Chinese and ethnic minorities. For example, their writing highlights examples of equality between men and women in the formerly nomadic nations. They avoid blaming the pastoralists for past wars, and seek to build a climate of partnership for the future. All this is as it should be. But what if we also need to talk about the effects of environmental degradation on human beings? Is it possible to discuss the social impact of desertification without blaming its victims?

Perhaps we can say that the people of the deserts are among the most tragic figures in history. Try as they might, they seldom were able to make their barren lands fully support them. Their history was often violent, with cutthroat competition for the means of life. Their ancestors often turned to raiding other nations in order to survive. When they invaded China, they became dominators, with all the psychological problems that involves. Any of us could have been born in their places. Perhaps their experience shows how much we all depend on a healthy environment for the most basic quality of life.

Chapter 10

Locusts and Desert Saints in North Africa

A LOCUST IS A KIND OF GRASSHOPPER THAT BURIES ITS EGGS IN THE SAND. THE eggs lie dormant, like the seeds of desert grass, waiting for water. If sufficient rain comes, the locust eggs and grass seeds both sprout at the same time, emerging as parasite and host. Desert locusts are large and aggressive insects that swarm in clouds and move downwind, devouring the vegetation before them more obsessively than goats. A few decades ago, naturalists discovered that desert locusts and common African forest grasshoppers are both of the same species. Their coloration is different: many forest hoppers are black, while those of the desert have sandy shades. Desert locusts also tend to be larger. But the main difference, which often blinds people to the similarity of these insects, is their behaviour. Forest grasshoppers live mainly as solitary creatures. They do not swarm, but quietly nibble on leaves in the forest. Desert locusts, on the other hand, rush about in swarms, as if driven by an instinctual panic to consume everything they can, as soon as they can. These vast differences in behaviour are probably connected to differences in environment. Desert locusts seem to be born with the expectation that there will soon be nothing to eat. And, sure enough, locust hoards leave virtually nothing to grow for another day.

The contrasts in appearance and behaviour between desert and forest people are not so striking. Humans are not as preprogrammed in their ways. They can make decisions that override habits from past generations. Therefore humans are open to far greater ingenuity, and stupidity, than other creatures. We justly see ourselves as a class apart from other beasts. But sometimes a similarity can be traced between the cultural adaptations of humans and other animals who live in the same environment. And in few regions of the world have the parallels between human and grasshopper behaviours been so striking as in certain periods of North African history.

Morocco's Migrants from the Sahara

In the early 1700s CE, the Ait Atta ("Atta people") of southeastern Morocco began moving north toward the Atlas Mountains. A series of droughts in the pre-Sahara had convinced them to leave their ancient homeland. To get a new home, they would have to fight. In preparation for their quest, the Atta clans elected a supreme chief. The young men formed war parties and began raiding other tribes. They rustled livestock, captured supplies of dates and otherwise

The Maghreb: North Africa's "Island of the West"

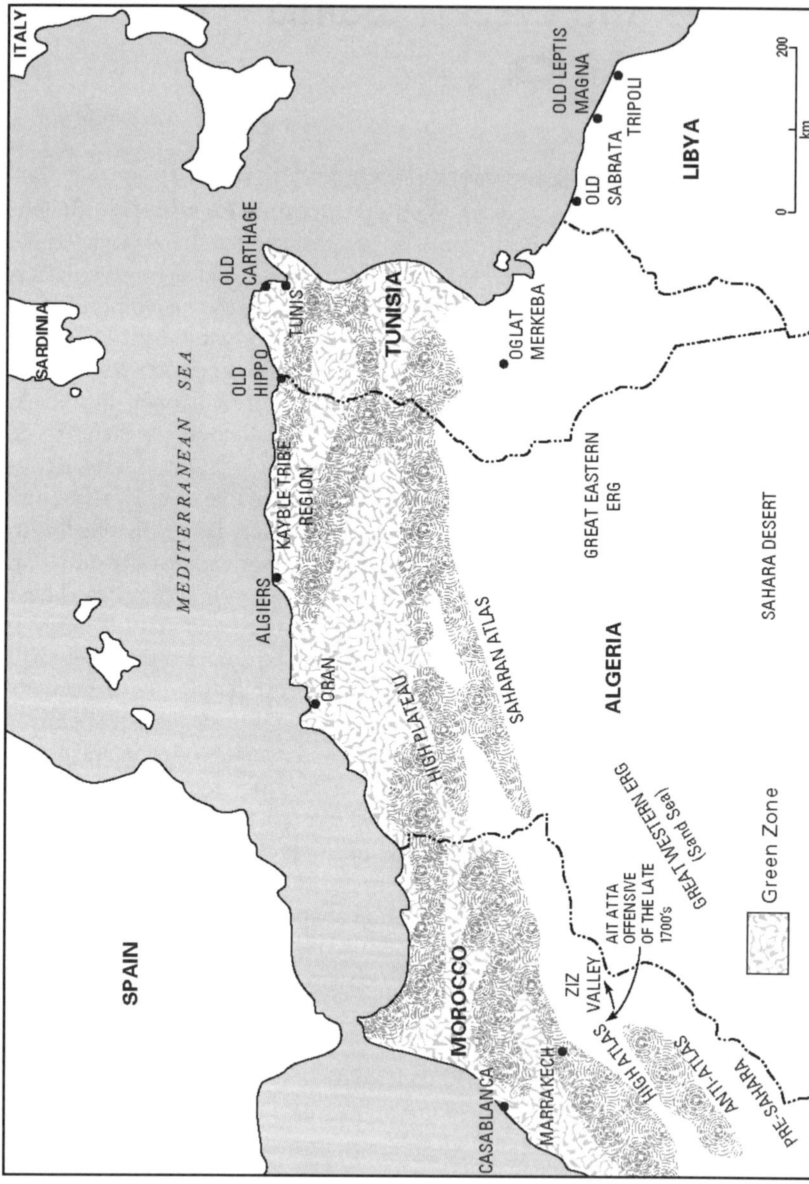

won glory for themselves. As they ranged northward, the Atta made a reputation for ferocity.

The Atlas foothill people prepared themselves for the onslaught by forming a defensive alliance. Several centuries before, the foothill people themselves had migrated northward from the Sahara to gain arable land. Now they called themselves the Ait Yafalman—"the people who seek peace." Against this alliance, the Atta warriors raided across a 150-kilometer front. They terrorized the villagers with their lightning raids but could not hold a captured mountain valley. Wherever the Atta intruded, the defenders slowly converged. Finally, the Atta had to admit defeat. They turned back, to prey on the more defenseless oases south of the mountains (Dunn 1972: 87–89).

The oases of the Moroccan pre-Sahara were basically watering holes, with date groves and some limited gardens. The *Ait Atta* conquered a string of such oases in the Ziz Valley around 1830. The terms of occupation varied according to the resistance the Atta received. Where the oasis overlords surrendered without a fight, they were allowed to keep their lands, trees and peasants; the new masters required only a heavy tribute from rents and crops. But in the Rteb area, the old Ait Izdig landholders resisted fiercely and lost. The Atta confiscated all their property and animals, and the surviving Izdig had to walk out empty-handed, hoping their relatives in other districts would take them in (Hart 1972: 55).

The Atta seldom quarreled with other pastoral groups, at least not over pastures. Their lands stretching off into the desert were so vast and sparse of vegetation that it was hardly worth marking or guarding boundaries. The Atta even made common cause with other pastoral tribes, such as the Bani Mhammad. Since both these groups shared common needs for food and water, they joined forces against whomever had these things. The two tribes made a formal pact, renouncing the right to exchange wives (Dunn 1972: 92–93, 97). They were, after all, brothers in arms. A brother in arms was more truly a brother than one simply born of the same woman; and, as the sister of a brother was a sister, brothers in arms did not exchange women.

It was mainly the oasis people whom the Atta and Bani Mhammad threatened with deadly force. The oasis dwellers had what the raiders needed—the dates, millet, melons and beans that made up an essential part of the nomads' diet (Dunn 1972: 104–5). The actual farmers of the oases were mainly Black people, known across much of Saharan Africa as the Haratin. These Haratin gardeners were almost never evicted from the land, because no matter who ruled the oases, someone had to grow the food. So the Atta made their traditional offer to the Haratin—military "protection" in exchange for a share of the harvest. If the Haratin tried to refuse, the Atta would simply cut down the date palms. The Atta had the power to destroy an oasis, and the power to destroy a thing was power over it. The Atta, however, saw their control as a service rather than a theft. They provided strong overlords, which could be better than weak ones. If one group of warriors was not powerful enough to monopolize local control, an oasis could be subject to extortion from more than one desert tribe.

The Haratin Farmers

Black farmers and former farmers are probably the largest ethnic population in southern Morocco. Many are descended from slaves brought from south of the Sahara. But some historians feel that many Haratin are descendants of people from the old Green Sahara who migrated north rather than south as the Sahara dried up (Adam 1972: 327). In southeastern Morocco they live in villages strung along the stream beds. Often they name themselves after their stream, for example, the *Ait Dra* ("people of the Dra River") (Hart 1972: 53).

In peaceful barter with oasis farmers, the nomads might offer animal products for fruits and grains. But the farmers were basically self-sufficient in food, and the nomads were not. To put it another way, the oasis gardeners were superior in producing the means of life, and the pastoral tribes were superior in the means of inflicting death. One type of superiority tended to overrule the other. As Ross E. Dunn says, "The Atta did not hesitate to use force ... when they wanted a larger share of the resources than they could conveniently secure through reciprocity" (1972: 105). The cultivators were sitting ducks whose assets could not be concealed or moved to a new location. They were shepherds of immobile plants. Everyone knew just where the Haratin could be found. Even if they put up a fight and drove the nomads off, the raiders could always return in the night to cut down the palm trees. Rather than lose everything, the Haratin agreed to pay tribute. And their submission to force proved to have a certain survival value. From a Darwinian point of view, the assertive farmers were systematically weeded out, leaving the more submissive ones to raise submissive children.

The Atta had to observe certain limits in their extortions from the Haratin. Once the Atta gained control, they terrorized the farmers only enough to spur payments of tribute. In the Ziz Valley, the Atta taxed the Haratin close to the edge of bare subsistence but not enough to kill them. Generally, the farmers had to give up four-fifths of their crops. As military control was the criteria of ownership, the Atta now owned the land; and, according to an old rationale, the owner of the land granted the farmers their soil, tools, seed and animals. These were four of the five elements needed for farming. The peasants were said to provide only the fifth element—labor. It was therefore deemed rational that the laborers should receive only one-fifth of their yield (Reudy 1992: 25). It could even be argued that the farmers were overpaid, because in addition to allowing the farmers their soil, seed, tools and animals, the Atta warriors also provided a sixth service—protection from additional raids by other nomadic groups, such as the *Ait Murghad* or *Ait Siddrat*. Since the Haratin were so low in status as to grub in the soil for food, they were lucky not to be charged more for their work.

If the farmers had to give up four-fifths of their crop, then they had to scrape enough from the soil to survive on one-fifth. So they were forced to maximize their yields and minimize their expenses in farming. Such overcultivation of the tiny oases was supposed to make up for a protein deficiency in the entire surrounding landscape—a landscape which was being steadily depleted by goat herds, the clearing of brush by fire and the drying climate. As the pastures of the pre-Sahara grew more barren, the pastoralists repeatedly tried to move closer to the moist Atlas ranges. "The effect over the centuries," Dunn says, "was a kind of recurring bumping action with both Arabic- and Berber-speaking tribes pushing one another ever closer to the northwestern coast" (1972: 85). In this movement, the oases farmers were often left behind, facing both the creeping desert and the remaining nomads, who were desperate losers from the race for northern pastures.

The Ancient Economy of North Africa

North Africa before the founding of Carthage was a relatively green country inhabited by sedentary villagers. Farming, of grains, figs, olives and grapes, may have started earlier on the Mediterranean's south shore than on the European side. The best farmland stretched from Tunisia to Morocco, in a coastal belt sheltered from the Sahara by the Atlas Mountains. The Arabs would later call this region "the island of the West," or *Maghreb*. The Arabs saw the desert as a sort of flowing sea, with an island of green floating in the northwest corner. Abdallah Laroui summarizes the archaeological evidence from this "island":

> everything in the monuments, furniture, weapons, clothing, and rites … suggests a sedentary population and nothing points to a life of nomadic shepherds. There are no offensive weapons, the clothing is worn without ornament; on the other hand, we find porridge bowls indicative of agriculture and large necropolises implying dense population. (1977: 60–61)

Starting in the 800s BCE, Phoenician shippers helped make Carthage one of the world's greatest ports. Here and in other coastal markets such as Sabrata, Utica and Hippo the western and eastern Mediterranean worlds met to trade. The North Africans sold their wealth of surplus grain, local crafts, plus gold and ivory from deep in Africa. By 500 BCE, at the latest, the Berbers were crossing the Sahara for trade. The comparative mildness of the desert at that time can be inferred from the means of transport. They hauled supplies from the Niger River to Carthage in horse- or donkey-drawn carts (Davidson 1991: 54, 17). Within a few centuries, driving such carts into the Sahara would become suicidal.

Around Carthage, the prehistoric forest belts were still largely intact. Even in the first century CE, the coastal lands of Tunisia and Algeria were described as "a country that was nothing but one shaded grove" (Davidson 1991: 56). According to Pliny, each grain of wheat planted in the African soil sprouted a stem of 150 grains (Attenborough 1987: 117). The rainfall was probably much like southern France. In the drier interior, deep-rooted trees proved hardier than grain and arboriculture was the main sort of farming. Already, foreign cities depended on Africa for food. And already, military rulers in Europe wished to avoid paying Phoenician middlemen for African produce. If paying soldiers would be cheaper than paying farmers, then the Romans preferred to take Africa's food by force. So when the Romans destroyed Carthage in 146 BCE, they eliminated the middle-merchants. They reportedly sold 400,000 Carthaginians as slaves to Roman landlords and took control of the farmlands. Then they took one-third or more of North Africa's crops as a tax.

Timetable of Conquests and Migrations Affecting the North African Maghreb

800s BCE	Formation of Carthagenian–Berber trading network
264–146 BCE	Punic Wars between Carthage and Rome
146 BCE–429 CE	Roman colonization
c. 100–700 CE	Christianization of North Africa
290 CE	Romans retreat to defensive walls before Berber attacks
300s–400s	Donatist "heresy" suppressed by Roman church and state
429–533	Vandal kingdom of North Africa
533–670 and 685–710	Reconquests by Eastern Roman Empire
670–711	Arab invasion, annexation to Islamic empire
740–789	Berber revolt against Arab rule, Kharjite "heretic" state
789–926	Shi'ite state, leading to the Fatimid empire after 910
900s	large migrations of desert Berbers into coastal Maghreb
1049	Maghreb throws off Egypt-based Fatimid rule, allies with Sunni Islamic caliphs
1050–1200s	Large migrations of Arabian tribes into North Africa
1537–1710	Algeria under Ottoman rule
1830–1962	French domination of Algeria
1952–62	Anticolonial wars in Algeria and Morocco
1976	Morocco and Mauritania attempt to partition the Western Sahara
1988	"Black October" riots against government austerity cutbacks

Rome's Breadbasket

The Romans saw themselves as developers of agriculture in Africa, and this was not entirely self-delusion. When Roman troops were not fighting Africans, they built viaducts and storage tanks. Their works tended to divert water away from native farmlands and towards the Roman towns and plantations. When Roman soldiers retired, the government gave them generous grants of African land. This too was development, because it employed Roman citizens. Besides, soldiers tired of war often made good gardeners. They industriously terraced the wadi beds or built catchment walls, allowing increased rain-fed farming. Many of their workers were poorly paid natives who had previously owned the land. The Roman methods of contouring fields to retain scarce rainfall were basically copied from the ancient North African jessour gardens.

The Romans, however, did not measure the success of their colonies by improvements to the land. For Rome, development in Africa meant increased supplies for the imperial center, ever more land awarded to successive generations of army veterans, and larger estates for Senators. By the first century CE, Rome was milking half a million tons of grain a year from Africa. Presiding over this enterprise were the Roman urbanites, residing in some six hundred towns or cities—Leptis Magna, Tripolitania and Sufetula each housed 50,000 to 100,000 people (Davidson 1991: 56). That was the height of ancient development. Within a few centuries there would be scarcely one town between Libya and Morocco.

When the Roman plantations expanded into ever more marginal lands, many displaced African farmers tried to cultivate the semi-desert. They in turn pushed the pastoralists to the borders of the Sahara, where their hungry animals stripped the desert's edge (Cloudsley-Thompson 1984: 12). Laroui feels that the bulk of the Maghreb's population was "renomadized," in "an involuntary return to proto-history"; the Roman walls against the nomads were "not so much a boundary of civilization as a ... frontier between the dispossessed, who were thrust out into the desert, and the needed workers, who were enslaved and crushed with taxes" (1977: 71–72, 54).

The Berbers found themselves caught between the advancing Sahara and the Roman legions. The desert, however, was harder to beat than the Romans. By sheer necessity, a Berber counterattack grew in strength on the empire's most desolate margin. First, nomad raids forced the Romans to build walls around their watered towns and estates. Then in 290 CE, the Romans made a major retreat, abandoning about a third of their former territory. They fell back to more defensible perimeters and gave up trying to dominate the pre-Sahara.

A Berber king, Jugurtha, tried to drive the Romans away soon after the defeat of Carthage. When his war effort failed, he offered to bribe the Roman Senate. Jugurtha attempted going over the heads of the colonists to simply buy a senatorial declaration of autonomy for his province of Numidia. "Rome," he remarked, "is a city for sale, and doomed to perish if it can find a purchaser" (Shaar 1971: 108). The Senators were eminently bribe-able, but in this case, taking the money would have been political suicide. To sell the Berbers their own country would have cut off or steeply raised the price of a third to one-half of Rome's food supply. It would mean that Rome would have to actually purchase all that food, rather than taking much of it as a tax. Besides, it would have been unpatriotic for the Senate to betray so many settlers and army veterans. For Rome to give up North Africa would have been harder than it was for France to leave Algeria in 1962.

The Romans' farms were mainly a series of monocrop plantations. Probably most estates invested little or nothing in fertilizer. Still, the land was so fertile to begin with that cereal harvests remained good for several centuries. When yields began falling, the officials at first treated it as a problem of tax evasion, so they tried to sustain their take of African wheat by tougher tax collection. But where good crops of wheat would no longer grow, the farmers increasingly turned to planting drought-resistant olive trees. The government then tried banning olive plantations, so as to force the farmers back to growing wheat. Finally, in the last stages of the empire, this ban was repealed. It grew obvious that many North African farms could produce olives and dates, but virtually nothing else. In some areas even the olive trees died (Attenborough 1987: 118). The Romans understood military threats and how to guard against them, but now they faced a larger, non-military threat. The situation resembled Ethiopia in the 1980s, where a strong and costly army stood ready to defend the country while the land itself degenerated beneath the soldiers' boots. Many explanations may be offered for the fall of the Roman empire, but in the view from Maghreb, it seems that Rome's African breadbasket dried up and the empire blew away.

The Vandals and Independent Berbers

As Rome caved in, Europe faced China's problem of barbarian tribes breaking through the walls. Some migrants from the eastern steppes pushed all the way through Spain to North Africa. Most of these invaders wanted jobs more than farms, and the work they knew was fighting. Their needs and abilities matched a growing Roman demand for mercenaries. The imperial army had an almost bottomless appetite for fierce warriors, and numerous Roman lords were hiring private armies. Sometimes they hired whole tribes, like ethnic corporations of mercenaries. So the Huns were invited to fight for one Roman lord after another, all the way to France and back to Hungary. The Romans seem to have viewed these traditional warriors as human mutants, warped into fighting machines for rent.

The Vandals traveled through Germany, France and into Spain. They had heard of North Africa as the richest breadbasket in the empire, and many of them decided to go there. They contacted the Roman governor of Africa, Bonifacius, and got an invitation to come and fight as mercenaries against the Berbers. So, about eighty thousand Vandals came to the Maghreb. On arriving, they quickly realized North Africa's reputation as a land of milk and honey was sadly out of date (Attenborough 1987: 125). The colonial officials had responded to their latest crisis by calling in a host of Eurasian mercenaries, but the colony could no longer afford to pay them. The Vandals therefore decided to collect their pay directly. In 439 they sacked Roman Carthage and made North Africa a Vandal colony. As the Romans fled from their own mercenaries, the Maghreb fell to men who knew raiding better than farming.

The Berbers by this time were largely children of the arid outback. As the Romans pulled out, these exiles from the desert moved in. Shepherds brought their flocks to graze on the abandoned plantations. Instead of being overfarmed, much of the country would now be overgrazed.

The Vandals concentrated on taking the remaining wealth of the partly abandoned Roman towns, and after exhausting this resource, they too began leaving Africa. Some went to Sardinia or Italy, which were now comparatively richer lands. As the warlords left, it seemed possible that the Maghreb people might finally be left in peace. The Roman empire, however, took one last swipe at them. In 533 CE, Emperor Justinian of Constantinople sent an army to seize Africa back. In his secret Anectota, the historian Procopius confided his opinion of this war:

> That Justinian was not a man, but a demon, as I have said, in human form, one might prove by considering the enormity of the evils he brought upon mankind.... For Libya, vast as it is, he so devastated that you would have to go a long way to find a single man, and he would be remarkable. Yet eighty thousand Vandals capable of bearing arms had dwelt there, who could guess their number? Yet more numerous than they were the Mauritanians, who with their wives and children were all exterminated.... Sending commissioners to value the province, [Justinian's appointed governor, General Belisarius] imposed grievous taxes where before there had been none. Whatever lands were most valuable he seized.... Wherefore mutinies arose resulting in the deaths of many. For he was never able to abide by established customs. (in Schaar 1971: 124)

Of all regions in the old Roman empire, North Africa suffered the most complete social and ecological collapse. Still, the fall of the empire was hardly the end of the Berbers, or of the Maghreb countryside. Once the Byzantine Romans withdrew, the surviving local farmers kept most of their produce for themselves. Much of the country went to seed, and the Berbers recaptured their orchards. By the 800s CE, travelers such as al-Ya'qubi reported groves of olives and dates stretching for a hundred miles in Tunisia. In place of the ruined Roman cities, the Berbers dispersed into farming villages. Laroui (1977: 122–23) says "they seem to have been independent, maintaining no regular ties with any suzerain and paying no taxes to anyone."

When Arabian cavalry units pushed into North Africa after 662 CE, they found few population centers larger than villages between western Egypt and Morocco. Of the six hundred towns in Roman North Africa, only partly-inhabited ruins remained. Arab commander Uqba bin Nafi established a town called Ifriqiya in 670, not far from the ghost city of Carthage and Leptis Magna (Willis 1971: 145). But Arab cavalry units riding into the Maghreb encountered a hornet's nest of resistance against all invaders (Laroui 1977: 69). Berber leaders like Queen Kahina held up the Islamic *jihad* for thirty-five years. Several times between 681 and 701, native Algerians drove the Arabs out completely. The Islamic empire met no fiercer opposition anywhere in its first century. When the Maghreb finally submitted to Islam after 708, it was at first only a military alliance in which Arab and Berber warriors united to raid Spain (Alport 1972: 144).

The Island of Rebels

To the outsider, North Africa presents an image of strict social conformity and religious orthodoxy. But within North Africa, a vast gap has repeatedly appeared between two different kinds of orthodoxy. The first is the orthodoxy of local communities, with their strong bonds and traditions. In the hands of these communities, both Christianity and Islam were mainly expressions of ancient values, such as hospitality and group solidarity. For these traditionalists, an injunction to love one's neighbour as oneself might seem an obvious piece of common sense. The moral was driven home in a thousand traditional tales: Once upon a time the ancestors knew hardship and lacked sufficient food and drink. Yet they shared what little they had and, behold, there was enough.

The second orthodoxy is that of the rulers, with their claims to absolute power over the people and resources of a desiccated land. This was a basically political and economic, rather than religious, orthodoxy. But it infringed on religion, in that it sought to identify the ruler's interests with the will of God and establish the ruler as an object of reverence. For example, when King Hassan II of Morocco spoke at the commissioning of a new army general in the 1970s, he said: "Anyone we entrust with a civil or military mission must reflect our mission, that of the commander of the faithful, obliged to serve as the divine shadow on earth" (Munson 1993: 127). At that time, the Moroccan army was engaged in the will of its master, annexing the former Spanish Sahara and exterminating the local resistance.

These two kinds of orthodoxy (of the rulers and the local communities) both claimed the same God and the same holy book as their sources of inspiration. Aside from this claim, the two orders of religion bore little or no resemblance. To the outside world, the orthodoxy of the rulers has appeared to be the "official" form of religion. This has been the outer vestment of Middle Eastern and North African civilization. When sultans like Hassan II spoke of their own power as the essential element in Islamic culture, the outside world often believed them. But within North Africa, the claims of rulers to godlike omnipotence were less readily believed. The divine right of sultans was not popularly accepted as an article of true religion. It was a belief of central importance only to the rulers themselves.

Of course, local people and their clerics seldom attacked the ruler's beliefs in public, and silence could be taken for consent (Munson 1993: 126–34). To challenge the divine right of kings often brought the death penalty more speedily than any other heresy. As King Hassan II announced to his country after a coup attempt in 1972, "God has placed the king on the throne to safeguard the monarchy, and to do this the Maliki school of Islam stipulates that he must not hesitate, if necessary, to eliminate one-third of the population infected by evil ideas to protect the two-thirds of the population not so infected" (Munson 1993: 134).

Naturally, the Moroccan people did not greet these sentiments with wholehearted enthusiasm, but they usually phrased any criticism of the ruler in indirect language. Statements of general religious principle could be made to affirm the communal values violated by the sultans. So a series of Moroccan proverbs have called down the wrath of God on unjust rulers: "The oppressor will not be helped by God"; "The oppressor is cursed by God and despised by the people"; and "There is no protection from the call of the oppressed to God" (Munson 1993: 143–44). On the side of traditional North African culture veiled to the outside world like the dark side of the moon, those who stood up against despotic warlords had been commonly revered as the greatest saints of Islam.

Over the past two thousand years, the North Africans have embraced two religions—first Christianity under the Romans, then Islam, initially under the Arabs. In both cases, the local people took the invader's religion and used it to express their own values. They turned both religions against those who brought them, repeatedly rejecting foreign rulers not only for their oppression but also for the betrayal of their own faiths. Islam, of course, evolved primarily within arid environments and has been strongly shaped by the high-pressure politics of scarcity. What is less well recognized is how deeply Christianity took its "orthodox" form within that same environment, before it ever became the orthodoxy of medieval Europe.

The Donatist Christian Revolt against Rome and Its Church

Before Christianity became the official religion of Rome, it had spread rapidly in the Maghreb as a popular cult. The Romans periodically repressed it, because Berber Christians tended to use Christianity as a justification for refusing to pay homage and taxes to the emperor. During this period of sporadic conflict and persecution, the Christian churches in Italy grew apart from churches in the colonies. For Roman Christians, there often seemed no real contradiction between loyalty to the empire and loyalty to God. In the colonies, the divergence of interests was more obvious.

The Maghreb was arguably the most hard-pressed of Roman colonies. There, the rationalizations of Roman Christians often appeared as nothing but self-serving injustice masquerading as faith. The Maghreb became a breeding ground for a "church of martyrs," in which refusal of loyalty to Rome was taken as an article of Christian faith. Death for the sake of that principle was proclaimed a guarantee of salvation in the next world.

After Emperor Constantine embraced Christianity (around 312 CE), the official churches of North Africa suffered from identification with the colonizing government (Reudy 1992: 11). When Roman or Latin church leaders gained high ecclesiastical posts in North Africa, indigenous Christians often rejected them. Perhaps most Berber Christians regarded any priest or bishop who compromised with Rome as unfit to be a spiritual leader.

In 312, the church appointed a new bishop of Carthage. The chosen candidate, Caecilian, was a loyal Roman. According to his accusers, he was so loyal that, earlier in his career, while Rome still persecuted Christians, he had allowed Roman officials to destroy Christian scriptures rather than defy them. Some of the lowly parishioners of Carthage presumed to veto his appointment, which would imply they had the right to choose spiritual teachers for themselves. As this protest movement, called the Donatist heresy, spread in the decades to come, Augustine, the Bishop of Hippo, saw the implications clearly. If the official church allowed popular veto of its appointed leaders, it might have no power over anyone.

St. Augustine (354–430), who perhaps shaped the Roman church even more than St. Paul, was both a North African and a well-educated member of the empire's upper class. During his youth, the plantations of wheat were failing, the plantation workers were rebelling and the barbarians of Inner Asia were thundering across Europe. During his last days, in 430, the Berbers were reclaiming their land, and the Vandals were hammering at the gates of Hippo and Carthage. In the conflict of Berber peasants and Roman masters, Augustine saw the empire as a force for God's order, and he felt it his duty to maintain discipline over an unruly flock. Just as many modern Moroccans feared the chaos that might result if King Hassan II was overthrown, Augustine placed his trust in a military dictatorship under God.

Theologically, Augustine argued that the Roman church priesthood stood above criticism from the laity. If the church hierarchy had appointed a man as priest or bishop, his qualifications were not subject to review by the people. The church's nomination defined a man as priest or bishop, not the consensus of the sinful parishioners. Why were the parishioners so incapable of choosing their own leaders? Because, reasoned Augustine, they had been in the grip of an inborn disobedience to ultimate authority ever since the first disobedient acts of Adam and Eve. The depth of this inherited evil could be easily observed. For example, if Augustine set his mind against desires of the flesh, such as desire for sex, his body still returned to such desires again and again. Despite his vow of chastity, Augustine dreamed of sex and woke with his penis erect. Obviously the body was tainted with disobedience and could not be trusted. The options, as Augustine saw them, were to give free reign to the chaos of disobedient passion (which the common people were obviously wont to do) or subject this disobedience to a superhuman authority. Christians, Augustine maintained, should recognize that they were born sinners, incapable of discerning or choosing between good and evil. They should give up relying on their own twisted common sense, and instead depend on a higher, external authority (Pagels 1981: 113–14). And who were those correct superhuman authorities? They were the heads of the established Roman church and state. Was not the church's succession of bishops and popes set in motion by Christ the king? And how could the rulers of this world have come to power save by God's will? If life was a battle of good versus evil, what else besides a powerful alliance of church and state could save the peasants from their own sinful ways (Pagels 1981: 125)?

This theological argument was most convincing to the empire's rulers and ecclesiastical heads. Those in high office did not usually ask why they themselves should be considered less prone to sin than their subjects, such as the peasants of the Maghreb. Therefore, Augustine's arguments against peasant revolt in the lands of scarcity were adopted as church dogma. A Christianity of the rulers was given its clearest formulation in North Africa, and then became the outer vestment of European civilization for the next thousand years.

As for the Maghreb, when the Donatist heresy was finally suppressed by armed force, the indigenous population quickly abandoned Christianity (Reudy 1992: 11). In this case, the religion of the rulers defended itself by destroying the popular religion.

The Maghreb's Muslim Rebels

In Islamic times, the Magreb's people resumed their rebellious ways. Supposedly, Berbers who joined the Arab Muslims as comrades in arms were to be welcomed as full members of the Islamic brotherhood. Many Arabs, however, felt that the Berbers lacked all the marks of higher Arabic civilization. To them, the Berber's conversion seemed an opportunistic gesture made to participate in plunder and evade the tax on unbelievers. Besides, Arab governors needed revenue with which to rule. Therefore they reimposed the unbelievers' tax on Berber Muslims. At this insult the Berbers rose to fight again, taking up the banner of an outlawed Islamic group known as the Kharjite sect.

The Kharjites did not recognize the hereditary Umayyad caliphs of Islam. They claimed the caliph should not simply inherit his position but should be chosen by a consensus of those learned in the faith. The leader should be the greatest exemplar of spirituality in each generation, be that person a prince or a slave. As the Quran said, "The most righteous among you is the most honoured before God" (49: 13). The original Kharjites fled persecution in Persia and Iraq by coming to North Africa. In meeting these exiled heretics, the Berbers found "other Muslims by whom they were treated and regarded as equals" (Alport 1972: 142–44). It was like a spark causing the pre-Saharan bush to burst into flames.

The Kharjite revolt of 740 drove the Arabs completely out of northwest

Africa. And aside from a later Arab reconquest of Tunisia, no Arabia-based empire ever ruled the Maghreb again (Reudy 1992: 12–13). Unfortunately however, the Kharjites' opposition to central rule left them prone to division: "Without the [autocratic] system to combat, " Laroui (1977: 96) says, "the Kharjites turned on themselves and devoured each other." In doctrinal disputes they splintered into sects. Their councilors were fallible, sometimes corrupt, and often ineffective as military commanders. For many in that time and place, such difficulties were enough to discredit primitive democracy.

A second Muslim Berber revolt was a Shi'ite *jihad* to remake the Islamic world. This was the Fatimid movement of the 800s, which rose first against the Arab rulers of Tunisia, swept the Maghreb and went on to conquer Egypt. The Fatimid Shi'ites also objected to the rulers of orthodox Islam, but for a different reason: they believed that the caliphate had been seized by the wrong family. With an old belief that holiness is inherited, the Shi'ites held that only a direct descendent of the Prophet through Muhammad's daughter, Fatima, could be a sinless, rightly guided head of the faith. The Fatimid's leader claimed to be just such an imam of the Prophet's lineage. Of course, his claim could never stand alone as the inspiration of revolt; he also had to inspire the hopes of the poor. Basil Davidson (1991: 132–33) says, "The impoverished people were the mainspring of revolutionary Shi'ism."

With an army largely made up of Berbers from the Kabylia region of Algeria, the Fatimids swept into Egypt in the 960s. The Fatimid caliph, Mo'izz, then built his new capital at Cairo and moved on to attack Syria and Arabia. The Maghreb found itself left behind and ruled as a province in an Egypt-based empire. Within decades, many Berbers viewed the Fatimids as just another foreign power, collecting taxes from the Maghreb's poor and giving nothing in return. In 1049 the North Africans rebelled again, this time claiming allegiance to the Fatimid's enemies, who were the Sunni orthodox caliphs of Baghdad. This reversal of loyalties involved apostasy against the Shi'ite faith, and acceptance of Sunni orthodoxy. So North Africa's rebels came full circle in the spectrum of Islamic denominations. However, this latest rebellion cost them more than any other. After the Berbers apostatized against Fatimid rule, the Fatimid caliphs took revenge. They invited certain Arabian allies such as the Bani Hilal and the Banu Sulaym tribes to come and take the Berbers' land. Starting in 1050, these huge tribes passed through Egypt with their vast herds of animals. On reaching the coastal strip of Libya, they fanned out over the countryside.

The Great Arabian Colonization

The first Arab invasion of North Africa in the 600s had been little more than an all-male military incursion. But after 1050, something like 200,000 Arabian men, women and children arrived as settlers, driving millions of sheep, camels and goats. The migrants moved east to west along the coast, officially sanctioned to plunder any Berbers who stood in their way. By 1150 they dominated the eastern half of North Africa. By the 1200s their frontier reached the Atlantic (Trimingham 1962: 19). In the 1300s, Ibn Khaldun wrote his judgment against them:

> After the Banu Hilal and Banu Sulaim [Sulaym] descended upon Ifriqiya and the Maghreb … three hundred and fifty years of struggle ensued during which the countryside fell into utter ruin.… Formerly the whole region between the Sudan and the Mediterranean had been well populated, as the remains of civilization, the debris of monuments and buildings, the ruins of villages and towns bear witness. (in Trimingham 1962: 19)

From his standpoint in time, Ibn Khaldun blamed the desolate condition of North Africa more on the Arabs than on the Romans. As the invaders were mainly nomads, their influx definitely upset the previous environmental balance between farmers and pasturalists. And though the Arabs may have been minor figures in a vast landscape, the degradation of North Africa grew increasingly serious. Laroui (1977: 216) comments, "It is certain that arboriculture suffered most. The vast wooded areas of earlier centuries as described by traveler-geographers seem to have gone out of existence."

Perhaps these migrants conquered North Africa more completely than the Anglo-Saxons conquered America because the Arabs almost fully converted the natives to their culture. By violence or social influence, the mass of Arabian nomads set their stamp on North Africa. When the Hilali Arabs became the cavalry and tax collection officers for rulers of the eastern Maghreb, they imposed their clan organization on the Berber taxpayers (Laroui 1977: 256). The Arabic language largely replaced the ancient Berber languages and scripts. Even for Berber speakers, the distinction between being a Muslim and a person of Arabic culture slowly faded away. Perhaps most Berbers came to claim Arabian ancestry. Their ancestors may have been pagans or "heretical" rebels against Arabian invaders (Geller 1972: 13–14). But most Tunisians came to view their ancestral saints much as Irish Catholics viewed St. Patrick and his priests—as bringers of the one orthodox faith from a distant Holy Land. Such was North Africa's loyalty to the civilization from Arabia. But within that great tent of cultural unity, the dichotomy continued between the religions of rulers and local communities. Of course, that same division appeared everywhere in the world, with the cultural canyon varying only in width and depth from land to land. But of all regions of the Earth, that division was possibly greatest in the Afro-Asian arid belt.

Political Evolution of Islamic North Africa

North Africa's journey between Islamic sects, from the Kharjite "heresy" to Fatimid Shi'ism and Sunni orthodoxy, was a quest of trial and error in search of solutions to the problems of scarcity, justice and power. The Kharjite, Sunni and Shi'ite sects served as both religious denominations and political parties. As Henry Munson (1993: 36) represents them, these sects stood for a spectrum of political solutions ranging from quasi-democratic to absolutist. And this quest for answers has continued down to the present time.

The semi-democratic Kharjites stood for rule by consensus of the learned. They were considered revolutionary because they rejected the very idea of rule by warlords, kings or sultans. For the Kharjites, such rule by a strongman was a violation of Islam. In fact, the Kharjites were often labeled as terrorists because they preached rebellion against any ruler who violated the holy law. But democratic principles did not fare well in North Africa, because the real-world competition of warlords and sultans tended to squash them underfoot.

The Shi'ites championed a seemingly opposite political ideal. Their hope was for a messiah-like king from the line of Muhammad's own family who would rule in perfect justice. Naturally, when such a sinless imam came, it was the duty of believers to fight for him. Next, he must be given absolute power over lesser men and women. But confidence in a perfect leader was easily dashed. Once in power, the Fatimid imams showed themselves quite capable of error and injustice. In the backlash of disillusionment, Shi'ism too was largely discredited in North Africa. The hope for a messianic king suffered, and the population turned to Sunni orthodoxy. But even within orthodox society, the basically Shi'ite idea of a kingly imam, or *mahdi*, survived. In recent centuries the hope for one who would right the wrongs of the world has risen again and again, especially when the rains failed and the peasants went hungry, yet the rulers went on feasting and building armies.

Sunni orthodoxy represented a compromise somewhere between the Kharjite and Shi'ite ideals. Like the Kharjites, the Sunnis believed that their rulers should be subject to religious law and should seek guidance from the learned clerics. Long experience had taught them that it was impossible to control how a warlord came to rule. But once a man gained power, he must be cautioned against violating sacred laws. Theoretically, this compromise created a sacred contract for justice in Sunni society. The right of warlord-sultans to rule was recognized, but their decisions should be guided by consultation, or *shura*, with the learned clerics. The sultans would be accountable, not directly to the ignorant public, but to the alumni, or *ulama*, who the people respected as learned and virtuous.

Probably a majority of Sunnis shared the Shi'ite belief that holiness was inherited, and they felt that the caliphs and sultans should be qualified by descent from the Prophet's tribe, the Quraysh. In this, they differed from the Shi'ites only in that the Shi'ites were more specific, holding that a true Islamic ruler should descend from Muhammad's particular family. Most Sunni rulers in the "Arab world" claimed to be *sharafa*, or men representing bloodlines from the Prophet's tribe. But in the hands of warlords, this real or alleged descent from the Prophet became a religious justification for autocratic power. Many Sunni sultans seemed to view themselves as almost incarnate, Shi'ite-style imams whose will was an expression of God's. For example, the great king Ibn Tumart (d. 1130) regarded himself as fiercely orthodox, yet called himself an imam and a *mahdi*, as if he were a Shi'ite messiah (Munson 1993: 22, 146, 49).

So, in Sunni tradition, the divine right of kings co-existed with a religious concern for social justice and a popular trust in guidance by the *ulama*. The tension between these different values might be compared to the balance among executive, judicial and legislative powers in many governments today. But in North Africa, the traditional division of powers was more an expression of hope than an enforceable system of checks and balances.

Sultans and Saints

In real life, the degree to which a ruler heeded the *ulama* was a matter of personal inclination. There were no checks on a ruler's power save armed revolt, or personal appeal to the king. The *ulama* could protest a sultan's decisions, but this took rare courage. Certain heroic members of the *ulama* sometimes dared to go before a sultan and accuse him of immorality. Such petitioners were generally killed. Afterwards they might be revered as martyrs. If the dead protester was widely hailed as a saint, his blood could weigh heavily on the sultan's hands. The ruler might then make a show of repentance. The system would then have "worked." Unfortunately, a holy man of almost suicidal heroism did not appear in every generation.

The leaders of the Almoravid and Almohad movements (in the 1000s and 1100s) combined religious denunciation of corrupt rulers with peasant revolt (Davidson 1991: 133, 186). But after that, Sunni clerics usually shunned violence. They even tended toward the otherworldly view that power must corrupt all it touches. So the saint Al-Hasan al-Yusi (1600s) emphasized a saying attributed to the Prophet: "The best of kings are those who visit the ulama and the worst of ulama are those who visit kings" (Munson 1993: 44–45). In that case, the sanctity of clerics depended on their remaining apart from the wielders of power. Perhaps their survival depended on it as well.

Al-Hasan al-Yusi shunned any official position for himself, lest he lend moral sanction to a despot. He did not, however, shun the challenge of accusing the sultan of immorality to his face. Al-Yusi confronted Sultan Isma'il of Morocco over his inhuman treatment of workers on the city's fortress walls. At least one injured worker was buried alive rather than slow the work. Al-Yusi, who was staying as a visiting scholar at the court, learned of this inhumanity and began smashing the dishes brought to his room, one per meal. Finally the palace ran low on dishes. When called before the court to explain himself, al-Yusi asked Sultan Isma'il: "Which is better—the pottery of Allah or the pottery of clay?" In this case, a religious leader won a promise of repentance from a sultan.

Decades later, in the early 1700s, Sultan Isma'il began building an army of Black slaves. This was a special army, accountable to the sultan alone, designed to allow him to act independently of any tribal group's military support. To build this army, agents for the sultan reportedly enslaved free dark-skinned Muslims. According to Laroui (1977: 272–73), "This method of recruiting ... was a tacit avowal that the ever-widening gulf between the state and society made it impossible to enlist a loyal army among the free population." When rumors of this abuse spread, Isma'il moved to undercut his critics. He called some of the *ulama* together and coerced them into issuing a proclamation (or *fatwa*) saying that nothing wrong was happening.

This was the kind of relationship between rulers and clerics which al-Yusi had warned against. Since it was popularly believed that rulers should act only in ways approved by the *ulama*, Sultan Isma'il simply forced the clerics to endorse his deeds. The very fact that the sultan had a *fatwa* proclaimed in defense of his policies underlined their questionable morality (Munson 1993: 2–8). In this way, Isma'il tried to use the clerics as pawns for overcoming the people's moral sensibilities. But to the sultan's dismay, one scholar, Al-Hajj 'Abd as-Slam Guessous, denounced the *fatwa* in public during Friday worship, exposing it as a coerced endorsement for a violation of Islamic law. The sultan had his soldiers strangle Guessous and dispose of the body in the night (Munson 1993: 52). Critics of the slave army were silenced and "recruitment" went on as before. As many slaves were dark-skinned Haratin farmers, the new army grew directly at the expense of Moroccan agriculture. But the vanished 'Abd as-Slam Guessous reappeared among the ranks of Morocco's saints, who were far better remembered than the sultans.

In treating the *ulama* as deputies of the ruling house, some sultans seemed to take themselves as the religious heads of a state church. Some of them even tried to replace respect for learning with respect for power. Traditional culture held a deep regard for education and for learned people. But for the autocrats, the learned community was a source of political competition to be controlled and discredited. In the 1500s, for example, as the grassroots movement of Sufi orders spread through rural Morocco, Sultan Muhammad al-Shaykh saw their growing network of local leaders as a threat to his authority. In 1547, he ordered leaders of the Sufi *zawiyas* (local circles) to report to him in Marrakech and had them massacred. Similarly, in 1965, as major protests by students and teachers swept Morocco, King Hassan II simply ordered his army into the streets, where it fired into the crowds, killing possibly several hundred people. The king then addressed the nation's students and teachers, saying, "There is no danger to the state as grave as the so-called intellectual. It would be better if you were all illiterate" (Munson 1993: 134). In a moral order upheld by the learned *ulama*, Hassan made it seem as if distrust of learning was an article of Islamic faith.

Autocrats and Tribal Elders

Centuries ago, Morocco's tribal leaders often had freedom to choose which overlords they obeyed. Many Berber tribes elected clan councils by a vote of all adult men. If the councils appointed a tribal chief, he typically served a one-year term. Down to the twentieth century, such semi-democratic local councils acted almost independently of any greater power. Their decisions were binding expressions of communal will. Among the Beni Mtir of Morocco, if a man openly disobeyed his clan councilmen, they might burn his tent or cut off his wife's hair (Vinograder 1972: 77). Modern urbanites might describe this as primitive tyranny, but it was a self-imposed tyranny by which small communities governed themselves.

The old tribal councils cultivated strict self-restraint among their people. Communal order depended on the kind of self-discipline that would stop a man who was burning with thirst from taking more than his share of water. This need for communal control reflected the urgency imposed by scarcity in the surrounding landscape. Old puritanical restrictions on sex provided a communal control over the individual's right to reproduce. That right was given only to those deemed capable of supporting children and raising them to be loyal, contributing members of the community.

The old communal tyranny was like an austere rationing system for hard times. But the hard times seemed to never end. Instead, the environment gradually grew worse. The strict rationing of goods and freedoms became a seemingly permanent necessity. Constraint and frugality turned into ends in themselves, rather than means to a different future. But however hopeless this may seem, the old morality was rooted in a great hope. It was a hope that if people behaved in correct ways, they might sustain themselves on their land for all generations to come. If the traditions of generosity were maintained, people would always be willing to share what they had. Even in times of hardship there would somehow be enough for all. It was a noble hope. For centuries it has been fading away.

In recent times, the clan councils still set some rules, but mainly for personal behaviour between clan members. "Higher," more centralized forms of authority gradually superseded the old communal ways. Where inter-tribal wars flared up in times of drought, the most successful war leaders tended to keep power permanently (Hart 1972: 30–31). Where small tribal groups feuded over the means of life, the greatest warlords forced a resolution in their own favor. The terms of peace might be grossly unequal, but they were often better than conflict. Many of the *ulama* reasoned that an unjust sultan was a lesser evil than chaos (Munson 1993: 39). Munson points out that though many popular Moroccan proverbs condemn unjust rulers, "These proverbs do not contest the idea that power should be concentrated in the hands of one man, they simply stress that this man should be just as well as strong" (1993: 144).

The old sultans commonly referred to their subjects with the word "'*abd*," a word combining the meanings of "subject" and "slave," as if there was no difference between the two. The term also implied that there were only two kinds of people—rulers and subjects—with no mention of other categories such as learned clerics or clan elders. The term '*abd* suggested that subject people were born to be the sultan's slaves, just as they were born to be slaves of Allah. And in most rulers' minds, this was not a degrading destiny for their people. On the contrary, only through total obedience to ultimate authority could subject people achieve the highest moral perfection imaginable for them. Through obedience they might become rightly guided in all their actions, faithfully following the path set for them. But such virtue did not come naturally to Allah's creatures. It had to be taught, often by forceful means. The common sense of it appears in another old Moroccan proverb—"If the slave does not taste the stick from Sunday to Sunday, he will say there's no one like him" (Munson 1993: 87).

For centuries the growing power of sultans and central governments was the wave of modernity. Only seemingly backward elements opposed it. Only a minority of nomads beyond the reach of the sultan's army ignored the central authorities, and only a few critics from the generally sheeplike *ulama* occasionally challenged the ruler's will. The local leaders grew dependent for their own positions on the overlords above them. Finally, the European colonists and later governments further centralized control of resources, forcing local leaders to become petitioners for dispensation from the center. So, both royalist Morocco and socialist Algeria grew as centralized as communist states. As John Waterbury (1972: 399) put it, King Hassan II "could turn the tap on and off at will in a game where ideas and programs counted for little and the competition for material advantage became everything." In such a system, political and economic authority was one.

Despite historical protests from North Africa's saints and rebellions of tribal leaders, the Maghreb remained a zone of autocracy down to the 1990s, a region of economic, cultural and political monopolies. The Moroccan constitution still vested supreme power in the late King Hassan II and, in Libya, Colonel Ghadafi occupied a corresponding position. The underlying similarity between the right-wing Hassan and the leftist Ghadafi was so great that at one point the two proposed merging their governments. In Algeria, the military-backed FLN party offered elections in 1992 but repudiated the results. Only in Tunisia was economic and social diversity becoming a centrifugal force for a less centralized system.

The stability of North African autocracy did not stem from any lack of competition for power. The autocratic establishments faced constant opposition from a wide range of social movements, political parties and military factions. But almost all the opposition groups were also authoritarian. The monarchists, socialists and Islamic revivalists, whether in or out of power, all agreed on the need for authoritarian rule (Munson 1993: 38, 51) and stood on guard against the worst dangers from the past. And in a semi-denuded land, the worst dangers were scarcity, chaos and domination by others.

Villagers of North Africa could not personally affect the rainfall, the locusts or the rise and fall of empires. It was easy for individuals to feel powerless in a fragile environment. However, they could choose to control themselves. They could respond to scarcity with solidarity or division, and they could choose between generosity or selfishness. But if solidarity was to prevail over chaos, it was generally presumed that people must submit their wills to a higher authority. What authority should that be? Out of North African history, two answers seemed viable. The autocratic elitists proposed avoiding chaos through obedience to one man's will, and the communal traditionalists advocated renewed conformity to codes of conduct inherited from the tribes of the past. Both kinds of authority claimed the same source in God, and both made obedience their cardinal virtue. Augustine would be gratified at his people's morality.

The Rise of Modern Fundamentalism

Sometime during the period of European colonial rule, the power of the autocrats started to wane. The conflict of interests between rulers and ordinary people led the sultans to first hire foreign mercenaries, then to seek credit from foreign sources and finally to form colonial alliances with the Europeans. It took little discernment for most North Africans to reject the foreigner's authority. But to challenge the native rulers, with their claims to descent from the Prophet, took both discernment and courage. Most common people had trusted the learned *ulama* to tell them right from wrong, and the rulers also claimed to be morally guided by these men. But the fraudulence of this claim became obvious when Yusif, the puppet-sultan of Morocco under the French, forced the *ulama* of his day to issue a *fatwa* in support of his administration. The *fatwa* called for obedience to the sultan, even though he was a tool in the hands of foreign interests. It quoted a saying attributed to the Prophet: "The sultan is the shadow of God on earth and it is with him that all the oppressed find refuge" (Munson 1993: 100–101).

Such abuse of religious authority now came under popular attack. Rightly or foolishly, many common people started taking the judgment of rulers into their own hands. As laypeople of rudimentary education, they tried to decide when the rulers were forcing the *ulama* to misrepresent the holy law. This was a challenge as daunting as any faced by the Protestant laypeople of Europe in the 1500s when they dared challenge the authorities' interpretations of Christian doctrine. In North Africa, however, the popular revolt turned against the state, and not against the church. The clerics were almost never blamed for the tyranny of rulers. Rather, if members of the *ulama* were intimidated, their intimidators were blamed. And as most commoners judged all matters by the standard of communal tradition, the main opposition to modern autocrats took shape as a fundamentalist Islamic revival. In such protest, the clerics could become real leaders again.

Starting in the 1800s, certain clerics in Morocco led opposition to increased taxes, which had been imposed to meet foreign "debts," claiming these taxes were "innovations" in violation of Islamic custom. Later, from the 1930s to 1950s, the clerics were again leaders in the war against colonialism. Even the sultan, Mohammad V, faced jail for his protests against the French. For a shining moment the sultan and the *ulama* stood side by side, risking themselves for their people. But then, in 1962, the greatest religious leader in that war, 'Arbi al-Alawi, broke with the Moroccan government in protest over the new constitution. This document contained two items which al-Alawi condemned. First, it vested supreme power in the king, who was to be chosen by birthright, and second, it provided for an elected legislature. Concerning his first objection, al-Alawi upheld the monarchy itself, but he objected that a sultan must be chosen for his religious virtue, by the *ulama*. The old sultan, Mohammad V, had opposed the French, gone to prison and earned the reverence of his people. But now, his son, Hussan II, was to simply inherit that sacred trust. Al-Alawi believed this was wrong—because the constitution left the *ulama* no say in the matter. Concerning his second objection, why did a champion of national liberation oppose an elected legislature? Al-Alawi opposed it for the same reason he opposed hereditary monarchy—because it gave the *ulama*'s old role of overseeing communal law to another group of people who were professional politicians rather than religious scholars. These two clauses combined to cut the *ulama* out of both executive and legislative power, leaving the clerics with only limited judicial power, mainly over personal and family matters. The same separation of the *ulama* from political power arose in modern Algeria and Egypt, and the same kinds of protests arose there as well. In Algeria, the Islamic Salvation Front called for constitutional changes to empower the *ulama*. In Morocco in the 1970s, a new leader of the Islamic opposition, 'Abd As-Slam Yasin, proposed a council of the *ulama* to determine the laws and the banning of all (other) political parties (Munson 1993: 110–111, 149–67).

For mainstream traditionalists, "Islamic government" meant a return to the ideal of *shura*, by which the *ulama* should have guided rulers since Islam began. But some extremists demanded more serious changes. In Algeria, and increasingly in Egypt, some so-called fundamentalists advocated violent revolution. They demanded, in effect, an end to the old Sunni-style compromise between religious guidance and military power. These fundamentalists would have a virtually Kharjite-style council of holy men rule the country.

The heads of government in Morocco, Algeria and Egypt reacted against the theocratic challenge in similar ways. They were willing to give ground to the fundamentalists in matters of family and personal law. Repealing the modern freedoms of ordinary women, for example, was an acceptable sop to the traditionalists. But like the shah of Iran, these rulers rejected any Islamic limitations on their own powers.

The Youth Wave

The people behind the new fundamentalist movements were mainly idealistic, underemployed youths. The grandfatherly elders of the *ulama* might serve as symbolic leaders but, as Fatima Mernissi (1987: xi) points out, most activists were young men from the towns and cities. The fundamentalists also drew disproportionate support from educated people and students.

The parents of Algerian fundamentalists had often come from rural villages. They had moved to the cities as colonial plantations took their land (by 1919 half of Algeria's native farmers were rendered landless), as colonial war destroyed their villages or as the urban economy expanded (Laroui 1977: 207). Their children had grown up in the postwar (post-1962) decades, in the days of climbing oil revenues and easy foreign credit, when job opportunities were then expanding in the public sector and in Europe. By the late 1970s, as the great oil age reached high noon, it seemed that modernization and money would steadily replace all the scarcity and oppression of the past. No longer would the new generation inherit a land of slowly dying pastures and farms. Instead they would live off new resources in ways their grandparents scarcely dreamed of.

The North African baby-boom generation that came after the colonial wars—twenty years or so after the West's post-World War II boom—was proportionately larger than the West's. But just as this post-colonial generation came of age, ready to ride the escalator of development, the oil age peaked and fell. By the sheer glut that came from pumping the diminishing resource, the price of Middle Eastern oil dropped. Between 1985 and 1987, Algeria's oil revenues declined from US $13 billion to $8 billion (Stone 1997: 96). At this, international credit dried up, leaving a bill of galloping interest on previous loans. The percentage of Algeria's foreign earnings needed to cover interest on foreign loans rose to 51 percent in 1986, 54 percent in 1987 and 87 percent in 1988 (Reudy 1992: 246). At the same time, Europe started slamming its doors on workers from Africa. And the bureaucracies of North African governments, which had previously soaked up masses of secondary-school graduates, began major cutbacks and layoffs.

By the mid-1980s, two-thirds of Morocco's young adults between 15 and 20 were both unemployed and out of school (Ponting 1991: 344). In Algeria between 1985 and 1991, 180,000 educated young people entered the job market each year; of these, approximately 100,000 found work. The rest commonly joined the ranks of the so-called *hittistes*, or "wall-leaners" (Stone 1997: 97). In previous generations, most people had lived in self-employed poverty on the land. But this generation faced not just underpayment or underemployment, but *un*-employment, urban style. If these young people could not find jobs, they did other things. The black market rose to fit their needs. Algeria's government heads felt they must control the growth of unlicensed or illegal businesses. They did so partly by trying to banish the urban unemployed. In the summer of 1983, government bulldozers swept through the shantytowns of Algiers, knocking down at least 200,000 dwellings. Police ordered the residents to go back to where they or their parents had come from.

In response to the debt crisis, the International Monetary Fund (IMF) demanded cuts to domestic spending, in favor of continued interest payments to foreign banks. The IMF made further credit to North African nations dependent on major cuts, such as cuts to education or to subsidies on the price of bread. When these policies took effect, the cities of North Africa exploded in riots more violent than any seen in North America during the twentieth century. In "Black October" of 1988, the governments of North Africa, which had previously stumbled onto unlimited progress, showed a more traditional face. Those who protested the ruler's control over scarce resources were gunned down in the streets. A new age of scarcity seemed to be dawning, and with it a return to the communal ethics of the past.

As an Algerian fundamentalist youth explained to Munson:

> In this country [Morocco] if you are a young man … you have only four choices: you can remain unemployed and celibate because there are no jobs and no apartments to live in; you can work in the black market and risk being arrested; you can try to emigrate to France to sweep the streets of Paris or Marseilles; or you can join the FIS (Islamic Salvation Front) and vote for Islam. (1993: 178)

The new generation had become explosive. It was now a mass of educated but unemployed people longing for an apocalypse of justice. Those standing outside the modern Algerian system often felt no less wrathful than those who had stood outside the French colonial system. And as psychoanalyst Franz Fanon had written during the French-Algerian war, "the [dominated] native is an oppressed person, whose permanent dream is to become the persecutor" (in Crossan 1991: 317).

For allies in their quest, idealistic young people tended to reach over a generation—past their progress-minded fathers and mothers—to join hands with their grandparents. The established bureaucrats, engineers and teachers of the oil-age economy, whose ranks were increasingly closed to their children's generation, began finding themselves surrounded. Both the young and the old seemed to reject the system they served. Many members of the middle class had opposed old-fashioned autocracy. Now many of them feared that only continued autocracy could hold back the fundamentalist tide (Munson 1993: 178).

In 1992 the Islamic Salvation Front threatened to sweep Algeria's elections, if elections were permitted. Instead, a seven-member High Executive Council took power, seemingly headed by the defense minister. According to reporter Francis Ghiles, the army blocked elections because it "had no desire to see its considerable influence questioned by Islamic fundamentalists" (in Butt 1997: 164). By the mid-1990s, the Islamic revolution was so pervasive that people spoke of the "government of the day" and the "government of the night."

The Western democracies offered financial and military support to "conservative" autocrats such as Hassan II. They did this in the interests of "stability," which was the same reason they had supported the shah of Iran—as if stability was the best they could hope for in North Africa and the Middle East. When Algeria suspended its 1992 elections, the West stayed quiet. It was obvious that democracy would lead straight to a fundamentalist landslide. But why were autocracy or theocracy the only serious options? Why did North

Africa have to swing throughout its history between autocracy for order and theocracy for justice? Why did the clash of these values always seem to produce some form of semi-theocratic dictatorship? Munson (1993: 170) says the autocrats and theocrats both saw North Africa's people "as a flock of sheep in need of a shepherd"; they disagreed "only as to the identity of the shepherd."

The Youth Wave's Female Half

During the decades of rapid economic growth, resources became available for something little seen in the past—education for women. This was generally counted a modern luxury, affordable after more important needs were met first. But in the great oil age, education for females was a luxury which ever-growing numbers of North Africans could afford. By 1977–78, Algeria had a million and a half girls and women in school. The number of women in salaried jobs was only 180,000, but this was double the number for 1966 (Reudy 1992: 229–30). After finishing school, most women returned to unsalaried work for their farms and families. Perhaps most people expected the benefits of women's education to be more social than economic.

In the 1980s, however, a new age of harder-nosed accounting arrived. Governments and lenders began measuring the benefits of women's education more strictly for monetary return. Of course the advantages of educating mothers were hard for accountants to measure, and women's production outside the monetary system did not count. The new cost-benefit analysis listed state-supported education in the expense column. If paying interest on bank loans came first, other things had to be cut, and the cuts had best fall on the least powerful. If the scaled-back centers of higher education had fewer openings, tradition indicated that girls should step aside and let their brothers be trained. In a time of scarce employment, women seemed to compete for jobs with men. In the zero-sum logic of scarcity, the rise of professional women seemed to offer no net gain. And if no obvious racial or cultural minority could be driven from the overcrowded job market, perhaps the minority of female workers would do.

The economic downturn compromised educated women perhaps more than any other social group. These people had prepared themselves for roles far beyond those of traditional arid-country women, but then the conditions and politics of scarcity had returned. In 1984 the Algerian bill on women's rights officially redefined women as legal dependents of their husbands or fathers, and made their right to work subject to a male caretaker's permission (Reudy 1992: 243). The educated women were caught in a crossfire between autocrats and theocrats. The autocrats compromised with both the fundamentalists and the International Monetary Fund. In that case, politics, economics and religion all seemed to require cutbacks affecting women.

So the brief era of women's advancement, save in Tunisia, seemed to be drawing to a close, except for one thing—the accomplishments already made by educated women could probably never be undone. A new kind of North African woman had already emerged. Human potential, once experienced, is seldom forgotten. The educated and professional women were perhaps the crowning glory of the oil-age economy, and that investment in human potential could still pay off in ways greater than all the investment in oil. By the year 2000, Tunisia's well-educated women were perhaps the outstanding factor in that country's modest economic miracle.

Possible Diversification of the Desert Economy

During the 1980s, Libya, Algeria and Mauritania each relied for over 90 percent of their national income on a single mineral product—oil in the cases of Libya and Algeria, and iron ore in the case of Mauritania. The Western Sahara, which was under Moroccan control, had a similar degree of reliance on phosphates. Perhaps it is the nudity of the environment which has always seemed to indicate a basically single-product economy, with a monoculture to match. But the modern switch to a mineral-based economy held at least one hopeful sign—it was a change. Instead of the seemingly endless overreliance on grass in the pastoral economy, till the scarcity of grass made the country a desert, people now relied on something else. After the grass and soil were nearly gone, people looked beneath the ground for something new to live on. And the resources they found there have sustained the richest economy the region has ever known.

Marching Down the Food Chain

In a long view of history, the people of North Africa have been slowly marching down the food chain. First they hunted the wild animals into scarcity or extinction. Then they domesticated a few species of animals and developed the dual economies of herding and farming, both of which relied on exploiting plants and topsoil, and slowly depleted the land. Next, people moved further down the chain of life, to the petrochemical remains of dead plants from the past. Beyond that, they traded inorganic minerals for food and supplies from other parts of the world. In a perverse view of history, it might seem that the North Africans (like the rest of us) have been eating their way down the food chain, destroying each link as they go. However, if they exhaust the organic minerals beneath the Earth, there may be nothing organic left to live on. And this will literally happen if people simply continue to move from one mono-economy to the next, consuming each resource until it is gone.

The change to a mineral economy was a shift from renewable resources, such as grass, to non-renewable ones such as oil. By that standard, the oil age was a step backward from sustainable living. But given the long exploitation of plants and topsoil, perhaps it was a matter of ecological life or death that the economy should turn to something else—almost anything else. Perhaps the mineral economy can be a breathing spell for plants, during which the crushing burden of supporting humanity can be partly lifted from the plant kingdom's back. If the desert is an environment with its food chain nearly cut, then reducing pressure on the damaged links might save the system. Besides, if the

Earth's mineral kingdom is much vaster and less fragile than its surface skin of plant life, then perhaps a mineral or chemical economy can have firmer foundations than one based on overharvesting plants.

Where tracts of the pre-Sahara are fenced off from direct grazing, the emerging cover of green bush is soon visible from orbit in space. If the pressure on North Africa's plant life can be reduced, perhaps the Sahara's borders will soon acquire that green tinge. With a carpet of scrawny acacia, mesquite or gum-arabic trees, the whole climate of North Africa might become noticeably more "Mediterranean." The landscape might then produce more fodder sustainably than the present desert produces unsustainably. No landscape is hopeless. Once, all land on Earth was bare, sun-blasted bedrock. Then the first organisms began breaking stone into soil. Now, with so much of the Earth's surface already organic, the process of soil-building can be many times faster.

The great oil age was a time when the Earth's stored-up capital of dead bodies from all previous life was dug up and burned. The income from this enterprise was either squandered, or else invested to build a diversified post-oil economy. In a simplistic big picture, Algeria and Libya seemed to be squandering their inheritance. As supplies began to run low, these countries seemed to be entrenching their dependence on one vanishing resource. But the percentage figures of gross national products may conceal a counter-trend of growing economic diversity. Growth for its own sake might not be a hopeful sign, but diversification probably is. And over the past century rapidly rising numbers of North Africans supported themselves in small industry, teaching, information systems, organized religion, entertainment and research. Tunisia, without the benefit of major oil reserves, started earning about 30 percent of its foreign exchange through tourism and more through diverse small industries (Borowiec 1998: 102). The Tunisian farmers made steady progress in exporting fruit from their sustainably expanding orchards. They re-terraced and reforested hill slopes to hold rain as in pre-Roman times, allowing orchards to grow in areas where the rainfall totaled less than ten centimeters per year. In the big economic picture, such things appeared as small steps, but at least they were steps in diverse directions. And some of these activities were actually good for the land.

Contrary Directions of Farm Development

For a close look at local trends in the countryside, the Oglat Merteba area in central Tunisia served as a focus of field research for the 1977 U.N. Conference on Desertification. The U.N. survey provided a snapshot of economic trends in this 20,000-hectare plot of the pre-Sahara. At that moment, the land-use breakdown was as follows:

- 150 hectares (0.75 percent of the area) were orchards or *jessour gardens* (traditional rain-fed desert gardens made possible by earth embankments, rock catchment tanks and rows of trees, allowing conservation of water).

- 2,000 hectares (10 percent of the total area) were being plowed for grain—in a zone receiving only 100–200 mm of rain per year.
- 3,800 hectares (19 percent of total area) were still usable as pasture for the people's animals.
- 14,050 hectares (the remaining 70.25 percent of Oglat Merteba) was desert. Perhaps surprisingly, the main economic activity in this area was firewood collection.
- The local people also operated a number of household industries, most importantly bee-keeping. The land occupied by these industries was negligible. (based on Walls 1980: 110)

The U.N. observer team calculated the total income from all these activities for the 1,320 residents of Oglat Merteba at 121,000 dinars. In 1977 this was roughly equivalent to US $268,000. The income was broken down by source as follows:

1. animal products from the 3,800 hectares of usable rangeland (37,000 dinars, or 30.6 percent of the total locally generated income),
2. garden products from the 150 hectares of traditional desert garden (30,000 dinars, or 24.8 percent of total income),
3. cereal crops grown on 2,000 hectares of semi-desert (25,000 dinars, or 20.6 percent of total),
4. home crafts and industries, including beekeeping and weaving (16,500 dinars, or 13.6 percent), and
5. firewood collected in the desert (12,600 dinars, or 10.4 percent). (based on Walls 1980: 110)

The growing of jessour gardens showed an ecological streak in North African culture. On the desert's border, where people remembered greener times, plant life could be all the more cherished for its scarcity. Traditional architecture housed lovely gardens within protective walls, and a wealthy family's villa might contain an atrium with a small grove of trees, flowers and a fountain of splashing water. In poetry and religion, such gardens were standard images of paradise (Attenborough 1987: 136). The science of desert gardening had evolved from ancient times, and this tradition was both alive and economically important in the jessour gardens. Along with beekeeping and crafts, these gardens represented the most intensive, sustainable and profitable features of the local economy. Nearly 40 percent of the area's income came from these activities, which required only about 1 percent of the land.

But just as Oglat Merteba was balanced on the edge of the desert, so its economy was split. Nearly half the income came through environment-enhancing work, but over half came from activities which further degraded the land. The herding of animals could be sustainable if the pastures were not overgrazed, but this was just what had already happened to the desiccated 70 percent of the area. And commercial collection of firewood from the desert was bound to hurt the area even more.

As a result of topsoil and plant-cover loss in Oglat Merteba, the water table was falling at about five centimeters a year (Cloudsley-Thompson 1984: 14). With that kind of water loss, even the date palms were dying in some areas. Trees on high ground and plants with wide, shallow roots were the most at risk. Good pasture perennials were also dying over wide areas, as the underground water faded away. James Walls (1980: 136) explains, "By 1974, an area [of wild pasture] which used to produce 1,500–2,000 kilograms of dried matter per hectare was producing 350 kilograms, with 80 percent of the surface naked of vegetation."

The two thousand hectares of cereal crops resulted partly from mechanized farming in the semi-desert. In the past, digging up the open range with a hoe was scarcely productive work. Traditional pastoralists had always planted grain in the pre-Sahara, but they simply broadcast seed on the ground, left with their herds for winter pastures and returned months later to harvest whatever had grown. Now, tractors were plowing up the best remaining pasture land. This trend, which was growing over much of North Africa, was almost certain to bring dust-bowl conditions, as in the Kazakhstan steppe of the U.S.S.R. in the 1950s (Walls 1980: 57). Andres Rapp claims that in many plowed areas of North Africa, "after two to three years of [mechanized] barley harvesting followed by grazing of the stubble, the entire topsoil is often blown off … down to … about 20 cm depth" (in Grainger 1990: 72).

This sort of "throwaway" farming, in which topsoil and reserves of underground water were treated as one-use disposables, was the most rapid method of desert creation yet invented. Yet just where desert expansion was the greatest danger—on the very edge of the Sahara—this kind of farming was expanding as if it were the wave of the future.

Scenarios for the Gap between the Desert and Sea

The environment of North Africa seems little changed from year to year. For most economists it appears unlikely that the Sahara, which has been crawling northward for five thousand years, could cross the Atlas range and close the remaining gap to the Mediterranean shore. It seems a remote possibility that the great Nile River could simply dry up, leaving about 75 million people stranded in a wasteland. But flood levels in the Middle Kingdom of ancient Egypt were sometimes nearly ten meters higher that the twentieth century average before the Aswan dam was built (Ponting 1991: 86). Since those times, most of the river's watershed has been slowly stripped of trees, and the river's flow is now a concern for the future.

Projections of disaster, of course, never reckon with the human spirit. The North Africans are survivors and will do whatever they must to survive. Perhaps they will grow omnivorous—able to rely on an ever broader spectrum of resources. If the world is like a tree, they, and we, can learn to eat the fruit of every part—without ever cutting a ring around the trunk. The oil industry of Algeria might then be replaced, not by utter collapse, but by reliance on far more plentiful resources. In 1994 a new design for solar cells allowed panels so thin that they could serve as roofing material. If exploiting that potential ever became a bigger priority than building armies, the use of firewood for heating and fuel might be reduced to levels where natural vegetation could again grow faster than it is cut. That alone would put North Africa on the road to a damper climate. Such are the hopes for averting ecological deadlines.

Chapter 11

Ancient Europe's Encounters with Desert Civilizations

IN HIS 1958 NOVEL, *DANCE WITH THE DEVIL*, GUENTHER SCHWAB GIVES playful yet vehement expression to German sentiments in favour of nature protection. At one point in the book, Schwab has Satan take a group of potential recruits on a time machine through Spanish history. Moving fast forward from ancient times, the passengers watch as forests melt into sheep pastures. One passenger remarks, "It is as if a dangerous bacillus were eating the skin of a living body." The Devil laughs, gloating over his progress. In describing the destruction of nature as a Satanic plot, Schwab counts on his readers to respond with anger rather than resignation. He presumes that the Devil's plan can be thwarted, even in Spain. God, in this story, is a sympathist for the forests, meadows and untamed rivers. Perhaps this is an unusual role for the God of the Bible, but Schwab assigns it to him anyway.

Like Schwab, many Europeans feel it obvious that their lands should be green and beautiful. Where this expectation is dashed, the locals usually cling to it anyway. Even residents of Berlin or London seldom accept the degradation of nature as "the way it is." They compare the reality before their eyes to images of nature in their minds. They may feel incapable of changing society but still pass judgment upon it. The judgments arise from comparing reality to dreams, and the dreams come from memories of the past.

Desert Influence from Two Directions

The deserts of Africa and Asia have never intruded directly on Europe but have approached it from two directions. From the south, the desert climate has reached over the water to Spain, Italy and Greece. From the east, an arid steppe has spread into the Ukraine. For over six thousand years, from about 4500 BCE to 1500 CE, the prevailing winds of migration blew from east to west—from the distant deserts or over the steppes to the lands of emerald green. The influx of people and cultures from Inner Asia and the Middle East came in trickles or floods and transformed Western civilization several times over.

The encounters of Europe's aboriginal people and its arid-country invaders unfolded on several fronts, with several main themes. The story can be presented like a play in three acts. The program might read as follows:

Act One: The Storm Plain
Europe's early farmers expand deep into the Ukraine. At first there is plenty of grass and land for all. But as the desert spreads through Turkmenistan, growing numbers of pastoral migrants follow the rain westwards. Under growing pressure, the nomads begin raiding the villagers, stealing their food and driving them out. The pastoralists then invade the woodlands of Europe, establishing themselves as a military nobility over the villagers there. This struggle between farmers and pastoralists replays itself for over six thousand years on the eastern storm plain.

Act Two: The Holy Empire
In an empire ruled by its strongest generals, the Roman armies increasingly rely on recruits from the eastern frontiers. By the third century CE, many of these men are believers in Near Eastern religions, including Mithraism and Christianity. By the early 300s CE, the Roman army's leaders are mainly Christians. Soon, Christian emperors set out to impose their brands of orthodoxy on the empire. That orthodoxy involves a wholesale adoption of Middle Eastern customs concerning women, marriage, divine kingship, nature and the structure of the cosmos. Over the next 1,300 years, Europe's old "pagan" cultures are nearly erased from living memory.

Act Three: The Ongoing Renaissance
Taking their long-established monopoly on learning for granted, church authorities are caught napping as translators of classical texts arouse public interest in "ancient wisdom" from the pre-Christian past. Inspired by the old philosophers, a new breed of experimental thinkers sets out to learn directly from nature, rather than from scriptures or other forms of sanctified authority. An ongoing age of natural science is born, which evolves as both a crusade against nature and a growing appreciation for how nature works. The religion of medieval Europe is not so much rejected as indigenized, till nature is commonly seen as an expression of God's love. Some old pagan religions reappear, like wildflowers no longer seen as weeds.

The following four chapters present this version of European history in some detail. But as the story's meaning depends partly on contrast with what went before, the tale must start with the culture and environment of Old Europe.

The Backdrop of Old Europe

Before about 8000 BCE, most of Europe was a treeless tundra, as cold as Canada's Northwest Territories. Then pioneering oak, birch or pine trees crept over the Balkans and up the Ukraine's river valleys. As the glaciers melted, summer floods surged south down the Rhone, Po, Danube and Dnieper rivers. The Baltic ice sheet cracked apart, and the warming ocean intruded all the way to Latvia (Dolukhanov 1979: 72, 116–17). So the zone of green crawled northward over the Carpathian Mountains. But as the tree belt advanced, it yielded territory to the south. The Ukraine's forests moved like a slow green wave toward the Arctic Circle, leaving a grassy steppe behind.

The Gardeners of 8000 to 4500 BCE

Neolithic farming in Europe could not start till vegetation prepared the soil. But almost as soon as environmentally possible, by about 8000 BCE, agriculture appeared on Europe's southern shores. Within a thousand years of the earliest known farm villages in Turkey and Iraq (c. 9000 BCE), such settlements spread to Greece (Dolukhanov 1979: 75). The first European farmers used crops and animals transplanted from the Middle East, such as emmer and einkorn wheat, two-row barley, lentils and mouflon goats (Gimbutas 1991: 5–7). The farms were probably just large gardens scattered in the woods or fields. Native European grasses, fruits, nuts and berries would be gathered in the semi-wild areas, and imported Middle Eastern plants grown in the gardens. This early Sesklo culture could be surprisingly prosperous. Marija Gimbutas' archeological team in Anza, Macedonia, found houses and barns from the 5000s BCE which were often larger and of better construction than village buildings of the 1960s CE. Gimbutas (1991: 64) suggests that the neolithic primitives enjoyed higher standards of living than many Balkan villagers in the war-torn twentieth century.

Gardening in Spain and France also started on the Mediterranean shores, probably in the 7000s BCE. The first farmers there also used cereals from North Africa or the Middle East, and their first domestic goats were Near Eastern mouflons, rather than the native *capra ibex* or *capra pyrenaica* goats. Later, farmers in Catalonia and the French Riviera began domesticating European plants, such as local varieties of lettuce, peas and bitter vetch (Gimbutas 1991: 3). With an expanding menu, these gardeners of the Chassée culture spread up the Rhone Valley.

The main stream of neolithic culture flowed up the Balkans to the Danube before 6000 BCE, up the Danube and Rhine to the North Sea by 5500, then into Britain, Scandinavia and Russia in the 4000s. In the lower Danube valley, hundreds of villages emerged like a new flora in the woods. Some were tiny hamlets, and others could house five thousand people. In northern Hungary, near the Carpathian Mountains, the Bükk culture of the 5000s BCE seems to illustrate the merging of prehistoric cave dwellers with neolithic farmers. The graves show a mixture of light-boned Mediterranean people and heavier-boned descendants of the Cro-Magnon population. The settlements were of one- or two-story houses, but caves were occupied as well. The villagers lived off their gardens, hunting and the mineral resources of caves. Caves were used as sanctuaries for traditional religion, but also as mines for flint, quartz, crystals and blades, and, by about 5500 BCE, copper (Gimbutas 1991: 43–46). Perhaps such a mixture of cultures, races and technologies gave rise to a mythology of forest elves, mountain dwarves and "true men." In the old Darwinian understanding of evolution, various races of primitive humans "replaced" each other, as the superior breeds eliminated their rivals. But remains like those of the Bükk culture suggest another possibility—that Neanderthal, Cro-Magnon and other people could have met and intermarried, breeding each other out of recognizable existence to produce the rather diverse population of Old Europe.

Around 4800 BCE, farming cultures from the Balkans moved into the Ukraine and found their most fertile fields. The Ukraine was a gardener's paradise. The threat of raiders from the treeless steppes had not yet appeared; the arid steppes were only starting to form in the far-off Volga-Donets region. The gardening civilization met no environmental limits except to the north. By 3500 BCE, a 9,000-square-kilometer area of the Uman district held 253 villages of the Cucuteni culture, some of which had thousands of houses. One site near Tallyanky covered five and a half square kilometers and could have housed ten thousand people (Gimbutas 1991: 105, 326). Since the land was vast and rich, there was probably no reason for conflict over any particular plot of ground. The methods of gardening and animal raising were a boon to all who learned them. The villages had no walls and stood in open fields beside the rivers.

Old European Culture and Religion

The Old European villages commonly contained special buildings apparently reserved for crafts. In Romania and the Ukraine, these buildings were often two stories high and held kilns for ceramic art. Most images found on the pots or statues were of women, pregnancy, babies or similar references to the regeneration of animals. The artwork suggested worship for the mother of all mothers, rather than the king of all kings. These centers for seemingly religious art stood among the residential houses and were probably run by groups of neighbours, rather than professional priests.

Many village art centers were also storehouses of animal artifacts. One such house in the Orkney Islands holds a large collection of animal bones, including skulls of all kinds, full skeletons of otters and eagles, and piles of fish or rodent bones. The hedge-ring monuments of ancient England held similar caches of animal bones (Gimbutas 1991: 207). These bone piles seem to recall the vast cemeteries of sacred animals in ancient Egypt. In Old European ceramics, the Great Goddess commonly appeared as a madonna of the beasts, cradling baby bears at her breast. Perhaps the European names for the constellations nearest the pole star—the big and little bear—suggest a cosmic mother of bears, with her children new-born from the womb point in the sky.

Burials of the dead in Old Europe suggest a culture which would have been unthinkable to many later Europeans. In the Balkan Sesklo, Starcevo and Karanovo cultures of the 6000s to 5000s BCE, dead women and children were usually buried beneath the floors of their homes. The men were buried elsewhere, perhaps in graves of men's societies outside the villages. The practice suggests that mothers were seen as central to their homes. It also suggests matrilocal marriage, with daughters inheriting homes from their mothers, and men being married or adopted into the family. After about 5000 BCE, a practice of group burials evolved over most of central and western Europe. Clusters of twenty or more bodies of both genders were buried at various times in common trenches. Hubert Ullrich performed tests on such a cluster of skeletons in central Germany and found them all genetically related. Perhaps this also means that mothers and their lineage of children were buried together, while men married from outside the clan went back to *their* mother's graves.

The megalithic monuments of Western Europe were also communal graves, developed to dramatize a whole mythology of life and death. In rocky areas these graves were built of mighty boulders to last forever. In woodlands they were often made of timber palisades and earth mounds, such as the rings at Quenstedt, Germany, or Windmill Hill, England. These contained ossuaries, or communal bone pits. According to local folklore, the dead were first buried in temporary graves until the flesh was gone. Then, in a seasonal procession marking the start of winter, the bones of the previous year's dead were carried to a common tomb. No markers identified dead individuals. Their bones were laid anonymously into the bowels of the earth.

The Old Europeans associated this return of the dead to Mother Earth with the winter retreat of life into the ground. The tomb in the earth was symbolically a womb, which would bring forth new life in the spring. The monument builders often aligned the grave tunnels so that on midwinter's morning a shaft of light shot down the passage to where the dead lay, signaling the start of regeneration (Gimbutas 1991: 305). These people apparently believed that their children sprang from the source to which the dead returned. It is unclear whether they also saw each individual child as the reincarnation of a particular ancestor; if they did, they still chose not to distinguish one ancestor's bones from another's. Perhaps the continuity of souls through the cycles of life and death seemed no different to them than the continuation of plant souls beyond the compost heap.

Gimbutas suggests that the megalithic monuments had further uses. In Malta, some stone monuments seem designed in the shape of an enormous pregnant woman. Inside one of these stone tombs lay a statue of a tremendously corpulent woman, sleeping on a pillow. She is called "The Sleeping Lady of Hypogeum." Gimbutas feels she is "almost visibly dreaming." But why is she sleeping inside the tomb? Gimbutas suggests "this represents a rite of initiation or incubation. To sleep within the goddess's womb was to die and come to life anew." The resemblance is very strong to what we know of the Greek mystery religions, and the dream-healing of Asklepios (Gimbutas 1991: 289).

The Advancing Steppe and Its People

While the glaciers of Russia were melting, runoff water filled the Black Sea, making it four or five meters higher than the present level. Later the glaciers disappeared, and the Black, Azov and Caspian seas started to shrink. By 500 BCE, the Black Sea was down to eight meters below present levels. The groundwater also subsided, till young trees in the southern Ukraine could rarely reach it. An archaeological dig at Kamennaya Mogila, near the Azov Sea, has recorded this changing environment like a series of time-lapse photographs. Around 5000 BCE the local settlement slowly turns from hunting forest deer to keeping cattle, goats and a few horses. The soil of lower levels is thick with pollen from birch, alder, lime and elm trees; but in the upper levels only grass pollen remains. The plant matter generated from each square kilometer of land fell approximately tenfold, from the 560,000 kilograms of growth per year in a typical broad-leafed forest, to about 50,000 kilograms of grass per year on the steppe (Dolukhanov 1979: 82, 88, 6). This crisis of aridity seems to have forced the locals into a life of nomadic pastoralism.

When the steppe reached the southern Ural Mountains around 1000 BCE, the local Finno-Ugric-speaking people divided into two main groups. Some moved north, following the retreating forests toward the Arctic. Others adapted to the steppe and turned to herding. These emerging pastoralists became the early Magyars, or Hungarians. As the climate grew dryer, they made ever longer journeys for pasture, till they were year-round nomads (Makkai 1990: 8). For hundreds of years they drifted south down the Volga basin. But there they encountered streams of aggressive migrants from dryer lands to the east. As the pastures were generally greener toward Europe, the Magyars were alternately lured or pushed westward toward the rain. Finally, around 750 CE, nearly two thousand years after leaving their forest homeland, they arrived on the Hungarian plain—partly as refugees, and partly as warlords.

The Horse Economy

In neolithic times, wild horses used to roam both the forests and steppes of Eurasia. People hunted them like deer. On the plains they ran in herds. Some admiring hunters watched them from afar and dreamed of getting close enough, not to kill them, but to jump on and ride.

Horse bones in ancient campsites suggest domestication when their numbers grow unusually large. And this seems to happen first on the steppes of Kazakhstan and the Ukraine. At Petropavlovsk, in northern Kazakhstan, bones from over 100,000 horses surround a settlement from about 4500 BCE. Perhaps the horses were first kept like cows, as edible, milk-giving animals. But horses were far harder to control than cows. Eventually, horses were found to be most useful, not as *objects* of herding, but as a *means* of herding other animals. A man on horseback could manage many times more sheep, goats and cows than a man on foot. He could run circles around the other animals. Mounted herders found it possible to run herds large enough to make themselves rich, and even large enough to, less beneficially, eat all the grass in sight.

A man on horseback could travel about five times faster than on foot. As Gimbutas points out, he might travel 150 kilometers in a day and be back the next night. Possibly no one would know where he went. This was intoxicating. Suddenly, the web of human contacts was vastly extended over the face of the

Earth. A horseman could have whole dimensions of his life little-known to his community. Because they had been based on the limits of travel by foot, traditional boundaries between tribes collapsed. With the new ease of transport, a homogeneity in tools and arts appeared across thousands of kilometers of the Eurasian steppes (Gimbutas 1991: 354). A shared vocabulary of horse and livestock words emerged, perhaps as the common core of the emerging Indo-European languages.

The earliest signs of a horse cult appear in the middle Volga region, where a boy's grave near S'ezzhee contains horse skulls, figurines of horses and a very long flint blade. The bones date to around 5000 BCE. A boy buried with such wealth would probably have been a chieftain's kin. This is perhaps the earliest grave in the world in which the attributes of nobility are fine horses and long knives. Other large graves of this so-called Early Yamna period were mounds of earth on the open steppe. Gimbutas explains that such a mound would not stand out in a forest; the personal burial mound (*kurgan* in Russian) was a marker for lords of the steppes (1991: 353, 356).

The biotechnology of horseback riding did not hit the Americas till after Columbus, which is probably the biggest single difference between the histories of dry lands in the western and eastern hemispheres. Riding horseback allowed large-scale ranching and long-distance poaching. It was the most devastating improvement in the means of war before gunpowder. This improvement in the means of violence unfolded slowly in Eurasia, compared to North America, where the natives obtained both horses and guns at roughly the same time. The horse armies of Eurasia took thousands of years to evolve, but political implications appeared from the start. Just as horses enabled riders to control larger numbers of sheep and cattle, so they enabled mounted warriors to control larger numbers of people. And if wealth could be gained by keeping ever-larger herds of animals, even greater wealth could be amassed by milking groups of human beings for tribute.

The steppe warlords first stand out in the historical record when their graves grow unusually large and rich. Nineteenth-century archaeologists were amazed to find the wealthiest graves of neolithic Eurasia out on the steppes, north and east of the Black Sea. These, it seemed, must be tombs from a higher civilization. A great Kurgan mound at Maikop on the Caucasian steppe (from around 1500 BCE) contained a king dusted in red powder, two more human bodies (a male and a female), some weapons of copper, precious stones, hundreds of gold objects and a throne "of which a king of France could have been proud" (Briard 1979: 23). The metal work especially seemed to indicate a highly developed culture. More recent evidence suggests that the steppe pastoralists learned their metal crafts from villagers in the ore-bearing mountains of Old Europe, Turkey or Iraq. Perhaps some of the grave objects were locally made, and others were stolen or traded from afar. Jacques Briard (1979: 24) says, "We should not insult these princes by believing their wealth to be simply the fruits of plundering raids outside the

Caucasus." Perhaps not. But over much of recorded history, plunder was central to the steppe economy.

Around 4400–4300 BCE, the Kurgan mound graves reached Moldavia, penetrating the regions of Old European Cucuteni villages. At first the horse pastoralists and village gardeners lived in apparently peaceful proximity. The villagers made kiln-fired ceramics as permanent artwork. In nearby camps, the Kurgan people hardened their clay vessels in open campfires. The Cucuteni villagers grew wheat, barley, oats and rye; the Kurgan camps contain grain gathered from wild grass. The villagers buried their dead in communal graves; Kurgan chiefs were buried in individual tombs, with their horses, weapons and wives. The Old Europeans seemed to believe that death meant return to the source of life in the earth. But the Kurgan lords behaved as if they would survive death as individuals; a dead nobleman took his wealth with him, including his women. As Gimbutas points out, here was no concept of death as a renewal for earthly life. For the Kurgan people, human lives were separate from the life of the earth. A steppe lord only visited this place, acquired whatever wealth and status he could, and presumed to take his holdings away (Gimbutas 1991: 363, 395, 400).

As the scale of migrations out of Inner Asia gradually increased, conflict between the Old Europeans and pastoralists escalated. The Cucuteni villagers started moving their homes up on the hilltops and raised defensive walls of earth and wood against the rising storm of nomadic raiders. But the farmers could not adequately defend themselves without abandoning their whole way of life. Many were probably captured to become farming slaves, and perhaps some adopted the pastoral or raiding ways of their enemies. By 3500 BCE, the storm winds of migration and war on the plain reached hurricane force. They ripped up the Ukraine's villages by the roots and blew them away to the west.

The Kurgan Waves and Battle-axe Wars

The archeological evidence of steppe invasions in Old Europe includes destroyed villages, Kurgan graves, fortified hilltops and sudden increases in numbers of horse, cattle, sheep and goat bones. On the Bulgarian coast, villages of the old Varna culture are suddenly scattered, and their farmlands turned to pasture for horses and cattle. The old villages often terminate in a layer of ash. Among the ruins often lie clusters of skeletons, many with their skulls smashed in from behind. In place of the old villages, smaller clusters of pit-huts resembling steppe encampments often appear. Crude, poorly fired pots replace the well-crafted pottery of earlier cultures. According to such evidence, the first "Kurgan Wave" arrived in eastern Europe around 4400 to 4300 BCE. It moved through the Ukraine, leaving most villages and towns intact but withdrawn into defensive positions. The horsemen then pushed into the Balkans, basically demolishing the old cultures there. The burial mounds of steppe warlords appear like giant's footprints up the Danube to the Rhine (Gimbutas 1991: 363, 364, 358–59).

As the first major invasions broke on them, the Vinca people of the central Balkans crowded westward against the Adriatic coast or into Austria. Gimbutas (1991: 352) claims that before this time, the Old Europeans possessed no weapons apart from hunting implements; therefore, "the agriculturalists were easy prey for invaders on horseback with long knives." Refugees from the middle Danube's Tisza culture fled north to the Carpathian Mountains or Poland. Old European civilization retreated to the north and west, and survived as a band of "Funnel-Beaker cultures" from Belorussia to Britain.

The clash of Old Europe and its pastoral invaders was perhaps the foundational trauma of European history. This series of events has been the subject of popular myths and folklore down to the present time, and these stories have themselves shaped history, by defining the origins and destiny of Europe. At least three myths merit our reconsideration:

1. The Invaders as Europe's Founding Fathers

Before World War II, many Europeans assumed that if the Kurgan, Indo-European or "Caucasian" invaders of Europe were militarily superior to those they conquered, then they must have been socially, technologically and racially superior as well. They must, in short, have become masters because they were a master race. Weapons of bronze appeared first in the hands of steppe warriors, and this seemed to confirm their technological edge in the war for the survival of the fittest. Now we know that the neolithic villagers of Turkey and Old Europe made jewelry of gold, silver and copper long before the Kurgans. In the early Bronze Age they began forging tools such as metal sickles for grain. But it seems they didn't think of making metal weapons for killing people. That invention required minds with different needs and goals. Gimbutas (1991: 394) claims that the Proto-Indo-European words for metallurgy were loan words from other languages, and the people of Kurgan Wave I had no metal weapons. In that case, the steppe migrants were perhaps superior in their use of animals and their means of violence, but little else.

2. The Siberian Source of Wild Barbarians

Another myth of the ancient invasions points to Siberia as the source of barbarian hordes. Obviously the pastoral migrants came from the east. And farthest to the east lay Siberia. That wilderness, it seemed, must be the original source of wild, dehumanized raiders. In popular Western imagination, Siberia became a polar opposite to the realm of civilization. Of course, most people today would find it odd to view wild nature as "dehumanizing"—as if wilderness was more a source human depravity than cities. And Siberia is no longer unknown to Europe and no longer serves as a mythic source of evil forces. The land now called Siberia includes vast horizontal bands of tundra, arctic forest, broad-leafed forest and treeless steppe. The tribes of warrior nomads came from the semi-arid grasslands, which lie mainly south of Siberia. The actual Siberians of the northern forests lived much like the Laplanders in Scandinavia and were seldom aggressive or warlike people. For example, the Tofa people of the Sayan Mountains (north of Mongolia) lived by herding reindeer and gathering forest plants down to present or very recent times. In their culture, fighting and games of mock battle were considered shameful, and Tofa folklore praised no war heroes. The Tofa elders believed that all things, even the stones, were alive, and that forces of nature would punish anyone who failed to pay other creatures proper respect (Sorin 1992). These people lived in an environment of Siberian tigers, bears, and mighty winters lasting from August to June. But they were never made cruel or savage by the wildness of that environment. It was not wild nature but the *degradation* of nature that drove people to fight "like animals" for the means of life.

3. Old Europe as Female, and Its Attackers as Male
According to this modern myth, the civilization of Old Europe had a female personality, while the Kurgan invaders were quintessentially male. This sexual role-casting has some justification. The Old Europeans did seem to honour mothers more than fathers, while the Kurgans gave highest place to male chieftains. Also, in their raids on farming villages, the Kurgan warriors tended to kill the enemy males and take the women as slaves. This suggests a dual lineage for Europe's future—a male line of Indo-Aryan warriors, and a subjected female line from Old Europe's age of the Goddess. This myth neglects two sets of actors in the drama—the Kurgan women and the Old European men. The Kurgans, naturally, were half female. We can be fairly sure that Kurgan women were co-creators of their culture and raised their boys to be good raiders. As for the men of Old Europe, they no doubt suffered many casualties but probably most survived. Their values also survived. In the future, countless non-martial European men would be artists, farmers and philosophers. The Old European heritage would be there for anyone, male or female, to claim.

The Troughs between Invader Waves

For about two hundred years the warrior bands of Kurgan Wave I raided their way across Europe. Then, about 4300 BCE, the eastern frontiers grew quiet again. Perhaps conditions on the steppes improved or returned to normal. The livestock drivers settled in Europe as minor lords or returned to their grassy universe of the steppes. The times of mass migration and heroic war may then have appeared as an aberration which had never come before and might never come again. Of course, the invasions did come again. Eight hundred years after Kurgan Wave I (c. 4400 BCE), the next mass migration, Kurgan Wave II, arrived in Europe, around 3500 BCE. Kurgan Wave III came five hundred years after that, about 3000 BCE. A "Battle-Axe Culture" then spread over Europe, reaching the British Isles. After Kurgan Wave III, the tempo of migrations stepped up. In the Bronze and Iron ages, invasions such as those of the Indo-European Celts and Dorians grew too numerous to clearly count. We must suspect that the steppe environment grew increasingly unstable.

Indo-European Culture

In popular nineteenth-century theories of racial evolution, the ancient Indo-European invaders of India, Persia and Europe were considered as the source of all higher civilization. The attention then lavished on Indo-European studies began to rival biblical studies. Linguists traced the relationships between Indo-European languages, such as German, Slavic, Persian and Sanskrit, in search of a common "Proto-Indo-European" language. Some scholars guessed that the original homeland for this race of races came from the Caucasus Mountain region—and the idea of a "Caucasian race" was born. Linguistic evidence placed the Indo-European homeland somewhere between the speakers of Finno-Ugric, Semitic and Caucasian languages, which would place the perpetrators of "civilization as we know it" somewhere on the steppes of Eurasia (Gimbutas 1991: 352). Some recent evidence suggests that the European population already spoke Indo-European during the 2000s BCE, and that the source of Indo-European speech was possibly Turkey (Barraclough and Stone 1989: 60). But this makes little difference to the history of steppe invasions. The people of Europe could have taken up Indo-European speech following Kurgan Waves I and II, by the mid-3000s BCE. And, most importantly, the history of steppe and desert invaders concerns environmental refugees, regardless of their culture or language. The victims of environmental disaster have no fixed ethnic identity. Whatever their earlier homeland, it seems that hosts of Indo-European speakers inhabited the semi-arid steppes during the early Bronze Age, and that many of these people migrated to India, Persia, Mesopotamia and Europe. And the indigenous people of those lands saw these migrants more as destroyers than bringers of civilization.

The Indo-European Family

In 1901, Otto Schrader claimed to have found the original Proto-Indo-European culture—in Serbia. At that time, the Serbian homeland was divided between the states of Hercegovina and Montenegro. The family structure there, Schrader said, was a clear reflection of ancient Indo-European society. Central to this was the zagruda, or extended patriarchal family, which was traditionally occupied in stock-breeding. The various local *zagruda* were aligned in clans called *rodu*, which claimed certain pastures, fields and forests as their own. Relations between the male clan members were usually close and had a military function. The men understood themselves to be pledged to mutual defense and blood vengeance against external enemies. The women were accountable for sexual loyalty, the betrayal of which could also lead to blood vengeance. These clans were loosely united in tribes, or *pleme*, which claimed tribal territory and had tribal heads called *zupan* or *starina* (Gimbutas 1971: 140–41). In Schrader's day this kind of extended tribal family was already dissolving into urbanized nuclear families. But much of the old order still survives in rural Serbia and the rest of eastern Europe. At least, the code of clan loyalty and blood vengeance seems alive and well in modern Serbia.

In traditional Eastern European families, a grandfather, his sons and their families, commonly lived together on the same plot of land. The boys built houses for themselves across the yard after they got married. Such clusters of houses dotted the countryside around Minsk in Belorussia, where Donvar-Zapol'skij described them in 1897 (in Gimbutas 1971: 134). The land, farming tools and animals were all collective family property. The boys worked with their father almost every day and ate their meals in the central house belonging to their dad. This Belorussian extended family was also called a *zagruda*. Gimbutas (1971: 134–35) says that in such old-fashioned Slavic families, "The house father is master of the *zagruda*, responsible for both its economic and moral welfare.... At home he is a judge ... settling all problems and quarrels. What he says is right." Women here were secondary citizens. In early medieval Russia, many men claimed two or three wives. Heads of families could even require sexual service from their sons' wives. If a woman left her husband's family, she had no recognized right to take any family property with her (Gimbutas 1971: 135–36). A father could discipline his wives or children in any way he saw fit, even by killing them.

Such strong extended families with military alliances between clans tended to evolve in hard environments. In places where a nuclear family might not survive on its own, the extended family tended to grow in importance till it quite overshadowed the individual. In the yet harsher environments of the Near Eastern deserts, the patriarchal extended family grew even stronger.

During the late Roman empire, Roman generals on the Danube frontier described the family values of invading Germanic tribes. Like the Aryans of northern India, these old Germanic Aryans executed women for adultery, paid dowry in marriage and buried widows with their dead husbands. Later, when Indo-Aryan tribes settled in Germany, they mixed with the Old European population. Their culture then evolved as a mixed bag of patriarchal and matrilinial traditions. Dowry was paid by the groom to the bride, and not by the bride's parents to the groom's family; a wife was then viewed as a prize to be earned, not a cost to be covered. Gimbutas (1971: 135–36) finds the most rigidly patriarchal ways tended to soften with time after the invaders settled in Europe: "The farther from the steppe and the nomadic life ... a people moved, the more of these cultural elements they lost."

The Religion of Heroes

Heroic ages tend to be times of conquest, with people on the move to capture new lands. The European invaders of North America had their own heroic age in the Wild West of the 1800s, and each Indo-European ethnic group has likewise had its own age of heroes. Those times of migration and war usually involved great suffering, but poets and songwriters tended to immortalize the moments of victory. In the characteristic setting of an Indo-European heroic epic, the tribe was ruled by a warrior king and his loyal soldiers. The men were faithful to their leader because he was the strongest in battle. Their heroism and virtue lay in willingness to do their lord's bidding no matter what the cost (in Kramer 1990: 223–24). This was the warrior's ethic, heralded in a thousand sagas by bards in the pay of warrior kings. If ancient tribes are compared to modern corporations, the sagas can be seen as forms of advertising. Songs of praise for warrior kings were sung to boost the followers' morale and to win a yet larger following. The heroic songs also held out a promise to warriors—if their deeds were sufficiently brave, they too might be immortalized in verse. "Immortality" for the ancient Indo-Europeans generally meant immortalization in a saga. Of course, such immortality was basically for men only. As women were meant to avoid the perils of battle, their chances of immortality were slim.

The social patterns portrayed in the sagas, with characteristic roles for heroes and commoners, and for men and women, lasted throughout the history of Indo-European nation-tribes. Kurgan lords and their armed hosts, Slavic "thousanders" (leaders of a thousand) and their *druzinas* (councils of noble warriors), Germanic princes and their *thanes*, Alexander the Great and all the emperors of Rome were subject to the ideal of a fighting prince with his band of heroes. These princes were expected to lead their men into battle regularly and join in the fighting personally. Their followers were expected to leave their families and serve their warlords unto death. The ancient bards and their epics worked this image of heroism into the Western mind so deeply that modern science fiction commonly presents bands of future warlords and loyal warriors battling in outer space.

The Kurgan Waves in Europe are recorded partially by the spread of warlord art. On pots, weapons and tombs, the old themes of regeneration, pregnant women, fruitful plants or multiplying animals were commonly replaced by images of the sun and lightning, or pictures of daggers and battle-axes. The images of weapons were so numerous and prominent as to suggest a cultural fixation on and reverence for their power. Herodotus confirms that the Scythian steppe people offered religious sacrifices to their symbolic daggers.

If weapons were objects of worship, they must have been crucial to life as the worshippers knew it. Weapons must have been central to their economy. Riane Eisler concludes that plundering and captive-selling were the steppe invaders' chief enterprises. If so, it was natural that they had greater esteem for instruments of death than for older religious symbols such as the sheaf of wheat or mother with child (1987: 49–50). Under invader rule, people still revered the holy things of Old Europe, but now these co-existed with a cult of death-dealing weapons. In the early Iron Age tombs of the Seine-Oise-Marne culture in northwest France, pictures of battle-axes are artfully contrasted with portraits of the holy mother of the universe. On one side was the cult of birth; on the other the cult of death. The earth still gave its fruits as always, as the mother gave birth to the child, but now the battle-axe could compel the earth and its farmers to yield their fruits to the warrior. Likewise, prowess in battle could gain women for a warrior, either as captives or through attracting women with the strength and wealth which weapons could bring.

The old steppe deities were lords of the majestic heavens, warlords in the sky, mirror images of the Indo-European kings themselves. "The Kurgan ideology," Gimbutas (1977: 288) says, "exalted virile, heroic, warrior gods of the shining and thunderous sky." The sky gods fought battles with giants or monsters in the realms above. The gods of war eagerly meddled in human conflicts, swinging the fortunes of battle in favor of those with the greatest reverence for them. In Slavic myth, the best remembered god was Perun, who was pictured as a mighty man with a copper beard, riding across the sky in a chariot drawn by a male goat. In his hand was a battle-axe or hammer, which he hurled to punish wrong-headed people, or to fertilize the female earth in springtime (Gimbutas 1971: 165). The weapons these gods bore were usually Stone or Bronze Age blades, axes or battle hammers. When a god threw his weapon, it appeared as a flash of lightning and thunder. Those struck by lightning evidently deserved the gods' wrath. As above, so below: those struck down by the earthly warrior's hammers deserved defeat. Victory in war came only to those favored by the gods. Therefore, trial by combat was held to reveal divine will more surely than human decisions or votes.

The most important sky god, naturally, was the sun. He reigned over all, seeing and passing judgment on everything below. The warrior kings could imagine no higher goal than to emulate such omnipotence by gaining comparable power over the earth. They dreamed of becoming "sun kings" over all they saw. This dream, however, was attainable by only one man at a time within each territorial universe. The means of gaining earthly omnipotence were weapons, the courage to use them, and the ability to organize hosts of men as armed units. If this was the highest aim to which a man could aspire, then life was a game of winner-take-all.

For raiders, war seemed genuinely productive. As the thunderbolts hurled by the gods seemed to compel the earth to conceive and give fruit, so earthly warlords forced the world to produce for them. Their powers to gain wealth seemed to surpass the powers of the fertile earth, or the mother with child. So a mythology arose in which the land's fertility depended on the virility of a warrior king, rather than the other way around.

The Forced Marriage of Gods and Goddesses in Greece

The full weight of Indo-European invasions came relatively late to Greece. The mountains to the north served as a protective wall against the early Kurgan invasions. In the Danube basin to the north, the old cultures were destroyed long before the age of written history or legend, leaving mute artifacts which merely suggest a civilization of the Goddess. But in Greece, the Achaean and Dorian invaders came in force only near the dawn of the Iron Age, around 1300 BCE. The memory of those invasions was still alive when the writing of myths, epics and histories began. The invasions of Greece, therefore, are the best recorded of early Indo-European migrations, and they were recorded critically by non-Dorian observers.

Actually, Kurgan Wave III penetrated to the Greek coast after 3000 BCE, but studies of graves from those times show the Kurgan warriors were a tiny minority of the population. This would account for the nature of Bronze Age Mycenaean Greece, with its small warrior elite, and its continuation of Old European-style villages. True Old European cultures, with unfortified towns and councils of elder women, survived mainly in the isolated Cyclades Islands and on Crete. But daily life in mainland Greek villages probably went on much as it had before.

Down to about 1300 BCE, the Mycenaean warlords ruled from various hill forts, including the acropolis of Athens. Clay palace tablets in Linear B script record amounts of flax woven, or barley, olives and grapes harvested. At Pylos, such tablets record only the previous year of business transactions before the record ends, because the clay was fired as the palace burned. The same thing happened in other palaces. At Thebes the firing dates to before 1300 BCE. Moving south, palaces were burned one after another. By about 1200 BCE, the flames reached Pylos, on the southwestern coast of the Peloponnesus. Before and during this century of invasion, the Mycenaean centers show signs of increased fortification. Water tanks appeared inside the fortress walls to enable the towns to withstand siege. A stone wall across the Isthmus of Corinth rose at this time, as a defense against expected invasion from the north.

The Greek Dark Age

In the wake of these invasions, it seems the entire population of Greece declined drastically. In southwest Peloponnesus, around Messenia and Triphylia, about 150 settlements are recorded for the early 1200s BCE. A hundred years later no more than fourteen are known (Olivia 1981: 36). Vessels of wood or leather largely replaced ceramic pottery. The invading pastoralists avoided pottery because it broke in the course of traveling. The system of roads between Mycenaean towns fell into disrepair, and written records disappeared for about five hundred years. Concerning the invaders, who were later called Dorians or Achaeans, H.G.L. Hammond writes;

> Some negative conclusions are permissible. The invaders brought no distinctive painted pottery or other mark of a developed civilization. They did not take to urban life. ... They may have had some superior weapons, but in the arts they were inferior to those they conquered. (1959: 75)

For the next several hundred years, Greece is mainly an archeological blank.

In Hesiod's explanation of history (in Olivia 1981: 115–16), various races came to Greece over the ages, and each set of newcomers was worse than the last. An original race of virtuous golden people was replaced by an invasion of somewhat tarnished silver folk. Next came the frankly base men of bronze.

Finally, the men of iron arrived to dominate the land, and these were complete barbarians. Under their rule, Hesiod complained, might largely replaced the sense of right. The web of human relations and culture that formerly bound people together came undone. All sense of modesty, decency and righteous anger against evil vanished. Next, Hesiod gloomily predicted, the family itself might be destroyed. Hesiod sounds like a grand old grouch complaining about the younger generation. But in his day, Greece was just emerging from the Dorian dark age, somewhat like Russia after centuries under the Mongol yoke.

The collapse of Mycenaean Greece roughly coincides with the fall of Hittite civilization in Turkey, the destruction of towns across Syria and Palestine and the attack on Egypt by "People of the Sea." Rhys Carpenter (in Olivia 1981: 33–34) argues that all this resulted, not from invasion, but from a cycle of major drought. Carpenter sees a pattern of recurring large-scale drought in world history, arriving approximately every 1,850 years. His theory is generally held to compete with or to replace the theory of armed invasion from the north. But what if drought, nomadic migration and military chaos all go together naturally? If so, Carpenter's theory of drought complements rather than contradicts the evidence of armed invasion.

The Dorians first took the flatlands and river valleys in Thessaly, Boeotia and the Peloponnesus. The aboriginal people, who were often called Pelasgians, retreated to less accessible areas, such as the mountains of the Peloponnesus, the rocky peninsula of Attica, or the Aegean islands and coast of Asia Minor. In such enclaves the indigenous people managed to stand their ground. When the hot lava of invasion cooled, Greece was a geographic patchwork of radically different cultures. Some areas were almost purely Dorian, such as Sparta. Some were basically Old European, such as the Cyclades Islands. Between these extremes, half-breed cultures emerged, mixing almost diametrically opposite traditions. The best-known example is Athens. There, Robert Graves (1960: 18) claims, a male military aristocracy combined with a female theocracy to produce classical Greek culture.

The Dorians pictured their gods and goddesses as warlords like themselves. They brought their local version of the Indo-European pantheon to Greece, including Zeus of thunder, Ares of war, Apollo of the sun, and Athena, a goddess of warriors. But in Athens, Old European deities of the earth, such as Hestia, Demeter, Hera, Artemis, Aphrodite, Pan and Dionysus, joined the Indo-European divinities on Mount Olympus. Joseph Campbell (1976b: 143–53) says the old goddesses and gods were not slain or demonized; instead, the mainly male divinities of the Dorians and mainly female deities of the aboriginals were said to have married. It may have been a forced marriage, but the heritage of Old Europe survived in the resulting Greek cultural family.

The Scattering of Pelasgian Greece

As the aboriginal Greeks retreated before the Dorians, those compressed into the Attic peninsula or scattered over the Aegean from there were known as the

Ionians. Another native migration pressed eastward from Thessaly into Asia Minor and to the Black Sea coasts. These were called the Aeolians. Herodotus identifies these groups with the original Greek Pelasgians. In these migrations, Old European society scattered across the eastern Mediterranean as an international diaspora. Homer probably lived in one of the resulting Pelasgian towns on the coast of Asia Minor. From that standpoint he told stories of pre-Dorian Greece and of adventures on the seas. His language was an inter-polis speech, transcending any local dialect and addressed to an international audience of Greek speakers (Olivia 1981: 106–7). Similarly, Aesop lived on the island of Samos, where he wove animal stories from across Asia, Europe and Africa. His island lay close to midway between those three continents, and Greek refugees made it a crossroads between worlds.

The Dorian and Achaean invaders generally treated other ethnic groups as enemies. But the diaspora of Pelasgian Greeks was a wide-flung network of traders. Their sailors operated like bees on the island-studded sea, cross-pollinating local cultures while acquiring what wealth they could. From trading in Asia Minor, Syria and Phoenicia, Herodotus claims the Greeks gained their skills in metalworking, the "Phoenician alphabet," the Phrygian and Lydian modes of music, various popular games, and the coining of gold and silver (Olivia 1981: 178). By incorporating the best of many cultures, the skills of exiled Greeks grew faster than they might through isolated tribal experience. Instead of focusing on their own tribal traditions or the military arts of defeating outsiders, the refugees prized learning about their trading partners. They valued knowledge of plants from many lands, sea creatures and the pre-scientific lore of ancient Egypt.

The Pelasgian Greek religious world became an amalgam of all eastern Mediterranean traditions. The pantheon of gods and goddesses in Hesiod's Theogony, for example, recounts successive generations of divinities, starting with the marriage of Uranus (the sky) and Gaia (the earth), much as in the creation legends of Asia Minor and Syria (Olivia 1981: 169). The Greek mystery religions came partly from Egypt. As Plutarch later explained,

> That Osiris is identical with Dionysus, who could more fittingly know than yourself, Klea? For you are the head of the inspired maidens [devotees of Dionysus] of Delphi, and have been consecrated by your father and mother in the holy rites of Osiris. (in Bernal 1987: 118–19)

Later, in the first century CE under the Romans, the Greek diaspora still served as a capillary network for religious and intellectual movements such as Neo-Platonism or Christianity. Between scattered communities of former refugees, a "cosmo-politan" culture emerged. So a saying about the fruitfulness of casting bread upon the waters gained currency among both scattered Greek and Jewish communities. Of all effects caused by nomadic invasions from the dry lands, this was among the most beautiful.

The Pure Dorians of Sparta

The Dorian Spartans reached the southern end of mainland Greece during the tenth century BCE. They claimed the Eurotas Valley, with its land and farmers, by right of conquest. The social system they established was like that in other Dorian-dominated regions, basically a system of force-backed apartheid between master and slave races. In Dorian Argos to the north, the conquered farmers were called gymnetes ("naked ones"). In Sicyon, they were *korynephori* ("those who should be beaten with sticks"). The Spartan word for peasant slave was *helot*. As Pavel Olivia (1981) explains, the helots of Sparta belonged to the land they worked, and were inherited or sold along with the land, like serfs in the later feudal system. Unlike serfs, however, the helots "were not part of the nation, but were natives enslaved by foreign invaders." While serfs were treated as very low-class citizens, the helots were treated as subhumans (Olivia 1981: 64–65, 68).

Plutarch said the Spartans whipped their helots at least annually, lest these slaves forget the threat of punishment and fail to produce for their masters. Regular intimidation of slaves was a function of government. The highest officials of Sparta began their year-long terms in office by declaring a temporary reign of terror on the helots. In this game, called the *krypteia*, young Spartan men went out into the countryside to stalk and murder slaves. The deaths were seemingly random, but any helots who had caused trouble in the previous year could be weeded out during these killing seasons. The ritual war was part reenactment of the national story, part naked intimidation and part a Spartan-style initiation into manhood (Olivia 1981: 69). All Spartan boys started their military training at age eight, and participation in the *krypteia* may have been their first baptism of blood.

Around 750 BCE, the Spartans expanded into Messenia, on the southwest coast of the Peloponnesus. To defeat the Messenians took two long wars. Then, as the Spartans tried to make helots of the defeated people, the locals exploded into revolt again. The fighting was so ferocious that the Spartans searched for new combat methods. The old duels between rival individuals gave way to closed phalanxes of shock troops (Olivia 1981: 78). The soldiers mutated into armored hoplites, whose metal helmets and armor created a terrifying, non-human appearance. So the Dorian economy of aggression became an ever more costly enterprise and conquest was only the beginning of a struggle without end.

On his portion of conquered land, each Spartan landowner faced a lifetime of vigilance against his helots. The Spartans may have been free from menial labor, but they did not have easy lives. Most Spartans were hated by their slaves and knew it well. Xenophon said the helots "would have liked to eat them [the Spartans], even raw" (Olivia 1981: 70). But this was the lifestyle the Spartans believed in. They were willing to take all the risks involved and felt it beneath their dignity to live without slaves. By some accounts, Spartan men kept their weapons within reach at all times. They furnished their homes with stout, boltable doors that could be slammed shut at a moment's notice. They took great care with weapons, lest any helot have chance to seize a sword or spear (Olivia 1981: 67–70). Thucydides (c. 460–c. 400 BCE) said the Spartans were hampered in their War of the Peloponnesus, because the slaves were an ever-present danger to internal security. On some Spartan farms, owners and slaves may have treated each other with unusual decency, but Thucydides expressed the prevailing logic in pointing out that justice was only an issue between people of equal strength.

In his enthusiastic *History of Greece*, published in 1900, John Bagnell Bury claimed: "The eminent quality which distinguished the Dorians ... was that which we call 'character'"; this "character" was maintained partly by "keeping their own Doric stock pure from the mixture of alien blood" (in Bernal 1987). The alien blood concerned was that of Pelasgian slaves. It seems that the Spartans treated children of mixed Spartan and Pelasgian blood, or any other "illegitimate" children, with considerable contempt. Around 700 BCE, the outcaste and illegitimate children of Sparta organized to leave their homeland. They took several ships to the east coast of Italy and built the town of Taras. Archaeological evidence from Taras suggests that this "colony" had virtually no trade with Sparta for the next hundred years (Olivia 1981: 79). Apparently, it took four or five generations to forget such bad blood.

The Dorians' concerns for purity went beyond sex and blood. They also practiced a broad censorship of non-Dorian ideas from other parts of Greece. Pavel Olivia (1981: 79) says this involved "an artificial isolation which had a deleterious effect on the evolution of Spartan culture"; after about 550 BCE, the work of Spartan craftsmen reportedly declined compared to more innovative crafts from other states, and "the same tendency can be seen in literature and in music," even in the Olympic games. However, in earnest admiration, John Bagnell Bury said that in Sparta, "the Dorian seems to have remained most purely Dorian" (in Bernal 1987).

The Hybrid Traditions of Athens

On the Attic peninsula, the record of ceramic pottery shows no serious break in style from the Mycenaean Bronze Age to the time of classical Athens. The "sub-Mycenaean" pottery evolved smoothly into a "proto-geometric" style. The local villages were not swept away, and no "archaeological blank" followed the Dorian invasions. According to legend, the hero Ion led four tribes of "Ionian" people into this seaward corner of Greece. Under Dorian pressure, these tribes (plus any people already there) squeezed into the land's end near Athens. The various tribal aristocracies then had to solve a problem of ranking among themselves. After some time, they formed a central council, first led by Theseus, which met on the old Mycenaean hill-fort (or acropolis) above Athens.

Like the Spartans, the Athenians needed military solidarity to survive in a hostile world. But the solidarity of multi-tribal Athens was different from the unity of a Dorian state. In Athens, the peasants and councilmen both came from the same tribes. In nearby regions, the Dorians had enslaved all Pelasgian people, and this was a common threat to Athenians of all classes. Perhaps this is one reason why, after 594 BCE, the Athenian council headed by Solon claimed to free all slaves. Previously, slavery had been the fate of debtors and prisoners of war. Solon announced a ban on the enslavement of Athenian debtors (but not of captured foreigners). How the debts were paid and what freedom really meant for former debt slaves remain unclear. But Athens apparently meant to claim a social distinction from the Dorian slave-states. The Spartans sometimes used slave troops in their front lines, who constantly threatened to turn on their masters. The Athenians hoped to achieve a stronger unity than that.

Of course, Athens was also influenced by Dorian culture and by the heritage of old Mycenaean nobles which descended from Kurgan Wave III. The *polis* (or "walled town") of Athens was a semi-military society. Even after the "emancipation" of debtors, many prisoners of war were still slaves, and women commonly lacked the most basic freedoms. The pressure of the Dorian enemy forced a relentless militarization. The goddess Athena herself was a symbol of the resulting hybrid culture. As in Old Europe, the city had its patron goddess, but, as Eisler (1987: 80–81) points out, this was a goddess, not of nature, but of war. She was not born from the Great Mother's womb but from the *head* of Zeus, the Indo-European god of thunder and rain. This creature of her father in heaven was a kind of military angel created to inspire the town's troops.

The Athenian state evolved in a crucible of conflicting pressures. On the one hand, the whole community needed to unite against its enemies. On the other hand, local rulers wanted aristocratic privileges for themselves. How could they get both things at once? The councilmen devised a way. In Solon's social order, the people of all tribes were divided into four classes, each with different rights and privileges. These were:

1. *Pentakosiomedimini*, or men with income of over 500 medimni (units of 52 liters) of grain per year. These were eligible to serve as city directors and were not required to fight in the armed forces.
2. *Hippeis*, or men making 300–500 *medimni*, who served in the army as horsemen. These men were also eligible for public office.
3. *Zeugitai*, meaning "those using oxen to plow their own land." These made 200–300 *medimni* of grain a year and served in the army as armored infantry, or hoplites. Such men could vote in the polis assembly but not hold office.
4. *Thetes*, who were landless laborers or anyone earning less than 200 *medimni* a year. In the army, *thetes* were the unarmored light infantry. Formerly, many had been debtor slaves. The *thetes* were allowed to come and sit in the polis assembly but could not vote. (Olivia 1981: 136–37)

In this system, the male citizens were ranked according to level of income. A noble-born man who did poorly in farming or trade could therefore fall in legal status. A successful *thete* or *zeugitai* could possibly rise. Solon and his councilors made power depend more on the means of production, and less on the means of coercion—in contrast to the hereditary and military sources of power among the Dorians. In Solon's model, the principles of ranking and privilege were retained, but the same standards for achieving rank were applied to all males.

Unmentioned in these social ranks were the Athenian women. Women in the 500s BCE had no political rights in the assembly, and no property rights independent of their male guardians. They were even legally barred from leaving their homes after dark, so far had Athens grown from Old European tradition. At least Athens still accepted that women could be artists, philosophers, teachers and priests.

Standing near the front line of battle against the Dorians, Athens slowly grew to resemble its foes. During the great Peloponnesian war with Sparta (431–404 BCE), Athens was almost ruined financially. It seized slaves from lands to the east and made them work the silver mines of Laurium. When units of slave-soldiers defected from the Spartan side, the Athenian leaders were split over what to do with them. Should the Athenians cast themselves as champions for the liberty of all slaves? Some of the defectors were accepted into the Athenian army, sent to fight on other fronts and later declared free men. But many Athenian officers were so involved in maintaining their own military discipline that they feared to reward desertion under any circumstances. It seems that most of the Spartan slave-defectors disappeared and were probably killed in secret (Olivia 1981: 70).

During the 500s BCE, a series of dictators took power in Athens. These leaders probably felt it obvious that Athens must beat the Dorians at their own game. Therefore the command of men and resources had to be centralized for the sake of security. The people must learn the values required for military survival. The tyrant Peisistratus (r. 546 to 510 BCE) proved a particularly imaginative engineer of Athenian culture. He patronized the arts on a major scale, but the sculptures or dramas he sponsored had an instructional value. His administration raised a limestone temple on the Acropolis. On one wall was portrayed the *apotheosis* of Herakles (Hercules), in which Herakles is transformed into an immortal god in reward for his mighty deeds. According to this state-sponsored message, the very strongest men sometimes rise above their fellow mortals and become gods.

The Peisistratus administration also supported an annual round of dramas at the forum of Athens. Certain dramas received classic status and were repeated every spring. Eisler (1987: 78–82) claims the chosen productions were a form of social propaganda, well calculated to promote the sentiments and loyalties favored by the administration. In the regularly performed *Orestia*, King Agamemnon goes to war across the sea. To secure his troops' safe passage, he offers his daughter, Iphigenia, as a human sacrifice to the lord of storms and winds. Years later, when Agamemnon returns from battle, his wife, Clytemnestra, avenges her daughter by stabbing her husband to death. The couple's son, Orestes, then reluctantly considers it his duty to side with his male parent and avenge him by killing his mother. This leads to the famous original trial by jury. The gods and goddesses arrive from Mt. Olympus to play roles as both witnesses and jurors. They must decide for the people of Athens if Orestes has done the right thing. At the culmination of the trial, Athena casts the deciding vote. She rules that Orestes did right and should not be punished. The goddess herself explains to the assembled people of Athens that a man is primarily the product of his father's seed and not of his mother's body. Athena herself was immaculately born from the head of her father Zeus and not from any woman. Therefore, Orestes owes primary loyalty to his father. This includes the duty of blood vengeance against anyone who kills his father, even if it is his mother. In this morality play, there was only one difference between Dorian and Athenian morality—the Athenians still needed to be told.

The Environmental Reduction of Athens

With the Dorian invasions, the Attic peninsula had absorbed a major influx of settlers, who cut the valley forests for cropland. Then they slowly consumed the hillside forests for firewood and lumber and used the cleared slopes as sheep and goat pastures. So most of Attica slowly turned to a hilly grassland. The forests of oak and beech gave way to a grazed-over "classical Mediterranean" environment of low bushes, olive groves, flowering grasses and scented herbs. This was the scenery later associated with the cradle of Western civilization, but it was a scene of desolation compared to the previous environment (Ponting 1991: 75).

Next the farmers expanded their fields up the slopes, accelerating the loss of soil. By 590 BCE, Solon urged laws banning cultivation of the steep slopes because of the obvious erosion. Later, the tyrant Peisistratus offered Attic farmers a bounty for planting olive trees. He argued that olives would grow best under Attica's increasingly dry conditions. The roots of olive trees could penetrate the limestone bedrock, which now lay close to the soil's eroding surface (Ponting 1991: 76).

By the time of Periclean Athens in the 400s BCE, the bulk of Attica's topsoil had washed away. As Plato explained in his famous account, in past centuries,

> the annual supply of rainfall was not lost, as it is at present, through being allowed to flow over a denuded surface into the sea, but was received by the country, in all its abundance, into her bosom, where she stored it in her impervious clay and so was able to discharge the drainage of the heights into the hollows in the form of springs and rivers with an abundant volume and a wide territorial distribution. The shrines that survive to the present day on the sites of extinct water

> supplies are evidence for the correctness of my present hypothesis. (in Claiborne 1970: 334)

When the land could no longer sustain its population by known methods, Athens began sending large companies of young people overseas to settle richer shores. These were "surplus people," whom the mother city could not or would not support. Perhaps their soil was eroded, or certain families claimed a lion's share of the village lands. Perhaps their family farms were too small to divide any further. Whatever the case, the results were similar. Members of the new generation grew up finding no place for themselves and made plans to go elsewhere. Their elders financed ships and provisions for them to sail away. So it has been in hundreds of overcrowded and depleted lands since. Perhaps nothing stronger can be said about the social effects of desertification.

On richer shores, the colonies grew like transplanted vines. They generally shipped their surplus food to their "mother polis" in Greece and received back the finished goods of Greek civilization. Step by unplanned step, the colonies of various cities grew into a tangled web of competing land claims and criss-crossing sea lanes. In times of war, the various Greek cities tried to cut off each other's food shipments with naval blockades, and colonial plantations or grain ships were vulnerable to attack. Without a strong navy, Athens could be cut off from its farming colonies and starved into submission. Thus the construction of wooden ships ate up much of the forest remaining near the Greek coastlines. So the Greeks sacrificed their homeland's capacity to regenerate itself—in order to fight for other lands.

As Athens grew dependent for food on its armed forces rather than its own soil, the professional army grew in political power. As Robert Claiborne (1970: 334) describes it, "Imprudent military adventures then dissipated much of the [state's] riches, and with them went much of the democracy—as shown ... by the condemnation of Socrates by a Committee on Un-Athenian activities." This was the Athenian democratic state which had held off the Spartans and the Persian empire, but then repeatedly fell under the rule of homegrown tyrants. By the 300s BCE, the political culture of Athens virtually duplicated that of its old enemies. Then, both Sparta and Athens were joined in a Persian-style empire under Philip II and his son, Alexander the Great. And these men were warlords on horseback from the Macedonian plains to the north.

The Mediterranean's Great Military Empires

When the Mycenaean Greeks invaded Crete (in the 1500s BCE), they redefined what the land and people of Crete were good for. The invaders decided what jobs and products were in demand. According to their values, certain traditional arts such as seal engraving, ivory carving and fresco painting fell from favor. The market was in weapons. According to Sinclair Hood (in Claiborne 1970: 327), "the inventiveness of the armorer ... remorselessly advances.... Swords are now better hafted and are shorter, adapted to cut as well as thrust, while elaborate bronze armor makes its appearance." In this economy, the sword and armourmakers were probably the best paid craftspeople.

With the progress of Greek military science, armies changed from bands of raiders to disciplined phalanxes of armored hoplites. Ships of adventurers evolved into armadas of galleys, sailing in formation. In such military units, individual fighters became cogs in a greater machine. Only the men of aristocratic families, who served as horsemen, remained freewheeling individualists in combat. Then, around 350 BCE, Philip II of Macedonia applied the principles of intensified war to horsemen as well as infantry. He turned his riders into phalanx-like units of cavalry, capable of charging like shock troops through the ranks of enemy soldiers.

The Greek King of Kings

In 338 BCE, Philip II turned his horse-army on the Greek states to the south. The generally fratricidal Greeks managed to unite against him, as they had against the Persians over a century before. In the climactic battle, Philip's teenaged son Alexander led a charge of two thousand horsemen through the Greek lines, cutting their phalanxes into fragments. This, evidently, was the best organized force of steppe horsemen to date. With it, Philip won a new kind of military control over Greece. This was not just another incursion of steppe warriors, but an Indo-European king building an empire in the Persian style of Darius the Great.

When Alexander inherited the Macedonian empire, the highest goal he could imagine was to take the place of the Persian king of kings. He therefore turned his armies to conquer the Persian empire, which comprised the whole semi-arid zone between Greece and India. In that direction also the country was mostly good grassland for horses. The empire Alexander achieved covered the whole Near East (except Arabia) with its African, Indian and European borderlands. This, for Alexander, was the entire "known world," that theater of the earth known and conquered by previous Indo-Aryan armies. Perhaps the Assyrian, Dorian and Indo-Aryan warlords of the past had each dreamed of a similar "universal" empire for themselves. The Persians had come closest to that goal, holding most of the region for centuries. Now the Macedonians and Greeks seized virtually all of it for a shorter time.

When all the warlords of Eurasia and the Near East were defeated by one fiercer than themselves, the result was a paradox. The kingdoms' feuding princes were cast down, and a single great emperor held a club over all. The resulting peace was so extensive, its like could not be recalled. Many believed that the unlimited greed and violence of the conqueror would lead to its opposite—a peaceful world society. They hoped that the free competition of military enterprise would lead to a monopoly on power by the final winner which would end the strife of the world. In Alexander's mind, the result of his great war would be a polis of the cosmos. And it might have, if the greatest warlord did not die like everyone else, leaving numerous followers fighting to take his place.

The Roots of Rome

The Romans were the next to approach Alexander's dream of one world, one warlord. But the Roman empire grew from deep inside Europe—out of the densest population belt, which then hugged the Mediterranean shore. With Rome, the idea that militarism has environmental causes meets one of its most interesting tests.

Italy before Rome was dotted with Greek colonies, including some of the largest cities in the world. But Rome was not one of these colonies. The Romans' language, Latin, was an Indo-European tongue, and Rome was an Old European Etruscan town seized by an Indo-European tribe. According to Livy's *History of Rome*, the Latins overthrew Etruscan rule because of the rape of Lucretia. One day, the poet explains, prince Sextus Tarquinius, the son of the local Etruscan king, came to visit the Latin family of Collatinus and his wife Lucretia. He stayed for the night and ended by making a beast of himself. After dinner, when all had gone to bed, young Sextus broke into Lucretia's room and raped her at knife point. This would be an outrage in any culture, but the Latin cultural response showed a particular moral logic as distinctive as a fingerprint.

As Sextus fled, Lucretia raised the alarm. The men of the offended family rushed to the scene. But before them, Lucretia seemed less angry at her attacker than ashamed for herself. She reportedly informed her husband, "Another man's traces, Collatinus, are in your bed." Then, as if the consequences were a foregone conclusion, she proceeded to kill herself. None of the men raised a hand to stop her. In their world, a woman was owned by her husband and was discarded if used by another. Their morality did not distinguish rape from adultery. The rape victim felt she must execute the penalty for adultery on herself—which in her culture was death. As a bride was accountable to the men of her family for her virginity, and might have to show them the blood on her nuptial sheets to prove it, so Lucretia had to prove her ultimate accountability to the *paterfamilia*. By her willingness to serve as her own executioner, she established her final obedience. The later Romans considered this an inspiring example for women.

After the rape victim was disposed of, there began a rather indiscriminate vengeance on the clan of the rapist. In the Latins' eyes, young Sextus had demolished the moral credibility of his father's ruling Tarquin clan. The Etruscans obviously were a pack of degenerate sensualists. No doubt their women were used to such immorality and would never think of dying for loyalty like Latin women would. In war council, the Latin men decided they could stomach no more of Etruscan rule. Their only acceptable choice was to seize Rome for themselves (Spivey and Stoddart 1990: 16). Livy's account is no doubt largely fiction and glosses over the early family ties between Etruscans and Latins, by which Collatinus was actually a blood relative of Sextus (Ward et al. 1999: 51). In Livy's patriotic retrospective, the revolt of 510 BCE was a struggle for Latin ethnic supremacy.

The Etruscans were either indigenous Old Italians or Old European migrants from elsewhere. They spoke a non-Indo-European language, and their families were matrilineal. The Etruscan kings of Rome were husbands of an hereditary line of queens. When their people wrote their names, they used a personal name and their mother's name. When the Latins translated Etruscan names into Latin, they added the father's names. According to Theompus, Etruscan women often had several lovers and raised the children regardless of who the father was (in Spivey and Stoddart 1990: 347–48). By Latin standards, they broke every rule in the book.

Even the partly Old European Greeks could be scandalized by Etruscan morals. According to inscriptions on Etruscan painted banquet scenes, the women portrayed were often wives of the feasting men. By contrast, in Greece and its Italian colonies, men commonly secluded their wives from such public revelry; instead, "professional ladies" called *heterai* provided female company at social events. This they considered far more ethical than the Etruscan practice of bringing wives and daughters into scenes of temptation.

Another divergence of morals concerns homosexuality. Nigel Spivey and Simon Stoddart (1990: 150) suggest that the Greek fashion of "higher love" between men reflected the low status of Greek women. If love between men was "higher," it implied that heterosexual love was lower. In other words, love between men and women was not love between equals. For men who felt this way, the stimulation of love with an equal might only be possible with another man. But Etruscan art seems to neglect the themes of homoerotic love. Perhaps the Etruscan playing field was more level, and Etruscan men felt women were every inch their match.

By contrast, the Latins upheld patriarchal traditions perhaps sterner than those of the Greeks. Latin custom gave fathers full powers over other family members, and the family "head" could exercise capital punishment on a disobedient child or wife. When a Latin child was born, it was placed at the father's feet. If he deemed the child too puny to be worthy of his lineage, he could order it killed. He could also reject the child for being an unwanted girl. Only if he picked the baby up and accepted it did the mother have permission to give it milk (Dudley 1970: 45).

The early Latins were one of several pastoral tribes that moved with the seasons between the coastal lowlands and the central mountains of Italy. Some of the other herding tribes were the Umbrians, Sabines, Marsi, Hernici and Samnites. All these groups tended to fight each other over pasturelands. They also had regular run-ins with coastal farming people such as the Etruscans. All these pastoral tribes spoke related Indo-European dialects and were probably related to the Apennine or Villanova people who appeared in Italy with horse-drawn chariots between 1300 and 900 BCE. Before that, speakers of "Early Italic" languages had been clustered in the northwest Balkans (Barraclough and Stone 1989: 84). Perhaps the Latins reached central Italy in the 700s BCE. The remains of a pastoral settlement are found on Palatine Hill in Rome dating to around 750 BCE. That would be just about the year, according to legend, when the father of the Roman people, Romulus, settled on the spot (Dudley 1970: 4–8).

In those days it was an economic law of life that when pastures were eaten up, pastoralists had to gain new grasslands in war (Dudley 1970: 25). In that case, every herdsman had to alternate between animal husbandry and hand-to-hand combat. War was a gamble, with close to all-or-nothing stakes. Even the winners seldom gained any lasting relief from their problems. All captured land had limited grass, and new enemies always stood beyond the last. Still, a tribal leader had little choice but to hope for one victory at a time.

The Heroic Age of Rome

When the Latins went to war, they often set the stage by forming alliances to tilt the odds in their favor. Some of their first alliances were with the highland Hernici and other Latin-speakers in the Alban hills. Every summer from 486 to 431 BCE, this "triple alliance" pushed against the surrounding nomadic tribes. Finally, they captured the mountain pass at Algidus, which stopped the Aequi from raiding down the Tiber to Rome (Dudley 1970: 21, 22). Such a victory might have assured the Latins sufficient land and security, but for them the law of life and the force of habit were one. Every conquest led to a new border with a new enemy. In fact, the Latin word *rival* meant "a neighbour who uses the same *rivus* (water source)," conveying the Latin assumption of hostility among neighbours (Hillel 1994: 36–37). It took thirty-six years (from 326 to 290 BCE) for the Latins to overpower the Samnites of the Liris River Valley to the south (Dudley 1970: 25). Each trial of strength could have only one winner.

As Rome slowly swallowed the western side of Italy, the Greek colonial cities tried to defend themselves, hiring experienced generals from Macedonia. Despite setbacks, the Romans conquered them one by one. By 266 BCE, Rome ruled the whole peninsula. A modern Italian might wonder how such an empire could be insufficient. But like a modern corporation that feels it must either grow or fall behind, Rome felt it must either win or lose with each new rival. The next neighbour beyond Italy was Carthage.

The three Punic wars between Rome and Carthage happened for many reasons, but it cannot have helped that many Romans saw the contest as inevitable. When that expectation became mutual, the rivalry passed beyond a matter of raiding, revenge, redrawing boundaries or saving face. It became total war, with the goal of each nation to exterminate the other from the face of the earth (Dudley 1970: 53). During the second war (of 218–201 BCE), Hannibal's African army was loose in Italy for fifteen years, trying its level best to scorch the land so completely that Rome's subjects would starve. Toynbee said the landscape of southern Italy never fully recovered (in Dudley 1970: 77). And the Romans later inflicted similar damage on Tunisia.

In the middle of the second war, the old Macedonian empire offered support to Carthage, and many ruling class Romans immediately concluded that the heirs of Alexander's empire must also be eliminated. After arranging a temporary truce with Carthage, the Romans turned on Greece. As usual, they first called for alliances, claiming to be liberators eager to help Athens and Corinth win freedom from Macedonia. Then the Roman army arrived with its nightmarish wealth of fresh combat experience. In three brief wars between 215 and 167 BCE, the Romans eliminated the Macedonian empire. After that, few were left able to oppose Rome in the whole Mediterranean world. The legions drove the Seleucids out of Turkey in 190 BCE. They took Spain as a captured Carthaginian colony. Egypt could be virtually adopted from the Greek Ptolemies. On the battlefield in Macedonia in 168 BCE, General Aemilius Paullus addressed his victorious men. His message was that the Romans must never rest from vigilance against all possible enemies:

> The successor of Alexander the Great, who attained the highest pinnacle of power, has in the space of a single hour fallen and been put beneath your feet.... Do you suppose that we ourselves have any guarantee from Fortune that will avail us against the attacks of time? Young men, abandon empty insolence and pride of victory. Humble yourselves as you confront the future, expecting a time when God shall launch against each one of you his jealous displeasure at your present prosperity. (in Dudley 1970: 87–88)

The Environmental Impact of Empire

In the Second Punic War, Rome conscripted hundreds of thousands of Italian farmers into the army. When the survivors returned home, many found their farms destroyed. Over much of southern and central Italy, the olive trees were cut and the farm buildings burned. The crops were torched and the animals slaughtered. Many of the veterans could not find their families. Those in the army had perhaps been in less danger than their women and children on the farm.

The prospect of returning to a dust bowl was not attractive for many veterans. They were mostly young men who had been soldiers most of their adult lives. Fighting was what they knew. Rather than start life over with almost nothing, many veterans chose to stay in the army. The empire still needed experienced soldiers, and the army still offered regular rations, camaraderie, adventure and the chance of supplementing pay with plunder. For a soldier from a small farm, it could easily be an economic improvement. Livy mentions a soldier by the name of Spurius Ligustinus who gave a speech in Rome about his twenty-two years of army service and thirty-one awards for courage. Ligustinus owned a tiny farm in the country on which he could have been an old-fashioned peasant, but through his army career he rose above all that. Ligustinus patriotically reported that his four boys were all looking forward to careers in the army (Dudley 1970: 78). It was the biggest business in the land.

The Roman government was concerned about the devastation of southern Italy. Obviously a great empire could not raise its expenses from depopulated, impoverished regions. The rulers offered northern Italian farmers larger farms in the south. Some families took the offer, but southern production still lagged far below prewar levels (Dudley 1970: 77). Therefore the government awarded large holdings of land to retiring army commanders or others capable of investing in commercial farming. In many areas, these big plantations replaced peasant family farms. The new plantation owners were often absentee landlords who treated their holdings as investments to support an urban lifestyle. For an absentee owner, the most maintenance-free use of land was ranching. Therefore, many new landowners surveyed their barren southern estates and decided to raise sheep and cows on the grass. To staff the ranches and plantations in the cheapest way, the owners brought in prisoners of war as slaves.

Between 200 and 150 BCE, about 250,000 slaves from the wars were brought into Italy and Sicily, most to work on the southern plantations and ranches. About 150,000 Molosians were sold by a Senate order. In 177 BCE, a flood of Sardinian prisoners hit the market. Then in 146, the entire surviving population of Carthage, reportedly 400,000 people, was sold to the highest Roman bidders. Donald Dudley (1970: 80) says, "The importation of slave labor was an enormous social disaster.... Immediately, it provided a new and sinister motive for Roman wars of aggression." Immediately also, the produce of slave plantations started to undercut the prices of food grown by family farms.

The slaves in southern Italy and Sicily formed an underclass of people whose self-interest lay in subverting the system in any way they could. While the bailiffs' backs were turned, bands of slaves became brigands. Stealing was no dishonour to those from whom the state had stolen everything. On many occasions, only the army managed to control these people. The slave revolts of Spartacus and the other "Servile Wars" were only the most enormous outbreaks. From 104 to 101 and 73 to 71 BCE, rebel slaves controlled most of rural Sicily (Ward et al. 1999: 162, 178–79; Dudley 1970: 80–81). The army had absorbed the original farmers and the empire used captured slaves to replace them. Now the army held the slaves to their tasks and exterminated any who rebelled.

By the 130s BCE, the social imbalance of this military economy reached a point of explosion. While the slave revolts escalated into the Servile Wars, the Roman Senate was tearing itself apart in heated debate over land reform. Tiberius Gracchus demanded a legal limit to the amount of land any one man could own. He proposed a maximum of five hundred *jugera* and argued that the state should redistribute any surplus land. This bill was actually passed into law, before Gracchus and hundreds of his supporters were murdered in a single night. This was partly a fight between ruling clans over power, and the land reform was not halted till 118 BCE. But when Marcus Livius Drusus tried to revive land reform in 91 BCE, he too was murdered (Grant 1992: 73–74). The slaughter of aristocratic reformers and rebel slaves bore a certain similarity. The level of violence needed to maintain the system had increased, but the military was able to meet the challenge.

During this period, a series of books emerged on plantation management. Cato's *De Agricultura* appeared about 160 BCE, to counsel farmers on cash-crop farming. Cato described the market opportunities for various crops, the most appropriate uses for slaves, and the latest methods of cultivation and animal care. He rated various crops in order of profitability as follows:

1. grapes
2. vegetables
3. willows
4. olive orchards
5. pasture for livestock
6. dryland grain crops
7. timber
8. coppice plantations (stands of trees from which one or two limbs at a time could be cut for firewood).

The grapes and vegetables brought the most money, but these crops required intensive management, access to water and nearby markets. Concerning the semi-arid plantations of the south, Cato's cost-benefit analysis indicated they were better used for pasture than grain crops. He also said cutting down whole trees for timber was more profitable than lopping off individual limbs at a rate the trees could sustain. Better yet—cutting down the forest for wood and then using the barren area for pasture would combine two uses of the land, each more profitable than a living forest (Dudley 1970: 79). These were the market values Cato would apply to all land the Romans could capture. Accordingly, dated cores of sediment from the lakes of central Italy show that pollen from the old oak forests faded away as the empire grew. The accumulation of silt in the lakes accelerated, and the lake levels fell (Spivey and Stoddart 1990: 32–33). The purpose for which the land existed had been redefined. As Dudley (1970: 87) points out, Cato was no rustic soul, speaking of family values or the good old days. His book "is written in the spirit of inhumanity that produces modern factory farming." The author, after all, was a patrician senator.

On large plantations the slaves appeared to work like free farmers, plowing, planting and weeding. Perhaps they seemed content, but they were not real farmers. A real farmer manages land and is responsible for its future fertility. A plantation slave's responsibilities were fewer than those of an assembly-line worker. As pay, the slaves had the right to eat some of the food they grew, as if existence was compensation enough. It would be a surprise if they cared at all for the land they worked. The plantation owners cared, but their usual concern was solely economic. They wished to keep a maximum spread between the profit of outputs and the expense of inputs. With such labor and management, many large plantations suffered serious soil erosion, and uncounted tons of topsoil silted up the ports of Paestrum, Ravenna and Ostia. The volume of erosion was so vast that it buried the beaches, pushing the coastline out to sea. By the late empire, the shores of western Italy stood one or two miles beyond the pre-Roman beaches (Spivey and Stoddart 1990: 24–25).

With so much topsoil flushed away, Italian agriculture was increasingly replaced by newer and larger plantations in the colonies. As the land distribution crisis racked Italy, the colonial plantations, or *latifundia*, of Spain and Africa started producing more, at less cost, than most Italian plantations. Even with the added costs of transportation, colonial wine and olive oil undercut Italian competition within Italy. So Rome grew ever more dependent on the colonies for its daily bread. By the 60s CE, dependence on African grain was so complete that Nero seized the most productive farmlands there for his imperial estates. Before that, according to some reports, a mere six landlords had control over most of the good land in Rome's African provinces. Nero either feared these men's potential stranglehold over the food supply or else reacted to actual blackmail from them. Whichever it was, Nero reportedly had them murdered and confiscated their estates (Dudley 1970: 213, 227–28). With that, he became the greatest absentee landlord on Earth.

In Italy, the first-century CE farmer-philosopher Columella blamed Italian estate owners for their woes. Italian farmers, he said, were ignorant concerning soil fertilization and needed to relearn basic human kindness toward their slaves, animals and fields. They should go back to the basics and learn first-hand from farmers of traditional experience, like Columella himself (Dudley 1970: 212–13). Other Romans also looked upon the depleted landscape and praised the old ways. Virgil, in his poetic work *Georgis*, spoke of farmers as the soul of the nation and hoped for a return of honour for their craft. With seemingly unrealistic sentimentality, he predicted that when farmers dealt fairly with the land and honoured its spirits again, then the good "reign of Saturn" would return: "That was how Etruria grew strong, how Rome became the fairest city in the world,... that indeed was the life Saturn led in the Golden Age, long before the trumpet had sounded for war, or swords groaned as they were sharpened on the anvil" (in Dudley 1970: 174–75).

"Paganism" in the Roman Empire

From the standpoint of the Middle Ages, "Roman religion" appeared to be a kind of monoculture before Christendom. The Romans, it seemed, all believed in a pantheon of gods and goddesses, with official priests and sacrificial altars. In orthodox Christian eyes, the gods and goddesses of Old Europe and those of the Indo-Europeans were all equally pagan. Actually though, most people in the Roman empire probably worshipped the deities of Old Europe. They still believed that farmers must keep good relations with the spirits of their fields. For them, Italy was as densely populated with spirits and deities as were the mythical countrysides of India or Ireland. The peasant's faith went straight back to neolithic worship for the female earth.

The official state cult of Rome, however, was of Jupiter, head of the gods—

and he was worshipped at the altar of victory in war. Triumphal processions of soldiers returning from their campaigns offered him tribute from their plunder. He was, once again, the Indo-European lord of thunder and lightning. Those who displeased him, he struck down with a bolt. Brave warriors who worshipped him became like bolts of lightning themselves, flashing forth in service to god and country. The official Roman view of life was powerfully simple. The rulers ruled, the subjects served and force decided who was which. Any fool could see that society would never yield its wealth to Rome unless forced. Compulsion was the law of life. To believe otherwise was to think as a child.

Within Roman society, the Old European and Indo-European traditions co-existed on different social planes. Among the peasants, the Old European ethic lived on. In the aristocracy and army, the cult of Jupiter and the goddess of winged victory prevailed. So long as each class left the other to its beliefs, there was a peace of sorts. When the Romans conquered new areas, such as France and Britain, they tried to honour the local religions as legitimate, but subordinate to the imperial cult. Local nature religions, after all, were appropriate for farmers and woodsmen. The Romans even recognized that most deities of northwestern Europe were the same as those of southern Europe, only with different names. This, it might seem, was real tolerance. The Romans required only that their imperial cult should be the crowning glory of local tradition. If that was accepted, all was peaceful.

But if local religious leaders, such as the Druids, rejected this, then the famous Roman tolerance tended to suddenly evaporate. The Roman army was unleashed to hunt down local cult leaders. The Druids' crime was to consider their religion at least equal to that of the conquerors. In that case, the two cults did not co-exist, one beneath the other, but conflicted directly as two ruling-class cults within the same territory. To enforce cultural control, Roman authorities outlawed all unauthorized fortune telling, Chaldean astrology, magic and necromancy. The rulers feared the influence of rebel fortune-tellers and tried to monopolize all prophecy in the hands of loyal state priests (Smith 1976: 10).

Among Roman farmers, the most sacred symbols of life included the mother with child, the full ear of grain and the bull. Many Italian peasants continued to worship the Great Mother Isis long after the empire was officially Christian. Later, of course, her name was changed to Mary, the Black Madonna. For other folk traditions and mystery cults, the cycles of grain and seed told the story of life. As for the bull, ancient people from Egypt to Spain saw this beast as the embodiment of virile potency. Villagers honoured bulls, festooned them with flowers and paraded them in seasonal festivals. Sometimes people played with bulls in the arena like the old Cretans, or ran with them through the streets, hoping to absorb some of the great creatures' sacred powers.

But under the Roman empire, bull worship took on new meaning. A new mystery cult from Persia spread in Roman society, especially in the army. This cult of Mithras made its central symbol a priest sacrificing a bull. David

Attenborough (1987: 111) says, "Although the bull was still acknowledged to be the fount of fertility, a human figure has now been elevated to stand alongside it and force it to release its power." The priest Mithras played a cosmic role, which many soldiers could identify with. The common soldiers too were heroes who had forced the sacred bulls of Egypt, Greece and Spain to give over their power. In Mithraic ritual, wonder at the power and fertility of nature was replaced by an idealization of the one who overcomes nature. The common soldiers were invited to share in that cosmic victory over the world.

Attenborough finds it no accident that the Romans converted a version of Mithraic ritual into popular entertainment. And perhaps no image of ancient Rome remains so firmly in the world's memory as the mass butchery of animals in circus arenas. Of course, major sports events always contain elements of a morality play. They capture the public imagination because they seem to dramatize the way life is. And the Roman circuses blatantly displayed a cosmic drama over and over, as if the thrill of it could never end. In this drama, a hero confronts the powers of bestial nature and overcomes them in the most literal way—not by taming them or communing with their spirits, but by killing them. To enact this story, with thousands of variations ad nauseam, required hundreds of thousands of beasts. To catch them, whole units of the Roman army were assigned hunting duties. In North Africa the army waged a kind of war on the animal kingdom for sacrificial slaves till it virtually exterminated the non-human population (Attenborough 1987: 113).

Many historians say the animal and gladiator games were a great diversion from the real social issues of Rome. Like modern sports spectacles, they were inherently passive and non-political: a spectator could not affect the outcome of a game one way or another. But perhaps these games showed another kind of diversion as well. At the empire's high noon, after Rome had subdued almost every enemy in the Mediterranean world, the role of the army changed. For the most part it did guard duty on distant frontiers. After numerous civil wars between contending Roman warlords, one general—the great Octavian Augustus—emerged after 27 BCE to rule them all. He reduced the government's expense, and risk of military coup, by demobilizing forty-two out of seventy legions. He settled 300,000 veterans as farmers on colonial land. It was during the following Pax Romana that blood sport in the ring reached a height of popularity. Perhaps these games served as a substitute for war, among people who were losing their appetite for real war. The warlike Latins had intermarried with the more peaceful descendants of Old Europeans for centuries. Meanwhile, out on the eastern borders, so-called barbarians from the steppes began filling the ranks of the frontier armies. And by the standards of these mercenaries from harsher lands, the Romans had gone soft.

The Army's Victory over the Romans

In 107 BCE, a new man to Rome's inner circle, Gaius Marius, became consul of the empire. His qualification for office was his hero status in a North African war and his popularity with the troops. As consul he repaid his men's loyalty with generous grants of African land. He removed any property-holding qualification for army service, so that landless men flocked to Marius in gratitude. The assembly of arms-bearing men (the *Comitia Centuriata*) elected him to five consecutive consulships (104–100 BCE), violating the tradition that a consul must rule for one year only and then not again for at least ten years (Ward et al. 1999: 161–62). This route to power was so natural in a military economy that it became prototypical for the whole future of Rome. In the new politics of international empire, soldiers looked to their leaders for rewards, and the generals looked to their troops for support as they stepped into politics. Dudley (1970: 98–99) says this "made for a *clientela* of a new type.... After Marius, nobody won supreme power in Rome without the army to back him."

If the most respected commander in the army regularly rose to be emperor, then Rome was a kind of republic with the ruler elected by soldiers only. But after the conquest of the whole Mediterranean world, the army was too far-flung for centralized leadership. Each main theater of operations, such as the Rhine, the Danube, the Near East or North Africa, required its own resident commander-in-chief. Thus, each major general had his own domain, tax base and army, and each was a potential candidate for emperor. If more than one theater-commander at a time aspired to ultimate power, the soldiers supporting different candidates would cast votes by killing each other.

The dictatorship of General Sulla (82–79 BCE) illustrated the growing split between Rome and its army. Sulla was a man of the troops. He came from an old noble family, but one which had long been out of influence. He was backed by soldiers but was treated as an outsider by the aristocrats of Rome. To enforce military-style respect from these civilians, Sulla simply murdered any patricians who criticized him (about 2,700 nobles), confiscated much of their wealth and used the money to reward his loyal troops (Ward et al. 1999: 169–73).

A few decades later, three famous generals contended in a classic power play. From Gaul and the Rhine frontier came Julius Caesar. Commanding the troops of the Near East was the mighty Pompey. On the Danube frontier stood the formidable general Crassus. These three, at first, made a show of ruling jointly, calling themselves the First Triumvirate. Then Crassus died in a war with Parthia. That could happen to the best of warlords. Julius Caesar marched his western army to Rome and proclaimed himself dictator. His opponents, however, killed him. The empire then exploded into civil war between generals Pompey, Mark Antony and Octavian.

Civil wars were a recurring headache in Roman "politics." Relief came periodically, usually when one warlord won decisively and temporarily united all the armies behind him. This is what Octavian did. He beat his opponents more decisively than any Roman warlord before him, or after. He then became the great Augustus Caesar, slightly deified, with power so well established that he scarcely needed to show his claws again. He united the army by decommissioning all the legions which had backed his enemies and paying them generously with other people's land. Seldom has the phrase "reward thy enemies" been put to better use. To symbolize his universal rule, Augustus brought the obelisk of solar worship from Heliopolis in Egypt to Rome. There it became an ornament of the emperor, no longer symbolizing the power of the sun over all the earth, but instead the power of the sun king (Dudley 1970: 248).

During the reigns of good emperors, who won loyalty from the whole army, the benefits of peace were obvious. Across the whole empire common laws, common currency and open lines of communication prevailed. The roads were first built for the armies and then happened to serve trade as well. On the roads and seas, merchants no longer faced a chaotic maze of local chiefdoms, each demanding new tariffs and bribes. The empire's vastness inspired travel and broad learning. The Greek scholar Posidinius (135–50 BCE) traveled over the Mediterranean world, studying the relations among plants, animals, climates, landscapes and human cultures. On the Atlantic coast of Iberia he worked out a theory of global tides in accordance with the cycles of the moon. His ethnographic studies emphasized the harmful effects of slavery and military oppression in the empire. He hoped society would learn from these mistakes and move onward to a *societas humana*, with equal rights for all (Dudley 1970: 131–32). Such were the economic and cultural flowers which grew in conditions of relative peace. But peace itself seemed to depend on the relatively unquestioned power of a *supremo*.

Naturally, the *supremo* sometimes took his fatherly role over society too seriously. Octavian Augustus felt he must promote social morality and control sex. He made adultery a felony in the criminal code, where before it had been a private matter settled between families. Accentuating the positive, Augustus offered privileges to good Roman families who bore three or more children. For childless couples and single people, he imposed financial penalties. According to Tacitus, the main effect of the sexual morality laws was to stimulate a crop of self-appointed spies who reported on other people's sex lives to the government (in Dudley 1970: 170).

After Nero died in 68 CE, once again the frontier generals marched their armies towards Rome for a military-style "election." Vespasian's Danube and Syrian armies from the eastern fronts prevailed, and a new ruling dynasty (the Flavians, of 69–96 CE) was born. This proved a taste of the long-range future. In 193 CE, when another furious war of succession broke out, the head of the Danube army, Septimius Severus, emerged victorious. The Danube army had grown into the largest regional force because it faced Europe's greatest threat—the slowly rising tide of migrants off the eastern storm plain. For the next three hundred years, commanders of the eastern frontier forces would normally control the empire.

To repay his men's support, Severus raised base pay for the army by a third. In addition, he increased the total manpower of the army to over 400,000. Many of the new recruits came from regions around the Danube frontier. As on the Great Wall of China, many recruits were direct migrants from the steppes, who came because they heard of regular jobs for fighting men. Such troops were often relatives of the unemployed migrants they were supposed to fight.

If the troops were increasingly frontier people, and those of talent rose through the ranks, then the commanders were soon easterners as well. The first of these native eastern generals to become emperor was Maximinus Thrax (r. 235–238), from Thrace. Thrax knew very well who he depended on for his position—he raised the soldiers' pay another 50 percent. With this, the military budget began to crush the empire (Dudley 1970: 259). Expense and income, of course, are always relative terms. What most Roman civilians saw as an expense was counted as a gain by the army. What, after all, was more valuable than what the army provided? Did not the empire depend for its very life on the army? What else besides the army of the Danube stood between Rome and the incursions of steppe barbarians? Thrax imposed his new military budget and the civilian population paid. But even this increased money and manpower soon proved inadequate. By 271 CE, the pressure from the storm plain was so great that the army abandoned Dacia province, retreating across the Danube to use the great river as a moat for a new defensive line.

After Thrax, the trend toward rule by eastern soldiers grew. Philip the Arab (244–249) came from the Mesopotamian front. Then came Claudius Gothicus (268–270), Aurelian (270–275), Probus (276–280), Carus and sons (283–284) and Diocletian (284–305), all of whom haled from Illyria on the Danube frontier. Diocletian tried to curb the recurring problem of rebellions by different army groups by forming a Tetrarchy (or College of Four Emperors) to rule over the empire's various quarters. Diocletian picked the other three emperors himself. All were from Illyria province and all were career officers in Diocletian's Danube army. Such a tight-knit team made for a more united government, at least temporarily (Dudley 1970: 264–65). But it no longer seemed that Rome was ruled from Rome; it was now ruled from the headquarters of the eastern front. The path to power no longer ran through Rome but through the eastern army camps, which were full of immigrants. The new ruling classes were largely easterners, not Latins, patricians or Italians. Italian Romans increasingly spoke of their rulers as unsophisticated barbarians. In turn, the emperors increasingly avoided Rome. They stayed with their men in the provinces, where they were understood and respected (Smith 1976: 35–37). Eventually, the new capital of Constantinople rose as the headquarters of the eastern command.

The man who moved the Roman capital to Constantinople, Emperor Constantine (r. 312–337), also had a primary hand in de-Romanizing the army. When faced with a migrating force of "300,000" Sarmatians at the Danube frontier, Constantine decided to open the gates and let them in. Some, rather racist, critics claimed that Constantine was bent on barbarizing the empire. But Constantine felt these Sarmatians would strengthen his hand. They would be grateful subjects and a fine new pool of army recruits. Their self-interest would lie in protecting their land grants and army jobs from too many more immigrants like themselves (Smith 1976: 66). Many ethnic Romans started speaking of the army as a force of foreigners for protecting the interests of foreigners. Army immigration policy was beyond civilian control. It was a policy of deals, bribes and threats. Each case was a special arrangement between semi-barbarian border guards and new hordes of steppe migrants. Largely by recruiting such migrants, Constantine and his sons expanded the army to half a million men. The main expansion was in cavalry units. Rome's chief enemies were steppe warriors on horseback, and men who knew this kind of war were needed on the Roman side (Dudley 1970: 272). By this point, the ethnic makeup of the forces had changed almost beyond recognition. Or perhaps the army was simply returning to its real roots, becoming a force of Indo-European warriors.

The Roman generals next started forming military alliances against each other, using foreign tribes as allies. For example, in 360, Emperor Constantius offered a deal to the Germanic Alamanni tribe. Constantius wished to eliminate his cousin Julian, the commander of Roman forces in Gaul. So, Constantius (who was in Constantinople) sent a letter to the Alamanni promising them they could keep any land in Gaul which they could seize from Julian (Smith 1976: 101). This particular attempt failed, because Julian's men intercepted the letter, but other such deals succeeded in the future. Soon, the eastern emperors were freely supplementing their forces by hiring whole migrant tribes *en masse*, as in the hiring of Alaric's Gothic horde.

By the late 300s, an ugly rift separated the empire's civilian and military populations. Many civilians viewed the army as a mass of greedy barbarians. Perhaps the civilian population of Constantinople felt this tension most directly. This city held a growing population of Roman merchants and bureaucrats, catering to the nearby eastern armies. As the great Hunnic invasions approached, masses of refugees filled the border camps, and the residents of Constantinople grew afraid. The refugees tended to join the army, but this was small comfort to the city dwellers. To them, the refugee troops looked disturbingly like Huns themselves. What, after all, was to stop the whole Hun horde from marching in, claiming refugees status and taking over the army? Yet the city would have no defense at all without its largely barbarian guards. In theory, the city residents should have been able to distinguish between hostile steppe migrants and those they employed to protect Constantinople. But given the circumstances, this was too much to ask.

In 373, the entire Gothic nation fled the Ukraine before the Huns. After much bargaining, pleading and threatening, the Romans let perhaps several hundred thousand of them cross the river into refugee camps south of the Danube (Smith 1976: 141). Some Goth leaders then claimed that Roman soldiers had stolen their weapons and food; many Goths therefore felt justified in raiding their hosts for plunder. Alarmingly, some were reported to be cooperating with the hostile Huns and Alans to the northwest.

In Constantinople, panic and rage over the immigration policy reached a boil. Emperor Valens had to break off his war with Persia and come back to deal with the Goth uproar. When he arrived in Constantinople, the crowds jeered him as a lackey of the barbarians. Some said he couldn't stand up to barbarian demands because he was not a man at all. Valens told them he was man enough to teach the Goths their place, and he was on his way to do just that. But he swore that when he got back he was going to flatten Constantinople. He might have done it too, but he never did get back.

The area south of the Danube was so chaotic that Valens' army met numerous delegations of Goths, each claiming authority to speak for at least some of their people. At the same time, different groups of Goths were raiding and killing some of Valens' men. The emperor set up a camp at Hadrianopolis to negotiate with some Gothic leaders. While they were talking, an army of Alan and Ostrogoth horsemen came charging over the hills, wiped out the royal bodyguard and killed the emperor (Smith 1976: 142). As far as most civilians could tell, the barbarian problem was completely out of hand.

It took a man like Emperor Theodosius to make a deal with the Goths. He hired them as mercenaries to fight rebellious Roman generals in western Europe. Once the Goths were offered a contract, they became a major military asset and helped Theodosius keep western Europe under the eastern army's control. That was how a Gothic chieftain named Alaric came to be one of the Roman army's main subcontractors. Alaric was determined to prove himself the emperor's right-hand man. He threw his troops into the front lines, taking terrible casualties in the west. But for all their sacrifice and loyalty, the Goths found themselves treated as second-class warriors, not on the regular army payroll. For the regular Roman troops, plunder was only a supplement to wages. But the Goths were *expected* to be raiders, and plunder was supposed to be their *primary* pay. That way, Theodosius could externalize some of his military expenses.

Alaric's men felt cheated. They rode to Constantinople in 396 and camped in protest under the city walls. To pacify the dangerous horde, the government coughed up a sum of gold. The Goths took their money and, seeing they were not welcome at the city gates, went away. Riding toward the setting sun, they remained a fully mobilized combat force. While waiting for the next war, they supported themselves according to the terms of their employment. They systematically pillaged the cities of Greece, unopposed by any other unit of the Roman army (Smith 1976: 188).

By 400, popular outrage over "barbarization" was so great that the Roman consul Aurelian tried to order all "barbarian-born" troops to resign from the upper ranks of the army. However, General Stilicho, the commander-in-chief of the western armies, who was a Vandal, did not approve of this order. Stilicho sent his men to brush off Aurelian like a fly and proclaim equality for barbarian-born soldiers. In the east, Constantinople reopened its gates to the Gothic troops who were camped outside as guards. Should they not be admitted as soldiers in service of the emperor? By what prejudice had they been excluded? In they came, mixing freely with the city residents. Many citizens of Constantinople expected to find the Goths disgusting, and did so. Familiarity was doomed to breed contempt. When the Goths held their own Arian-style Christian worship services in the city center, many orthodox Christian Romans felt their city was being defiled by both barbarism and heresy. Next, a comet appeared in the sky, and some orthodox Romans proclaimed it a sign from heaven that barbarians were intolerable to the Lord. In the ensuing race riot of July 12, 400, crowds of Romans raced through the streets, trapping and killing every Gothic soldier they could find.

The army units camped outside the gates could have taken revenge and flattened the empire's capital. Instead, they simply turned their backs on the city. As the citizens of Constantinople watched from the walls, the army broke camp and marched away, leaving the city almost defenseless. If the Huns or Persians heard about this, they would come as fast as their horses could run. Overcome by fear, the city sent a delegation of repentant civic and church leaders hurrying after the offended Gothic troops. The delegates offered apologies and begged the mercenaries to come back and accept their pay. However, Alaric took his troops to Italy, looking for more grateful people to guard (Smith 1976: 200).

In 405, Alaric got his golden opportunity. That year an estimated 200,000 Ostrogoth warriors poured over the eastern frontier and rode for Italy. This was a real steppe horde, the likes of Kurgan Wave III. They were first-generation invaders, fresh off the steppes of Inner Asia. They could not yet imagine anything better to do with Europe than plunder it and cart off the loot. Alaric moved in to become the savior of Rome. Along with regular units under Stilicho, Alaric's Goths chased the plunder-laden raiders away (Smith 1976: 203). But did the Romans show any gratitude? Did Alaric receive anything like the recognition he deserved? Once again, he got nothing of the sort. The Roman aristocrats treated him as a dangerous boor. They distrusted his motives for military service, just because he was a steppe warlord. Alaric knew very well that his background qualified him eminently to meet the military needs of Rome. He knew his ambitions were no different from those of all previous Roman army commanders, but in him these same ambitions were treated as vices rather than virtues. Alaric finally got the message. He was indeed both a hero of the army and an enemy of the Roman people. The army lived off the people, and it always had. The confusion of goals in his mind disappeared. In 410, he and his men rode to Rome, broke through the gates and sacked the city to their heart's content.

Chapter 12

The Age of Desert Civilization in Europe

AFTER 300 CE, THE STREAM OF MIGRANTS FROM THE EASTERN STEPPES became a torrent, and finally the greatest flood of invaders in European history. The numbers of Huns, Avars, Ostrogoths, Franks and Vandals were exaggerated by contemporary writers, and many Europeans were confused between one steppe tribe and the next. Some Romans thought the migrants were really all one tribe and only called themselves by many names to confuse others. A detailed account of all the hundreds of migrant groups which fought and displaced one another is probably impossible to reconstruct. As N.H.H. Sitwell (1984: 50) says, "A single tribe cannot be systematically traced through history any more than a single cloud can be systematically traced through a thunderstorm." Still, the dominant tribes kept their own accounts of exploits and the genealogies of their royal families. The patrilineal heads of tribes moving west in this great influx of people known as the *Volkerwanderung* became the "founding" families and nobility of almost every nation in Europe.

Climatic Factors in the Great Migrations

According to some climatologists, the *Volkerwanderung* and the "fall of Rome" happened during a climatic downturn, as the weather grew colder and dryer. The Inner Asian climate intensified and encroached into Atlantic Europe. In general, European weather has seemed to oscillate between warm and wet Atlantic periods and cold and dry Inner Asian periods. The effect of cold and dry climate swings on old Inner Asia is fairly clear. The growing seasons shortened, drought blighted the summer grass, and the long strings of good years, in which the herds grew fat and multiplied, drew to an end. Past successes in expanding the herds turned into a liability, as more animals ate the grass, which now grew more slowly. Whole tribes of pastoral people struck camp and tried to follow the rain. In the great migrations, the men rode ahead on horseback to secure the way and drive the herds, while the women and children came behind in wagon trains. As the animals quickly spoiled dry pastures, the tribes were constantly on the move (Attenborough 1987: 122). So the steppe people converged by the hundreds of thousands on the eastern approaches of Europe. As they crowded each other, disputes over grass and water flared into wars of blood vengeance.

David Attenborough (1987: 122) says the cruelty of the steppe people in war shocked even the Romans: "After a fight, they flayed their victims and slung the bloody skins over their horses as trophies." The warriors hoped to win through intimidation rather than attrition. A truly mighty hero should have to simply growl and the enemy would flee. Unfortunately, the real boys and men of these tribes were seldom so horrifying and had to do far more than growl to drive each other away. Their ferocity was a mask over their desperation. Those who rushed into battle were running from something they feared even more. So the hordes of the Avars who attacked Europe in the 600s CE were reportedly fleeing from man-eating griffins on the steppes behind them (Golden 1990: 257). When such fleeing migrants fought each other, the winners and losers usually pushed each other still further west, in a general flow toward the lands of green.

The earlier Kurgan Waves, or migrations of Indo-Europeans, had seemed to originate from the western half of Inner Asia. But in the *Volkerwanderung*, huge tribes such as the Huns, Avars and Turks arrived all the way from north of China. And as these speakers of Altaic or Turkish languages moved westward, they pushed perhaps most tribes between Mongolia and the Ukraine before them. During this period, the populations of Inner Asia seemed to move toward Europe *en masse*. And as they rolled into eastern Europe, previous migrants such as the Franks were pushed forward into western Europe.

The *Volkerwanderung*

The steppe barbarians, according to Roman historian Ammianus Marcellinus,

> are no good at all at fighting on foot, but perfectly at home on their tough and ugly horses. They even ride side-saddle when they relieve themselves. It is on horseback that each of these people buys and sells, eats and drinks and, bent across the neck of his stead, takes a deep sleep. (Attenborough 1987: 122)

To Marcellinus it would probably have seemed absurd that these horsemen would ever become the nobility of Europe. Imagine his surprise if he could have moved ahead in time to the 1800s, to a university-town salon in France, Germany or England, where upper-class ladies and gentlemen discussed theories of racial purity. At first he might have felt at home. Then he would realize that when these people spoke of racially pure Europeans, they meant people of Indo-Aryan blood, related to the ruling classes of *post*-Roman Europe.

According to popular history, the driving force behind the *Volkerwanderung* was the great Hunnic horde. Marcellinus described these people as "an unknown race from the far ends of the earth [who] moved like an avalanche and crushed everything they encountered on the way" (Gimbutas 1971: 75). Like the Mongols, they rode in units of tens, hundreds and thousands. Their economy had changed from mainly pastoral to mainly military. "There is no indication," Denis Sinor (1990c: 197) says, "of any Hun desire to change a pattern of life in which warfare provided the principal means of income"; their economy required "the constant lengthening of the action radius—no place can be plundered indefinitely." On their way from eastern Asia they raided deep into Persia, Armenia and the Near East. They devastated the partially agrarian tribes of the Ukraine and then arrived at the borders of the Roman empire. In Europe the Huns conducted major raids, demonstrating their famous fighting abilities. Various Roman officials promptly offered them jobs as mercenaries. The Huns then served for pay or plunder in various Roman conflicts, or in wars to control other refugee tribes which they had pushed into Europe in the first place.

The ferocity of Hun soldiers pleased their employers. But along with the soldiering went merciless looting, which outraged every local population. With each mercenary contract they served, the Huns made more sworn enemies. And, as different groups of Huns took contracts for different local rulers, the Hun confederacy lost its unity. Finally, after Attila died, the Huns' many enemies turned on them from all sides, chasing their scattered units back to the steppes. Later, some Hunnic chieftains regrouped on the plains and became part of the Bulgar migration, which established the kingdom of Bulgaria.

The Ostrogoths, Visigoths and Vandals were all steppe tribes that crowded into Europe partly due to pressure from the Huns. Many of these people were already farming villagers, whose ancestors had settled as farmers in Central Europe long ago. Around the 100s CE, many Gothic farmers migrated east again, to cultivate relatively empty land in Romania and the Ukraine. In times of peace they were farmers encroaching on the nomads' domain. They slowly filled the Prut, Southern Bug and Dniester valleys with farmstead villages (Gimbutas 1971: 70–71). Even the local steppe Scythian and Sarmatian peoples increasingly turned to settled farming for several hundred years. Then in the late 300s, the plains farmers came under attack. Like the Cucuteni farmers of four thousand years before, the villagers found themselves stationary targets for ever larger units of mounted raiders. First the Ostrogoths retreated, pushing the Visigoths before them. Then in 371, the Gothic kingdom of the steppes dissolved. The Visigoths east of the Dniester broke and ran, trying to ford the river by night. The Huns slaughtered almost everyone they caught on the east bank. The uprooted settlers converged on the lower Danube, carrying little save their weapons and some farm animals. In becoming refugees, they reverted to what their ancestors were—Indo-European armed tribes on the move, trying to survive by any means necessary.

The 200,000 or so Goths who crossed the Danube stretched the upper limits of immigration Rome could tolerate. But still the steppe people kept coming. Next, new groups of Sarmatians and Quadi appeared on Rome's frontier. They appealed to the border generals with pleas and threats—insisting it was impossible for them to survive on the steppes. Roman emperor Valentinian decided to see the situation himself. His armed column rode through a seemingly empty countryside northeast of the Danube. The migrants' stories, Valentinian believed, were fabricated lies. He accused the Quadi and Sarmatian leaders of being the very raiders they claimed to flee. They were legions of thieves, trying to worm their way past Rome's defenses by the novel means of sheer deceit. The emperor found his predicament impossible. His responsibility was to stop these hundreds of thousands of invaders from swamping the empire, but he couldn't keep them all out, or control them all if he let them in, or believe anything they said. With all else failing, Valentinian tried to overpower the tribal leaders by sheer verbal abuse—as if he could puff himself up and make himself more fearsome than the whole Hun horde. But no matter how loudly he roared, nothing would make the migrants go back. Valentinian screamed at them till a blood vessel burst in his brain (Smith 1976: 140–41).

In the following decades, the Goths, Visigoths, Ostrogoths, Sarmatians, Longobards and Vandals all moved deep into Europe. First they filled the ranks of the Roman army or served as private mercenaries. Later they took political as well as military power over the western empire. They spread through Germany to the Rhine, pushing the Franks into France. The Ostrogoths and Lombards moved through eastern Europe into Italy. In 406 the Alans, Vandals and Sueves crossed the frozen Rhine into Gaul. The Visigoths settled in southern France and Spain; and the Old European Basques retreated to the hills of northern Spain. The ethnic map of Europe as we know it was falling into place.

Constantius, a co-emperor with Honourius in the early 400s, tried to solve the problem of settling these intruders without causing another war. His solution was to allow the Gothic and other armed tribes to serve as military overlords for the farming estates. In exchange for protecting the estate operators from further disorder, these strongmen would receive proprietary rights, with one-third of the estate's income as rent (Smith 1976: 223–24). So the invaders of Europe were recognized as overlords, much as nomadic tribes were overlords over oasis farmers in the Middle East. Actually, the new overlords didn't really need recognition from Constantius; the Visigoths and Burgundians in Gaul simply commandeered one-third of all the Roman estates they conquered, and the Ostrogoths did likewise in Italy (Ward et al. 1999). So the medieval social order took shape, including the ethnic composition of its classes. Until this time many European villages had retained the old communal ownership of land, sharing out plots among local families according to need. But now a new set of military retainers gained lordship of the land. The Old European farmers became peasants belonging to an estate, while armed men on horseback oversaw and "protected" them. At night these noble warriors retired to their forts, which eventually grew into castles.

In 493, Theodoric the Great, king of the Ostrogoths, took the mantle as Roman magistrate in Italy. To validate this inheritance, it remained only for the Gothic historian Cassiodorus to write an epic history of the Goths. He gave his people a unified patrilineage to rival the biblical Hebrews and surpass the antiquity of Rome. According to Cassiodorus, the Goths had worshipped Mars, the god of warriors, long before Rome was born and had educated Rome in the

Martian cult of victory. Herwig Wolfram (1985: 22, 38) explains the enormous political and psychological importance of this saga: "Preeminence over the Romans and also the innate community between them and the Goths was thereby established." Almost every migrant ruler had a similar ethnic saga composed. The invaders from the steppes thereby proclaimed themselves the true custodians of European civilization.

After the Hunnic horde, the western Slavs moved into Poland and Bohemia. The southern Slavs pushed into the Balkans, settled there and raided what was left of Greece. The land was relatively vacant, since the Huns had driven many people from the eastern European farmlands. In the late 500s, a new steppe horde known as the Avars approached, probably originating from around the Gobi Desert. They invaded the Balkans, pushing more Slavic people before them. The Avars were equipped with stirrups on their horses, and Persian-style light riding armor. They forced the Wendic Slavs to serve as slave-troops in their front lines. Finally, a man called Samo rallied his people to successful revolt, and the southern Slavs escaped Avar control (Szadeczky-Kardoss 1990: 212–13). Meanwhile, other Slavic people fled the Avars by moving north into the cold forests of Russia. Marija Gimbutas (1971: 86) explains, "In times of attack everyone fled to the forests and swamps, thus luring the enemy into the treacherous forest as a trap."

The migrating steppe tribes tended to pile up in Germany, steadily raising the pressure on the Rhine frontier. This was the premier age when throngs of Indo-Aryan warriors dominated Germany. Later, such warlords and their ancestors would claim themselves to be "true" or "pure" Germans. Germany drew its name (Alimania) from a major confederation of these tribes which fought the Romans—the Alimanian alliance, or the alliance of all (real) men.

The pressure in Germany next produced an Anglo-Saxon invasion over the sea to England. After the resulting series of wars, the Old Europeans and Celts of the British Isles retained independence only in the hinterlands of Wales, Scotland and Ireland. This was the heroic age in Britain, but it produced two different lines of heroic sagas. The aggressive Anglo-Saxons and Jutes produced their odes to warrior thanes, such as *Beowulf*. The Celts and Old Europeans produced another body of legend, which could perhaps only arise on a frontier where the *Volkerwanderung* came to a halt. The original Arthurian legends concerned a Welsh king holding back swarms of Anglo-Saxon attackers in the 400s. The later Arthurian sagas presume a stance of protection for non-warriors, and their heroes do not glorify conquest or plunder. In their moral universe, might and right are not the same. But the strongest sign of Old European influence in these legends is that women are treated almost as goddesses.

When the smoke of the *Volkerwanderung* cleared, the ethnographic map of Europe stood roughly as we know it. Little remained to complete the picture besides the Magyar incursion in the 800s to establish Hungary.

The Near Eastern Cultural Conquest of Europe

During the late Roman empire, according to Donald Dudley (1970: 247), "The most striking advances were made by the Oriental religions, especially those of the Great Mother, Cybele of Asia, of Isis and Osiris from Egypt, and of Mithras from Persia." In the time of Augustus the Roman patricians had regarded all these faiths as dubious "foreign cults." Later, Judaism, Christianity and Manichaeism would emerge as yet more influential foreign faiths. And by the year 400, these various "Oriental religions" virtually had overrun the Roman empire.

After Rome conquered much of the Near East and Egypt, the religions of those lands became indigenous to the empire. Their believers were members of Roman society and increasingly acted the part. St. Paul, for example, used his rights as a Roman citizen to organize Christian churches all over the eastern Mediterranean. But why did these religions gain pre-eminence over the ancient religions of Europe? Christian historians point to the spiritual inadequacy of paganism and to the obvious superiority of the true faith. But those who do not find this obvious require more than a theological claim for an answer.

All historians describe the late Roman era as a time of mounting hardship and insecurity. According to John Holland Smith (1976: 3), the trials of daily life in the third and forth centuries were neither just "a challenge nor even a background to daily events fatalistically accepted; they were a mountainous burden, so that even free men felt themselves to be the oppressed slaves of fate." The extractions of the military grew so burdensome that many people abandoned their fields or workshops to escape the tax collectors. The rulers responded by legally binding farmers to their land and workers to their father's vocations. This would keep the cows in place to be milked. In centuries past, the army had won spoils of victory which benefited the Roman people. Now the standing army grew at the expense of almost every civilian. Naturally, the Roman cult of victory lost appeal in the eyes of the ruled.

Civilians in the vast empire tended to look for salvation in cultural traditions kindred to their own. Italian farmers adopted the Egyptian farm festivals of Isis and Osiris, or the Greek mysteries of Demeter and Adonis. Both sets of myths complemented the nature religions of western Europe. Both featured goddesses of the earth whose sons or brothers had died but risen again, showing the great mystery of eternal life. This was cultural osmosis between local religions of the sacred earth.

But Rome's military men on the eastern borders absorbed eastern culture in other ways, and for other reasons. The troops on the Persian frontier confronted a Zoroastrian theocracy. In the air of that war zone, they picked up Persian ideas and Persian cults like those of Mithras and Mani. The cultural influence of ancient Iran was enormous. All the surrounding peoples of western Asia were affected by Persian monotheism, including the Jews, Christians, Arabs and the soldiers of the eastern Roman army. All these people increasingly believed in one ruler, one God above him, one true social contract, and a war of good versus evil to exterminate falsehood. In early Christianity, this series of myths could be taken in several ways. These beliefs could hold promise for the salvation of the weak or for the victory of the strong, depending on who the believer identified with.

Between the religions of farmers and the military classes in the Roman empire, the Christian movement was able to come up the middle. For lower-class followers of nature religions, Christianity was a new mystery cult of the madonna and son. For the military, it was a cult of one universal ruler in the war of good against evil. With its many-sided character, Christianity could win both the poor and the military, leaving the old Roman aristocracy isolated. First, a series of Christian movements spread through the southern and eastern margins of the empire, among the peasants of North Africa and the Near East. The first officially Christian state was Armenia. The Gothic tribes of the steppes were largely Christians before they migrated into Europe to become mercenaries for the empire. By the early 300s, most leaders of Rome's eastern-based armies were Christians.

Christian Barbarians

At first the Roman establishment viewed Christianity as a hostile cult of the rebel Jews. The Jewish Zealots of Palestine had risen in furious revolt in the 60s CE, sometimes battling the Romans to the last man and woman. The most fanatical Zealots maintained that one ruler must rule the world, and he must be a Jewish king; one god must be worshipped, and he must be the god of Israel. To most Roman aristocrats this seemed an egomania taken to the point of suicidal war on the world. Most Romans could not distinguish Christians from Jews. After all, the Christians swore allegiance to only one god, namely the Lord of Israel, rather than the god of Roman victory.

But after the destruction of Jewish Palestine in 70 CE, some influential Christians were less critical of Roman values and were rethinking their place in the empire. The Roman government still held them suspect by association with revolt. But writing in the 200s CE, the theologian Origen justified the empire as a necessary stage in God's plan:

> A peace was prevalent which began at the birth of Christ [the Pax Romana]. For God prepared the nations for His teaching so that it might not, because of lack of unity between nations due to the existence of many kingdoms, be more difficult for the Apostles to carry out the task laid on them by their Master when he said "Go and teach all nations." (in Dudley 1970: 250)

Irenaeus likewise pointed out the spiritual value of a good empire: "Earthly rule, therefore, has been appointed by God, and not by the Devil, for the benefit of nations ... so that, under fear of human rule, people may not devour one another like fishes" (in Pagels 1981: 116). Some Christians spoke of a day when strife would cease; when one Lord and one empire would rule all the world. If both the Christians and the Roman commanders envisioned one world order, one law, and themselves as the agents of victory over evil, then they might well be comrades in arms.

The persecutions of Christians came at a time when Rome's emperor-generals were fearfully concerned to enforce loyalty among their subjects and prevent mutiny among their troops. Emperor after emperor had died at the hands of rebellious ethnic groups or during mutinies of their own soldiers (Smith 1976: 22). The prevailing political culture fostered a ruthless scramble for power over the dead bodies of rivals. Some politician-generals hoped to change this context and foster a sense of deeper loyalty to one overlord (hopefully themselves), as in the earlier Pax Romana. And an obvious solution appeared in the east. The Roman eastern frontier troops were immersed in a world where monotheistic religions were taken as the cultural bulwark of monarchy. This was true of the Persian Sassanian empire, the would-be empire of Israel, and the new Christian kingdom of Armenia. The idea before the noses of the Roman commanders was that a Near Eastern monotheistic religion could be a ruler's best bet to inspire loyalty. Perhaps the Roman god of victory was not the best religious basis for a world empire.

In one modern version of church history, the early Christians brought a message of peace and were persecuted by the Roman army. This account presents the Christians as ancient pacifists, and the army as a bastion of militant paganism. This version of history ignores the spread of Christianity within the armed forces and neglects the role of Christian soldiers in transforming classical Rome into the Holy Roman Empire. Actually the army was a hotbed for religious movements. Masses of young men were thrown together in communal camps, facing deadly danger and having the formative experiences of their lives. They sat around campfires, occasionally talking in youthful earnestness about their beliefs. Good evidence is available of Christian proselytizers in the ranks. For example, the evangelist Aurelius Prudentius Clemens is recorded as appealing to soldiers and disparaging the pagan goddess of winged victory: "Why soldier, if you lack faith in your own powers, arm yourself with the feeble solace of a woman's shape? Never yet has an armored legion seen a winged girl guiding the weapons of the fighting men" (Smith 1976: 157).

The Christians adopted the term *pagani* for their enemies. This term was formerly used in the army to mean cowards who would not fight, or farmers who tried to stay at home and evade their duty. In the Christian context, it meant someone who would not fight in the army of Christ (Smith 1976: 81). The image of early Christians as pacifists may make army and church seem incompatible. But this was almost never true in armies of the later "holy" Roman empire. In fact, the early Roman church and army tended to fit together like a new cult of victory. The church's geographical and hierarchical organization was basically that of the army. For example, around 90 CE, Clement, the bishop of Rome, appealed for order in a divided congregation (in Corinth) as follows:

> Let us then serve in our army, brethren.... Let us consider those who serve as our generals.... Not all are prefects, nor tribunes, nor centurians, nor commanders, or the like, but each man carries out in his own rank the commands of the emperor and his generals. (Pagels 1995: 53)

General Constantine, the first Christian emperor, was a man of partly barbarian blood who had grown up in army camps and come to power in the usual way. In 305 when Emperor Diocletian retired, the various regional commanders moved their troops into position for a military-style "election." Constantine was commander on the northwest frontier, and his immediate rival was General Maxentius, who was based in Rome. The war started with a propaganda campaign, with each general grandstanding like a modern politician. Even where might was right, a politician had to win supporters, or nobody would fight for him. Constantine proclaimed that his opponents could do no right, and he would do numerous good works should he win the emperor's crown. While Maxentius appealed to his constituents in Italy, Constantine courted the frontier armies. Perhaps this was his biggest advantage, but he sought religious advantages as well.

Neither politician wished to antagonize potential supporters, but Maxentius appeared hostile to one religious minority, namely the Christians. He had announced tolerance for the growing Christian community, but then he had to maintain order among squabbling Christian sects. When two rival Christian groups each elected a pope, or bishop of Rome, their partisans fought in the streets, excommunicated each other and raided each other's places of worship. Maxentius felt he must separate the combatants, and so he exiled both bishops to islands in the Mediterranean. Next, both bishops died in exile. Maxentius was left looking like an enemy of all Christians. Constantine picked up whatever points he could by announcing that he stood for religious freedom (Smith 1976: 40).

In Constantine's mind, Christian support was an important edge in his victory over Maxentius in 312. Possibly many Christian soldiers felt that Maxentius opposed their faith. Still, this factor was important only in retrospect. At the time of his first victory, Constantine simply proclaimed "to Christians and everyone else the right freely to follow whatever rule of faith they choose, so that whatever divinity [is] enthroned in heaven may be well disposed ... towards us" (Smith 1976: 49). He had played the Christian card and won. If it was helpful to him, he would play it again.

In 323, Constantine moved against his last rival for power, commander Licinius of the Near Eastern armies. In his opening propaganda salvos, he accused Licinius of persecuting Christians. Constantine claimed that his victory over Maxentius in 312 had been foretold: before the battle a cross had appeared in the sky as an omen of victory for the pro-Christian side (Smith 1976: 50–60, 47–48). This was the story he spread to the armies and civilian populations of the Near East, which were by now heavily Christian. Constantine knew the politics of frontier regions better than his more aristocratic opponents. He was offering Christians of the Near East not just freedom from persecution, but equality in status with the imperial cult of Rome. This time the Christian card was definitely important, and, whatever tipped the scales, Constantine won again. Now he was sole emperor of the Roman world. Now, not only were semi-barbarians ruling Rome through its frontier armies, but also the cultural core of Greco-Roman civilization was being co-opted by a frontier religion.

Soon after Constantine finished executing Licinius, his family and key supporters, he turned to building a new capital city in the east. Rome was no longer central to the rulers, either militarily or culturally. Constantine stripped it of its official status as first among cities. His new city, Constantinople, was also to outstrip Rome in grandeur, luxury and art. To adorn it, Constantine commandeered art treasures from pagan temples and official buildings all over the empire. The new capital contained more churches than Greco-Roman temples, and the churches were more richly endowed. Old patrician Romans saw all this as cultural treason, but many Christians proclaimed it a glorious victory over heathenism (Smith 1976: 64).

The Christian emperors of the fourth and fifth centuries were a dominant interest group, but still vulnerable to rejection by the old aristocracy or the pagan majority. The rulers sought allies to strengthen their own positions. As mentioned before, Constantine allowed an estimated 300,000 Sarmatian migrants into the eastern empire and saw this as strengthening his own constituency of semi-barbarian mercenaries (Smith 1976: 66). Later, Emperor Theodosius let in the Goths and employed them for the same reason. The Goths were mainly Christians (of the Arian sect) by 373, before they migrated off the steppes. Later, they formed a cavalry army within Theodosius's Christian forces, which crushed the pagan rebel commanders of the west at Flavia Frigida in 384. Theodosius had outlawed divination at pagan temples, but the ancient temples of Greece and Rome still stood, drawing their faithful. Alaric's Goths did what they could to change that. According to Claudius Clāudianus, the Gothic mercenaries so devastated the sanctuaries of Greece that in Eleusis hardly one stone stood atop another (Smith 1976: 187, 189). And that was just the start.

A pattern of alliance between the eastern armies, border tribes and Christian militants was well established by the time of Constantine's son, Constantius. When a pagan army commander in western Europe rebelled against eastern rule, Constantius offered to give Christianized Germanic tribes land in Gaul if they would cross the Rhine to attack these pagan Roman forces. Constantius had consulted his bishops on the issue, and they had told him to support Christian barbarians before pagan Romans. Alaric's Christian Goths felt the same way when they sacked Rome. Alaric's men saw their own gain as paganism's loss. With relatively good conscience they helped themselves to the wealth of Rome's Latin aristocrats. However, Alaric ordered his troops to leave the Basilica of the Apostle Peter untouched. Christian churches were the only undesecrated temples left in the city (Smith 1976: 85, 217).

Councils to Establish One Faith

While establishing his new capital at Constantinople, Emperor Constantine moved to hold a great church council at Nicaea. This was a state-sponsored conference, endowed with legislative power to set the official form of Christianity. Constantine felt some urgency to settle this matter. He had risen to power partly with Christian support, but then inherited Maxentius's problem of civil conflict between Christian sects. As Constantine explained it:

> I think it contrary to the divine law to overlook these controversies and quarrels [between Christians], for they may rouse the Highest Divinity to anger, not only against the human race, but against myself also, to whose care He has by His Heavenly Will given the government over all earthly things. (in Dudley 1970: 271)

So Constantine decreed that hundreds of bishops come together from across the Roman world to settle their differences. After this council there would be one emperor, one church and one Nicaean creed. The emperor would eliminate his political rivals; the true church would eliminate its religious rivals; and the peace of a new world empire would begin.

The church council of Nicaea, however, did not end dissension among Christians. Naturally, many complicated issues of diversity remained. To deal with all the issues of an emerging orthodoxy, the church held a series of high councils. These established certain Christian books as canon scripture and banned others. The secular authorities then sometimes imposed the death penalty for possession of the banned books. Most of the formative councils were held fairly close to Constantinople, in Turkey, Greece or Lebanon. Nearly 90 percent of the participants came from the Near East or North Africa. The decisions were taken by majority vote, with some instances of vote-rigging. The majority of the representatives assumed that their own attitudes reflected God's will. Being ancient Near Eastern people, they commonly took Near Eastern values for granted in deciding the official content of Christianity. As when Mohammed's followers added Arabian traditions to their prophet's message, these Christian men added their Levantine cultural universe to the sayings of Jesus. They rejected the most narrowly ethnocentric interpretation—that true followers of Christ must follow the entire body of Jewish traditions—but the broad background of Near Eastern culture still formed reality as most of the creed makers knew it.

As mentioned before, people in the desiccated Middle East commonly saw nature as their enemy. They sought God's help in a hostile world and often hoped for salvation *from* that world. "Salvation" for many of them was not so much a psychological state of wholeness in life, as a destiny to paradise in another world. Of course, only the virtuous, or those made virtuous by God's power, would merit such salvation. And the virtue most generally endorsed as meriting reward was obedience to traditional social mores. In that case, salvation depended not on right relations with nature, but on proper relations to the community—whose traditions came from God. In the arid lands, most ethnic communities believed that their own social code was the standard by which God would judge them as fit or unfit for salvation. But as this religious common sense had often led to extreme ethnic exclusivity, the early Christian bishops generally tried to be more inclusive than their ancestors had been. Most of them wanted to broaden the basis of Christian community. They proposed looser, more cross-cultural codes of morality and claimed that these simplified codes were the new universally valid requirements for salvation.

So the mainstream Near Eastern context of life was written into early church dogma and then imposed as official orthodoxy in Europe. In applying the council doctrines, most European bishops took Near Eastern standards as orthodox, and judged Old European nature religions as devil worship. They taught that European women must assume a proper Near Eastern-style submission to male authority. The men in turn must submit their wills to the one true socio-religious authority. Most orthodox Christians of the late Roman empire could see nothing in the teachings of Jesus to contradict these views.

The War on Nature Religion

After the Second Church Council of Braga in 572, Abbot Martin of Dumio, in northwest Spain, received a letter from one of his bishops. The bishop wrote to humbly ask Martin how he should do his job. This honest soul admitted that the Basques and other villagers of Galicia had not truly been won to Christ. Most often the *rustici* of the rural areas simply added Christ to a list of many deities they worshipped. Then, if the locals felt that invoking Christ brought them no benefits, they promptly forgot that name. What could be done, the bishop asked, to make them understand God's message, when they were so set in ignorance (Smith 1976: 238–39)?

Abbot Martin took the question seriously. He did not merely exhort his bishop to more vigorous efforts, or speak of the mission in an abstract way. He wrote out in full a sample sermon as he would give it to the unbelieving rustics. In part, this is what he wrote:

> Most beloved brethren, we want to tell you in the Lord's name something which you need to hear at least a little about.... When God first made heaven and earth, in the spiritual domain he made spiritual creatures, that is angels, who were in his presence and praised him. But one of them, the first made of all the archangels, seeing himself glowing with so much glory, did not give honour to God his Creator, claiming that he was himself as good as he. And for his pride he was thrown down with many other angels who thought as he did from that heavenly domain into this lower air below heaven; and he who had been the first archangel lost the light of his glory and was made a dark and horrible devil....

> Men, [meanwhile,] forgetting the divine creator of the world, began to worship created things. Some adored the sun, some the moon or stars, some fire, some deep waters or springs of water, believing not that these things were made by God for man's use, but that they were themselves gods in their own right. So then the devil and his servants, the demons, who had been expelled from heaven, seeing ignorant man forgetful of God his Creator and going astray with regard to created things, began to display themselves in a variety of forms and to speak with them and to desire of them that they would offer sacrifices to them on mountain tops and in leafy woods, and worship them as divine. One of them, assuming the ways of evil people who live out their lives in all sorts of crime and wickedness, called himself Jupiter: he was a magician and involved himself in such immense incestuous adulteries that he had his sister, who is called Juno, for a wife, and corrupted his daughters Minerva and Venus, and filthily fornicated incestuously with his granddaughters and all his relatives. And another demon called himself Mars: he was the instigator of contention and discord. And yet another demon called himself Mercury: he was the inventor of all theft and fraud.... And yet another demon took upon himself the name of Saturn: it was he who, living in the utmost barbarity, used to eat his own newborn children. And another demon pretended to be Venus, a whore, who played the prostitute not merely in innumerable adulteries but also with her father Jupiter and her brother Mars. So that is how men in those times were lost. They worshipped ... devils ... as gods ... building temples ... making statues and altars....
>
> And there was yet more. For many were the devils expelled from heaven. So some ruled in the seas, or rivers, or wells, or woods.... And those in the seas were called Neptunes, and in the rivers Lamias, and in the wells Nymphs, and in the woods Dianas: and they were all malign devils and nefarious spirits, who hurt and troubled unbelieving people who did not know how to protect themselves with the sign [of the cross]. (in Smith 1976: 239–40)

In this version of mythology, Lucifer and hosts of evil angels were not cast down to hell for their sin of pride, they were cast down to this earth. This world became their abode. Those who loved this world and took nature as their teacher and mother were under the spell of the fallen ones. Each soul was born into this realm of sin and depravity as a trial of faith. The test was for the soul to be plunged into evil surroundings, yet to remain faithful and obedient to the one true authority, who was not of this world. And the soul which passed this test would find its reward. In a reverse of the operation by which the evil angels were cast down, the faithful souls of this world would be raised up, to a better life in heaven.

What was required for the soul to win release from this world? The answer was complex, though the preachers tried to simplify their instructions to the rustics. First, the people had to believe in a Near Eastern prophet named Jesus. Next, they had to reject their ancestral European religions as evil and follow a social code imported from the Middle East. Third, they had to obey the priests and rulers claiming to mediate between the one true God in heaven and this fallen world. To reject any of these things would mean expulsion from the one true community and damnation for all time.

In pagan Italy, the villagers believed their fields were densely populated with gods, fairies and spirits. The general word for these was *numina*. So the spirit of Proserpina presided over stems, and Nodutus over foliage; Flora was the spirit of grain in its flowering stage, and Lacturnus possessed the grain as it filled with "milk." The terms for animating spirits were nearly as numerous as the words for the plants and animals themselves. Country people saw nature as a community of fully conscious entities, with whom they must trade favors to get by.

All this concern for spirits was an utter abomination in the eyes of orthodox Christians. St. Augustine described the *numina* of European pagans in great detail, explaining the intricacies of local nature mythology "with frigid contempt" (Dudley 1970: 48). Of course, Augustine was an educated man and isolated from life on the farm. Perhaps it befit his position to dismiss rustic folklore as superstition, and as belief in things which do not exist. Possibly the *numina* were few and far between in Augustine's North Africa, and it was fashionable in Hippo to not believe in them. But if Augustine felt that *numina* simply did not exist, why did he feel it was so *evil* to believe in them? In North Africa, the locals believed in nature spirits also, but these were often identified with desert winds, dust devils or sandstorms. The nature spirits there often seemed decidedly hostile. The later Muslims called such spirits *djinn* and believed them to be agents of Satan. Those who conjured nature spirits were wizards and devil worshippers. Augustine made a similar charge against nature worshippers in Europe.

The early Christians commonly blamed natural disasters on devil worshippers. Witches were people known for having close relations to nature spirits and were therefore suspected as traitors to their own kind. If lightning struck a church steeple, a witch might be blamed for calling down evil nature spirits against God's community. As Martin, the abbot of Dumio, carried on with his model sermon in 572:

> How could any of you who have renounced the devil and his angels … revert to the cult of the Devil? But what is the lighting of wax lights at rocks or trees or wells or crossroads if it is not worship of the Devil? Divination, and omens, and observing the days of the idols, what are these if they are not worship of the Devil? Observing the Vulcanalia [23 August] and the Kalends, decorating tables, wearing laurels, taking

> omens from footsteps, putting fruit and wine on the hearth, and bread in the well, what are these but worship of the Devil? For women to call upon Minnerva when they spin, and to observe the day of Venus at weddings and to call upon her whenever they go out on the public highway, what is that but worship of the Devil? To make evil incantations with herbs and to invoke the names of demons in those incantations, what is that but the worship of the Devil? (in Smith 1976: 241–42)

The banning of pagan religion advanced in stages. First, Christianity gained equal status with the Roman cult of victory, as one of Rome's officially supported faiths. Then the bulk of state patronage went to the Christian churches. By the 390s, the government was actually penalizing public practice of pagan rituals. Later the authorities closed the ancient Greek academies and dismantled the altar of victory in Rome. Christian officials cut off public support for the Vestal Virgins and let their sacred fires go out. Only private and family pagan practices remained unchallenged, so long as these were done without publicity.

When pagan temples lost the support of the state, the cost of their upkeep fell entirely on private contributions. Many temples fell into disrepair. But it was still a crime to destroy them, until the state withdrew its protection of pagan property. The message to Christian fanatics was that if they destroyed non-Christian holy places, they would no longer be punished. So began the period of state-tolerated vandalism of the classical past. An account concerning Bishop Martin of Tours was typical:

> Once at a village, when he [Martin] had destroyed a very ancient temple, he undertook also the cutting down of a pine tree standing close to the shrine, when a priest of the place and a crowd of pagans began to resist him. And these people … would not allow him to cut down the tree. And he was careful to point out that there is nothing sacred in a tree trunk. They should rather serve the God whom he himself served: the tree must be cut down, because it was dedicated to a devil. (in Smith 1976: 176)

Smith compares these Christian zealots to the mobs of Red Guards in the Chinese Cultural Revolution. Of course most modern Christians would never dream of destroying someone else's place of worship and would be aghast if they heard anyone actually propose such a thing. But modern common sense might not affect their inherited beliefs about the Christianization of Europe. According to well-known church histories, the destroyers of pagan temples were saintly warriors for Christ who liberated Europe from the forces of darkness. The views of their victims were almost completely purged from the historical record. Only in recent years have documents of pagan protest come to light. For example, a petition called "In Defense of the Temples" written by a pagan man named Libanus dates from 386. In part, he says,

> [The monks] say they are making war on the temples but their warfare is a way of pillaging what little poor unfortunates do have, the produce of the fields and the cattle they feed. [After their raids] ... the attackers withdraw, carrying off the goods whose surrender they have compelled. They grab people's land, claiming the place is sacred.... They who (as they say) give honour to their god by fasting are getting fat on the wretchedness of others.... If they hear of a place with something worth raping away, they immediately claim that someone is making sacrifices there and committing abominations, and pay the place a visit—you can see them scurrying there, these guardians of good order (for that is what they call themselves), these brigands.... In the city of Beroea there used to be a statue of Asclepios ... in which art surpassed nature.... But this statue, which ... was the work of a great genius [Phidias] has been smashed. (in Smith 1976: 166–647)

While classical academies still survived, well-educated voices contested orthodox Christian claims. The pagan philosophers seldom attacked the words of Jesus but opposed other beliefs incorporated into Christianity. The pagan philosopher Celsus (c. 180 CE) felt that most Christians saw nothing good in the world outside their own community. He accused them of thinking "God has ... disregarded the vast earth to give attention to us alone" (in Armstrong 1993). The Neo-Platonic teacher Plotinus of the third century "saw no proof that the material world is ruled by evil." Plotinus argued, quite the contrary, that everything in physical, spiritual or mental reality came ultimately from a single source, which he called the One and the Good (Smith 1876: 15).

With similar common sense, the British holy man Pelagius (c. 360–420) argued that human sin did not infect nature. Joseph Campbell (1976b: 465) summarizes the points of his "Pelagian heresy" roughly as follows:

1. Adam's sin of disobeying God was not the cause of bodily death in the world. Death is natural and Adam would have died anyway.
2. When Adam sinned, he hurt himself alone. He did not infect the human race or the natural world with any taint of inherited original sin.
3. All newborn babies are in the same pure state in which Adam was born. If they die in this state, they will have the eternal life they never lost. People, therefore, are *naturally* saved unless they go wrong in their lives—instead of *naturally* damned unless saved by the church.
4. As the human race does not die because of Adam's sin, so it does not all come back to life because of Christ's resurrection.
5. Jews who follow their own Old Testament law (and others who follow their own traditions) may achieve eternal life without becoming Christians.

6. There were morally perfect people in the world before Christ, and so there are ways to salvation other than through the church.

Most likely, the vast majority of British people living in recent centuries would feel these points make perfect sense. In fact, Pelagius seems to have made sense to the British of his own time, and his teaching won wide acclaim among European Christians. For a short time his ideas were even declared orthodox by the pope. Then his opponents made their counterattack. St. Augustine led a delegation from North Africa to Rome, appealing for the excommunication of Pelagius. Augustine argued that if Pelagius denied the fallen state of man and nature, then he must be speaking from that fallen state himself. Pelagius, Augustine argued, had taken sinful nature as his standard of truth, rather than God's supernatural grace and revelation from above. Without higher guidance, nature must have led Pelagius to worship the things of this world. How then could this voice of fallen nature be adopted as the church's message of salvation? What if others followed his example and relied on their own sinful natures to decide what was right, rather than relying on guidance by the church? Most likely the flock would go to the wolves and the shepherds would lose their sheep.

Pope Zosimus, who had declared Pelagius orthodox, had to face Augustine's angry delegation. Riots over Pelagius broke out in the streets of Rome. The bishops of North Africa and the Near East (where the large majority of Christians lived) vented their outrage (Pagels 1981: 129). Finding his own position precarious, Zosimus backtracked, declaring Pelagius a heretic in 417. With this, the church overrode the last major philosophical challenge to Near Eastern-style Christianity for the next seven hundred years.

Soon after the Pelagian controversy, the serious elimination of nature religion in Europe began. A church council at Toledo, Spain, in 585 declared

> that every priest shall make most careful inquisition in his own area, together with one of the magistrates of the district, for survivals of the sacreliges, and shall not hesitate to destroy whatever may be found. And those people who have fallen into such errors shall be coerced from this danger to their souls by whatsoever warnings may be required: which, if any neglect to do, let them understand both parties are in danger of excommunication. And if any lords omit to extirpate this evil from their holdings, or refuse to prohibit it among their people, they too shall be driven from communion with the bishop.(in Smith 1976: 242)

Within a few centuries, this kind of general inquisition by both priests and nobles established the cultural climate of medieval Europe. In Christendom, rural peasants were suspected of harbouring treasonous thoughts against their lords, and also against the ultimate Lord. The rural areas were described as "God-forsaken places," akin to the desert in which Satan tempted Christ. But in the center of town, where the lord's palace and the church stood together, people could feel close to God. The authorities painted the nature spirits of the old country religion in tones of darkest evil. The old pagan lord of the animals who danced with horns on his head was now the very image of the Devil. Campbell (1976a: 629) claims that under the medieval order, "Every local deity was a demon—every natural thought a sin."

Europe's Women in a Desert Civilization

The army of Rome was an effective machine for indoctrinating conquered males. It sucked up hundreds of thousands of men and drilled them into fit instruments of the empire. After about 325 CE, this mighty machine became the strong arm of Christendom. The other arm was the all-male priesthood. Through these institutions, sufficient men were recruited to uphold the Christian Roman empire. But Europe's women were harder to co-opt. The all-male army and priesthood fell short of indoctrinating women in the ruler's values. As if in another cultural state, the women often continued living by their own traditions. Such a gender partition of the cultural world may have seemed normal in classical times. In pagan Rome, the domain of exclusive male power scarcely extended beyond the ranks of the army. But under Christendom, the realm of male control was supposed to expand—in religion, property rights and family life—till it matched the degree of male control known in the Holy Land.

Women's Ways in Old Europe

Under Christendom, the traditions of Old Europe were systematically suppressed from public memory for over fifteen hundred years, and only fragmentary reports remain concerning pre-Christian women. The evidence appears in folklore, some written records and in "pagan survivals" of custom within living cultures. Even before Roman or Christian times, Indo-European invasions had changed the status of women over most of Europe. Still, a common tapestry of evidence remains scattered from Basque Spain to Scotland and from non-Latin Italy to the Aegean islands. All these regions record matrilineal rules of inheritance and succession. In the Scottish highlands, married men moved to the house or land of the bride's mother, and this was common custom into the twentieth century (Gimbutas 1991: 344, 348). Among the old Celts of Ireland and the Etruscans of Italy, women had power equal to men in initiating marriage or divorce. Actually though, these legal terms for friendship and parenthood hardly applied. Most women of Old European traditions probably lived without marriage at all, at least not as the church would recognize it. Pagan women were commonly in charge of their own households and had as many male or female friends as they wished. When a man and woman lived together, by no means did the woman's property pass to the man's legal control. No man "gave" a woman in marriage to another man. Friendships and love affairs between consenting people were not officially authorized, made eternally binding or made mutually exclusive by any higher authority. Women had ways of preventing unwanted births and raised children they chose to have with the help of their friends. A mother had charge of the children she birthed, not the father, especially since it was sometimes unclear who the father was (Gimbutas 1991: 348). Therefore there was no such thing as an "illegitimate" child. The idea of casting such contempt on an infant, or the "legitimacy" of the love which gave it birth, would probably have been either shocking or laughable.

Corresponding to these freedoms and powers over private life, the women of Old Europe were commonly active in public affairs. Gimbutas distills the following characteristics of Old European political and social patterns from descriptions by classical commentators such as Herodotus (Greek, fifth century BCE) and Strabo (Roman, first century CE):

1. Inheritance through the female line—including succession of royal families, with queenship passing from mother to daughter.
2. Endogamy—matrilocal marriage, or group marriage, combined with common ownership of property.
3. Metronymy—naming of children by their mother's family name.
4. An important role for the queen's brother—who often ruled jointly with the queen, with the queen's spouse often serving as a consort only.
5. A generally high social status for women. (1991: 349)

In religion, Old European women played roles as priestesses, oracles, healers or counselors. The Basque priestesses were oracles, practicing in their mountain cave sanctuaries down to the 1900s. Often their authority was greater than that of the local Catholic priests (Gimbutas 1991: 343). Obviously, Abbot Martin of Dumio had trouble with the women of northern Spain.

The Problem of Controlling European Women

Spiritual and political power for females was an anathema to Near Eastern morality. At first St. Paul enthusiastically encouraged female leaders in the church. Then, in the face of rising persecution of the church, he tried to remove the offense of female leadership, in effect asking Christian women to sacrifice their leadership roles for the good of the church's reputation. Words to this effect were raised up as canon scripture, overruling any implications in the Gospels as to how Jesus honoured women. With little more than a flourish of indignation, Paul and other like-minded churchmen managed to banish female power from the Near Eastern churches. But when Christianity came to Europe, the problem of powerful women arose on a far larger scale.

For most European Christians, the social mores of the Near East stood in stark contrast to those of their homelands. Basically, the church took Near Eastern customs as those of God, and European customs as those of the Devil. Therefore it was the church's mission to make European women conform to the social position of their sisters in the Near East. As Paul supposedly proclaimed, "For the husband is the head of the wife, as Christ is the head of the Church. As the Church is subject to Christ, so let the wives be subject in everything to their husbands" (Ephesians 5:23–24). With this concern firmly in mind, the message of God's love to men and women became a message that love must be strictly controlled, lest it take a form unapproved of in the Holy Land.

In the land of the faith's birth, control of women was the essence of male honour. For the Near Eastern majority of bishops filling the early church councils, local tradition, Hebrew mythology and the Gospel of Christ reinforced a common message for women. Women were weak, relatively helpless creatures, more subject to sin than males. It was therefore the sacred duty of men to protect their women from temptation, lest they bring dishonour on their family names. Tertullian gave this logic its standard theological baggage, teaching Christian women:

> You are the devil's gateway.... You are she who persuaded him whom the devil did not dare attack.... Do you not know that every one of you is an Eve? The sentence of God on your sex lives on in this age; the guilt, of necessity, lives on too. (in Pagels 1981: 63)

Middle Eastern women had lived with such attitudes for hundreds or thousands of years. European women, however, were a tougher case.

To control love, the church claimed a monopoly on the sacrament of marriage and stigmatized all relationships not consecrated by its priests. The priests judged traditional European families who had formed their own bonds independently, or lived in patterns different from church norms, as living in sin. For the church, lovers who separated and found other lovers were "bigamists." Homosexuality, that great classical tradition, was a criminal or even a capital offense. The word "faggot" comes from the practice of burning homosexuals. In Christian marriage, men were legal watchdogs over their women's conduct. The church's rule of monogamy applied to men as well as women, but the code was more rigidly enforced on women. They, after all, were the ones subject to bearing an "illegitimate" child. All these rules were now presented as a "standard" part of European tradition. But for all the strength of this church- and state-backed agenda, it took hundreds of years for Christian marriage to prevail over traditional lifestyles. In Scotland, "Celtic secular marriage" remained the most common family arrangement until the 1600s.

Divorce was allowed until after the year 1000, as it is allowed in Islam or Judaism. But after a divorce, or even the death of a lover, the church tried to ban remarriage. If remarriage was condoned, it would allow half-brothers or half-sisters to be born, which would lead to family conflict over estates. The church lent its moral authority to prevent any such thing. But despite priestly threats of excommunication, remarriage remained common, and people continued rebuilding their lives by celebrating second or third marriages outside the church. Finally, in the eleventh century, the church authorities reached a consensus for drastic action. They would cut off the supply of divorcees seeking remarriage by banning all divorce in the first place. Now not only was remarriage banned, but, in addition, lovers once joined could never be legally separated.

Before this ruling, Christian couples had reason to fear losing each other because of their own bad behaviour. Now the church attempted to put that fear to rest. According to canon law, the "international law" of medieval Christendom, partners were now bound together no matter how badly they treated each other. But just as the rules of Christian marriage were difficult to enforce, so additional hundreds of years passed before the ban on divorce was widely accepted.

When women could initiate divorce, it made little sense for a father to give his daughter in an arranged marriage because the daughter could overturn this arrangement at any time. But after divorce was abolished, a girl's marriage could be more predictably arranged. A tradition common to the Near East then arrived, of arranging betrothals for child brides. Where this became a standard medieval practice, women lost both the freedom to initiate marriage and the legal option to end it. Marriage could then be an economic arrangement between male family heads, using their daughters as bargaining chips. The church imposed its seal of approval on these arrangements and threatened eternal damnation for women who tried to escape (Campbell 1976a: 53).

If such rules were supposed to teach women a moral lesson, the lesson was *not* that women should make their decisions based on "right" moral principles. Instead, these rules implied that women should not make choices at all. The correct morality for them was to follow whatever their men desired without question. This accorded perfectly with the moral doctrines of St. Augustine, who taught that humankind (and especially womankind) is contaminated by sin and therefore cannot trust its own mind. In that case, women had best not try to think for themselves but should only follow the instructions of those set over them by God.

As women lost control over marriage, their rights to property also withered away. Old European customs had commonly recognized "married" women's property as separate from that of their husbands or lovers. But the Near Eastern custom would place all women's property in the hands of male family heads. This ideal also took many centuries to impose and was never fully accepted in many areas of Europe. Only in 1366, more than a thousand years after the Christianization of Rome, did the Statutes of Kilkenny end Scottish women's rights to own property in their own names.

As for roles in religion, the old Celtic Christian church at first presumed it should have both male and female priests, in accordance with local tradition. Then the female priests were excluded as a condition of merger with the greater Roman church. After that, the closest thing to female leadership in the church was a priest's wife.

It may be a well-forgotten fact, but for the first thousand years of Roman church history, the priests could marry. Members of monastic orders lived chaste lives, but common priests were almost always married. If they didn't marry, it was often assumed there was something wrong with them. Only after centuries of struggle, and under the determined leadership of Pope Gregory VII (d. 1085), did the church manage to close all legal options for marriage to its priests (Ranke-Heinemann 1990: 108–9). A kind of sacramental apartheid was then imposed, completing the separation of church and women. Now not only were priestly roles reserved for men alone, but the sanctity of the priests depended on segregation from females. This was a drastic break with Near Eastern tradition, and it happened only in western Europe. The Greek Orthodox, Slavic or Eastern Orthodox, Coptic and other churches of the East retained marriage for their ordinary priests. What corruption did this measure prevent in the West?

In the Near East, Jewish rabbis or Muslim mullahs were expected to marry. They were also expected to uphold local morality to an exemplary degree, including the morals of male authority over women. In many cases, the more holy a man aspired to be, the more strictly he controlled his women. Otherwise they might expose him as a man unable to rule his own roost. The Near Easterners commonly controlled female power by imposing segregation of women from public life, or *purdah*. Where men dictated the terms of women's lives, this could work. But in the strong-woman's world of western Europe, the priests were in no such position to dictate the parameters of women's lives.

When western European priests married western European women, the old tensions between equals remained, even though church doctrine held that the priest must be the social and spiritual head. The priests' wives often made a lie of the doctrine. Where local people commonly turned to the village wise women rather than the priests for advice, a priest's wife could easily eclipse her husband's influence. If she had greater personal force than her husband, she might shape his views more than the bishop did. If she was a representative of local women's traditions more than a supporter of the all-male church, she would be subverting the priesthood to paganism. This must have been a danger to orthodoxy from the first. Perhaps the problem slowly grew over time, or the problem of controlling priests' wives showed no sign of going away. Clearly, the patience of the western church hierarchy ran out.

According to some church historians, the rule of chastity for priests came about due to concern that hereditary dynasties or powerful families might rise to control the church. In this view, the policy of celibacy was merely a prudent choice, allowing the church to remain a society of volunteers, without children or inherited position. But why was this measure not needed in Islam, Judaism or Greek Orthodox Christianity? Why did the church of western Europe alone feel need of this wrenching reform? Perhaps it was the high status of women in western Europe which seemed to require this countermeasure. And perhaps it was the weak social position of women in the eastern lands, which made it easier for priests, mullahs and rabbis to have both wives and doctrinal purity.

The church in western Europe decreed a kind of *purdah* in reverse. Instead of secluding women from public life, the priests had to segregate themselves from the women. Of course, vast numbers of priests subverted the new rule for centuries. When ordered to put away their wives, many priests protested that without support from their wives they would go hungry and naked (French 1985: 157). Eventually, however, the doctrine of purity from sexual pollution prevailed. The church required of its priests a superhuman sacrifice. They were to cut themselves off from womankind, sexual love, children and family, all in the interests of preserving the church from the influence of strong women. The segregated priests then increasingly viewed women as a separate camp, or another species, more prone to sin than the male. The women went on with their own cultural lives, more apart from the church than before. They therefore seemed more opposed to the church and more identified with the Devil. A few hundred years more of this, and the priests and pastors of various Christian churches would declare war on that enemy camp.

Christendom's Women in the Early Modern Age of Scarcity

At the close of the Middle Ages, Europe passed through its first series of homegrown environmental crises. Between the 1300s and 1600s, Europe was a continent stalked by starvation and subject to all the social pressures of serious natural scarcity. The population peaked and fell twice, first with the Black Death, then in the famines and plagues of the Little Ice Age. By the early 1300s, Europe's population had risen past 80 million. And given the non-diversified nature of the economy, this squeezed resources in many villages close to the limits. General malnourishment rendered the population ripe for a plague so serious that it would take 250 years for the population to recover. Then, as the Little Ice Age set in before 1600, the death rate climbed again to match or surpass the birth rate. This time famine was more prominent than plague. The methods of farming had improved since the 1300s, and Europe could have supported more people, but the climate undercut the farmers. The growing seasons shrank with late and early frosts, and the crops sometimes failed for several years running (Ponting 1991: 97). According to David Sabean (1984: 7), all known records of wage and income scales for Western Europe in the 1500s show a dramatic fall accompanied by "great waves of mortality." The people of northern Europe from Muscovy to Scotland were the worst affected. In Scandinavia some regional famines brought a mortality rate of up to 50 percent.

As usual, scarcity motivated the most powerful people to consolidate control over resources. Where the expansion of farmlands approached natural limits, leaving only isolated patches of forest, the pressure for economic expansion turned inward. The most powerful local men tried to take the communal lands. Land ownership grew more centralized, while the poorer families had to subdivide tiny plots. The pressure of scarcity worked its logic, not only between peasants and landlords, but also within peasant communities. Competition among farming tenants increased, raising the temperature of suspicion and hostility. Women who owned land seemed to prevent men from having enough.

In this crisis, the surviving property rights of women tended to collapse. Scotland's Statutes of Kilkenny had come in response to conditions leading up to the Black Death. More sweeping changes in women's legal status came in the 1500s and 1600s. A series of reforms in several countries rendered women as legal minors, subject to their husbands or fathers. Most cities in Italy banned estate inheritance to a female (French 1985: 171). The French provincial law codes, concerned almost entirely with property and inheritance, also reformed women's economic standing for the worse. In some regions, within much of Germany, for example, women's legal rights to property remained, but even there the feeling grew that it was antisocial of females to insist on those rights. Widows who tried to hold onto their land, or who lost it and grew hostile, were commonly suspected to be malicious witches. By no accident, the accusers and prosecutors of witches were commonly paid with the property of the convicted.

For all these reasons and others, women across Europe were rendered economic dependents in an age of hunger. Where food was scarce, sons got more to eat than daughters. Single women, especially older women past child-bearing age, were commonly the first to starve. In former times, the older women would have been respected "clan mothers." To cut them off from property and support would have been a sacrilege to most Old Europeans—like starving one's parents would be to a Chinese Confucianist. To justify their actions, the winners in the time of scarcity needed a new code of ethics.

Premodern "animistic ethics" revolved around the ideas that nature is alive and people must not take more than they give. A greedy taker would be rejected by the community and the spirits of nature. Greed came before a fall (Schneider 1991: 185–87). The greatest evil imaginable was to hoard food while others starved. This violated every instinct of decency and parenthood, especially motherhood. In the premodern age of scarcity, it commonly seemed inevitable that for one family to "get ahead," others must be deprived. It also seemed inevitable that those whom the greedy pushed aside would take revenge. If any local man tried to maximize taking and minimize giving, then the spirits of the fields, forests and streams would harass him like pack of gremlins. The evil eye of jealousy would curse him, and those less well off would feel justified stealing from him. This was the tyranny of "animistic ethics" in the semi-pagan countryside (Schneider 1991: 184).

When the church overthrew the gods and goddesses of pagan Europe, the peasant's animist ethics survived. The militarized nobility claimed to preside over the land, collecting dues and allocating plots; otherwise they generally left the peasants alone. In that benign neglect, the old village world lived on. The peasants and nature spirits dwelt together in a delicate balance of many lives.

The peasants took what they needed from nature. They did not practice self-sacrifice for the sake of plants and animals, but they did deem it dangerous to harvest things faster than they grew. And they feared that if they became enemies of nature, nature would destroy them, rather than the other way around.

The landlords were less afraid of offending nature and their neighbours. As times got hard, they moved to increase their take from the fields. They intervened in the peasant's world, sticking their inexperienced hands in to extract more. They ordered their peasants to clear the woods and plant the fallow fields. Many landlords reduced the old three-field, crop-rotation system to two fields: one for grains and the other for pasture. Thus the fields were pressed into harder service, bearing crops, then feeding animals, with no real time for rest. As both the fallow and marginal lands came under continuous use, gross production rose—but the average yield per unit of land fell. Sabean (1984: 7) says, "The expansion eventually led to a destruction of the ecological balance—soil exhaustion, harvest failures, depleted resources … the agrarian product of the feudal classes was continually threatened." At least the fallow time of winter remained, which was more rest than most farmlands in the Middle East got.

The peasants couldn't name chemical nutrients in the soil or measure the rate of humus depletion. They only knew that the land was alive with spirits, and those spirits would take revenge on those who hurt them. Perhaps many farmers saw the Little Ice Age as nature's judgment on human greed. In passive and active ways, they resisted demands for accelerated production, both out of self-interest and semi-pagan conviction. Their reluctance to work the land harder had to be overcome by pressure from above. To their masters, the peasants' reluctance seemed to stem from general laziness and superstition. The landlords opposed the laziness, and the church opposed the superstition. Both opposed the property claims of village women.

Passive traditional landlords saw other landlords wringing more from their estates. Clearly many landlords were willing to incur hatred from their peasants by enclosing communal property. The peasants, who had always used their surrounding fields and forests, were now threatened with daunting penalties for trespassing on the newly private property. They were now supposed to live entirely from their own small plots. Next the landlords might jack up the rents. In Spain and Britain, whole townships of peasants were evicted as absentee landlords found it more economic, and less bothersome, to use the land for sheep. As a result, much of Spain was overgrazed till it became almost indistinguishable from North Africa.

In the late Middle Ages, the nobility and church took up to half of all village production, if the accumulation of rents, taxes, tithes and requirements for donated labor are all added together (Ponting 1991: 97). To protect their claims to this income in hard times, the nobles had to stand on military alert against peasant revolt. Feudal lords called on each other like brothers when their peasants turned ugly, and sent regional militias to put down local uprisings. Where the peasants resisted enclosure, eviction or high taxes, the landowner's armed forces intervened. And the mobs of protesters suppressed by the soldiers were often mainly women. Many village men were better conditioned to obey their lords. Most clergymen taught the villagers to obey. In peasant revolts or eviction struggles, the clergy almost always supported the landlords. In many recorded incidents, local clergymen responded to evictions by reminding the peasants that misfortune was a sign of God's displeasure; the poor should look into their own hearts for the reasons God had turned his face against them.

The church had its own needs to protect in an age of hardship. Both Protestant and Catholic clergy defended those interests through a working alliance with local rulers, and through a slowly escalating round of crusades against heretics or unbelievers. During the sixteenth-century Reformation, as Western Christendom began disintegrating into factions, each church had to compete for tithes. Each sect faced external opponents as well as a home front of semi-pagan parishioners. Therefore the churches escalated their interventions in the peasants' world. Was not peasant failure to support the church due to unbelief? Was not unbelief paganism? If idolatry for nature spirits remained at large, it was high time to root it out.

According to Jane Schneider (1991: 190–91), the message of Reformation preachers, both Catholic and Protestant, involved a renewed appeal to "salvationism": "Having as their central characteristic the project of individual redemption in another ... life, these religions charted a departure from ... earlier spirit beliefs." The clergymen taught that honouring nature spirits was antithetical to salvation. Instead of worrying about offending nature spirits, the farmers should worry about their fate in the next world—which depended on a proper relation to the church and the other authorities. The desired upward-looking attitude was one of service to the social hierarchy, with greater indifference to the environment (in Schneider 1991: 195). The peasants were to become a more obedient and productive workforce, more willing to cut the trees, dam the streams or dig the mines. Instead of hesitating to take more than they gave from nature, they should accommodate their master's desires. This would show a faithful and obedient attitude that would be rewarded in the next world.

Sabean (1984) traces the "massive inroads" against the old peasant ethics made by the early Protestant pastors in Germany. As a keynote of their message, these preachers emphasized the sinner's obligation to forgive enemies. Before partaking of communion, Lutheran ritual required the worshippers to publicly recite the Lord's Prayer, including the words, "forgive us our trespasses as we forgive those who trespass against us." But some villagers refused to say these words. Given the state of their relations with their neighbours and local authorities, they would not say what they could not feel. And if they could not recite the prayer, then they felt obliged to abstain from the rest of the ritual as well. The pastors admonished such people for their presumedly stubborn and blasphemous attitudes. If these abstainers were unwilling to share in the sacrament which bound the people of the Protestant state, they could be presumed to be dangerous heretics. If their pastors' warnings went unheeded, they might be taken in for questioning and imprisonment. At this time the village commonlands were being enclosed all over Europe. In western Germany only a minority of children could inherit land, because further subdivision was a problem. Growing numbers of villagers were becoming landless laborers, especially the single women and widows. In this situation, conflicting claims to land were endemic, and the strongest parties usually won. In the old village ethics, opposition to greed had been the main weapon for justice and balance. But now the religious authorities were teaching people to put the shoe on the other foot. Now it was those who bore resentment against the advantaged who were accused of being evil. The weakest were now told to forgive the trespasses of the strongest. According to Sabean, "stigmatizing ... resentment was an early example of the now familiar western cultural pattern of blaming the victim" (1984: 47–52).

By such a combination of economic and religious forces, traditional communities could be wedged apart. The old sensitivity to balance between village and environment faded away like a mist. Traditional morality was turned around, so it was not the greedy or the hoarders of food in times of famine who were now the enemies of society. Instead the lords and clergy turned the spotlight on their own opponents. The stereotypical face of evil was repainted. Rather than a fat, greedy man who would eat his own people, the image became that of a spiteful old hag, casting spells on upstanding householders out of jealousy (Macfarlane 1985, in Schneider 1991: 204–5). The most numerous critics of the authorities were the local people dispossessed of property. And over much of Europe the entire female population could be counted, at least in legal terms, among the ranks of the dispossessed.

The efforts of churches and states to reform their peasants might be compared to those of modern corporations to train their "human resources." The reformers tried to use positive means such as persuasion, evangelism, retraining and incentives, but very often these methods were judged inadequate. Old loyalties to nature spirits and neighbours remained in force, like an undercurrent of rebellion. The reformers of peasants grew frustrated. They tended to approach their subjects with a certain fanaticism, recalling the idol smashers of early Christian times. As when St. Augustine had denounced the Roman peasants' worship of *numina*, it was not just a matter of paternalistic contempt for peasant superstition. There was, in addition, an almost obsessive hatred, as if for a real and deadly enemy. To uncover and destroy that foe, the authorities adopted harsher means—including eviction, imprisonment, excommunication, the rack, the gallows and the stake. With such a conjunction of ecological crisis, economic pressure and Near Eastern morality, village women and their traditions could be almost totally suppressed. What followed was an attempted "final solution to animism," directed largely against the real religious leaders of traditional Europe—the village wise-women.

Obviously many leaders in the medieval church feared witches. Witches were not simply dismissed as silly old women with untrue beliefs. Most any local priest could see that people often took their troubles, questions or illnesses to the village wise-women. Certainly the services of "witches" competed with those offered by the church, and the church took its claims to monopoly in cultural matters seriously. But more than that, many clergymen really believed that witches had supernatural powers. Witches were known for conjuring the clouds, changing one substance into another, causing animals to change shape, and giving or taking away other people's sexual fertility (Gimbutas 1989: 210). Such powers, at least when attributed to women, were taken as deadly threats to God's order. In Christian myth, all forces in the world had to have supernatural causes. The world was basically dead and inert, like the Sinai Desert, with all movement caused by impulses from beyond. The witches' powers therefore had to have a supernatural source, and that source could not be God (O'Neil 1987: 89). In the dualistic view of the universe, there was only one other source of power, namely Satan, also known as "the prince of this world."

Perhaps the world views of priests and wise-women came most directly into conflict over the shared concern to heal the sick. In the biblical view, illness and health depended mainly on the patient's relationship to God. Health was defined not so much by bodily vitality, as by adherence to biblical teachings. Any illness might be a chastisement from God for nonconformity to Christian morals. In that case, the services of a priest were required to hear confession and advise the proper penance. Otherwise, the malady might come from the Devil or his legion of evil spirits. In that case also, a priest was required to cast the demons out. And if the patient was close to death, a priest had to perform last rites, lest the soul go to hell. So the medieval church presented itself as the true source of healing for both body and soul. The social demand for healing and survival, which in modern times makes medicine perhaps the biggest business of all, was to be channeled into an endless appeal for ministrations from the church. There should be no need to call on any other source for healing, unless the sick lacked faith.

Simply by offering their traditional remedies, the village wise-women seemed to presume themselves better than the church. Their self-assurance reminded churchmen of the Devil's arrogance. Mary O'Neil (1987) describes the dilemma of traditional healers in Modena, Italy. These women had long been told their work was a sin. At church they confessed to accepting patients, and the priests generally said, "You are forgiven. Go and sin no more." But during the Renaissance and Reformation, churches all over Europe grew more aggressive in demanding their own brands of orthodoxy. The priests began threatening healers that if they practiced their arts again, they would not be forgiven. The town drunk might be forgiven his innumerable sins so long as he came to confession, but not a witch.

Naturally, the local people continued bringing patients to the female healers. They pleaded for remedies on behalf of sick loved ones. What should the healers do? Some yielded to the requests and tried to help. Others refused to practice healing any more, even when the patients' families offered "to take the sin onto themselves." Those who continued to practice did so with increasing secrecy, allowing the opportunity for blackmail to emerge. If a healer's remedies failed and the patient died, the disappointed relatives might report the offending healer to a priest. It would be the medieval equivalent of a malpractice lawsuit. On the other hand, if a healer declined to help the sick, she could be suspected of refusing out of malice, and the prospective customers might seek revenge through an accusation of witchcraft. Some healers had little or no land and needed their healing practice as a means of living. To avoid being reported, they sometimes required an oath of silence from their patients. To make the oath stick, a healer might insinuate that she had ways of knowing if the oath was broken, and that harm would come to the patient if it was. Of course, such precautions could easily backfire. The healer could be reported for extracting oaths of secrecy and threatening people with magical harm (O'Neil 1987: 91–95). Fortunately, in most of southern Europe the death penalty was seldom used against witches. The women were merely detained or made to do penance. But even such "lenient" penalties were enough to make many traditional healers take their knowledge of herbs and treatments with them to the grave. The prominent male doctor Paracelsus decided to burn his own text on pharmacy in 1527, because he had "learned from the Sorceress all he knew" (French 1985: 163).

The early modern age of scarcity involved a swelling in the ranks of single women. There were more widows who were formally forbidden to remarry, and more poor young women who couldn't afford to marry in the first place. In Spain, a woman's family had to pay dowry to marry her off, but in the 1500s and 1600s growing numbers of farm families could not afford it. One alternative was to give the girls to a convent. But the convents also required dowries for girls to be left there. Without marriage or the convent as viable options, many young women tried to make it on their own. As they were increasingly pushed out of farm ownership, many turned to spinning and weaving. Almost all nunneries in Spain, and perhaps most independent women too, relied on silk or cloth weaving. The word for an unmarried woman became "spinster." Next, the early modern economy virtually demolished this trade as well, with currency devaluations, cloth imports and mechanized spinning. Many nunneries were left virtually without income and were then seen as an economic drain on the church (Perry 1987: 152). The Vatican passed periodic restrictions forbidding any more nunneries or female religious orders. The nuns might have taken work in the towns or fields, but the Council of Trent re-emphasized strict rules of enclosure for nunneries. The church's women had to live in *purdah*.

Under such conditions, what could self-supporting women do? Many tried to hold on to their land and were abused for keeping it from male relatives. Some tried to specialize in healing and were denounced as witches. Some saw no alternative to starvation other than prostitution, for which they were damned as well. But most embarrassingly for the church, large numbers of women tried to live as independent Christian saints. In the first population crunch before the Black Death, numerous laywomen's religious movements spread and sought official recognition. In some cases, as among the female Franciscans or in the Dominican convent under Hildegard of Bingham, the spiritual leadership of women was accepted almost as fully as Jesus accepted it. But more often the religious authorities tried to slam their doors in the women's faces, as when the Beguine movement was excluded from the church in 1312.

In the second population crunch of the 1500s and 1600s, Spain experienced another rise of independent female religious vocations. With no credentials or authority other than their own devotion, numerous women won followers and reputations as local saints. They were called the "*beatas*." As they were not under church control like nuns, the *beatas* were not enclosed in nunnery walls. They acted out their vocations in the streets, usually organizing help for the destitute. They sometimes set up women's shelters for abused wives, prostitutes and homeless women. The records of their deeds survive mainly in the Inquisition archives, as many of them were tried for heresy in the early 1600s (Perry 1987: 152).

One of the more prominent trials of a *beata* was that of Catalina de Jesus of Seville in 1627. As Mary Perry explains, Catalina had a disturbingly huge following in the city. Even some ordained priests considered themselves her disciples. Her possessions became holy relics to those who loved her, and her works of charity were more prominent than those of the cloistered nuns. But of course a woman in such a position eventually came into trouble. The Inquisition took its time gathering its case against her. She was reported for urging abused wives to leave their husbands. She spoke in public, quoting scripture and explaining its meaning, which was a function forbidden to any but a priest. She lived on gifts from supporters, for which the priests accused her of corruption (even though church-imposed tithes were enforced by threat of excommunication). The Inquisition cast doubt on Catalina's claims to chastity. How could she have chastity if she had no man or enclosed convent to protect it? And if she could not have chastity, how could she have sanctity?

Catalina was not accused of witchcraft. Instead, the Inquisition presented an argument believable to the city where Catalina was popular. In a public hearing, priestly psychologists carefully discredited her delusions of grandeur. They presented her as a victim of feminine mental weakness: "To present them [the *beatas*] as witches or sorcerers would have acknowledged that they possessed special powers" (Perry 1987: 158). The treatment given to Catalina, and many other *beatas*, was a lifetime of penance in seclusion, under the close supervision of a confessor. No further visions, prophecies, sermons or contacts with the public were allowed. Catalina was to work in a house of seclusion for her food so as not to be a burden on society. Perry (1987: 147–48) explains, "Prosecution by the Holy Office, therefore, served to reintegrate the *beatas* into the traditional gender order." Apparently this was the mainstream church's imported vision for the spiritual development of women.

Chapter 13

From the Book of Desert Wisdom to the Book of Nature

IN THE MEDIEVAL ORTHODOX CHRISTIAN CONCEPTION OF THE UNIVERSE, God dwelt in heaven, and he was the cause of all good things on earth. The only science that really mattered was the divination of God's will. But since mortal eyes and ears could apprehend only earthly things, they could not reveal God's design. The true causes of things could be seen only by revelation, and revelation was a difficult and politically dangerous game. Competition among those claiming to possess the correct revelation could be fierce. To deal with such competition, the medieval church claimed a monopoly on revelations from the Lord. All unauthorized visionaries were suppressed, and all higher education was placed in the church's hands. The subjects of higher education were theology, church dogma and law—in other words, the discernment and enforcement of God's will. This was the intellectual project that had replaced classical philosophy.

In Europe, however, philosophy was never quite subordinated to revelation. Even leading churchmen were not content to explain that God makes all things occur but humans cannot know the mind of God. The theological seminaries of Europe stood on mental foundations from the old schools of inquiry, and certain habits of thought remained. Instead of simply asserting a revealed truth and ending inquiry with that, the schoolmen found themselves constantly debating just *how* God made things happen. For each effect on the inert earth, the scholastic theologians postulated a chain of causation and command. Orders were passed from God to the archangels, then to angels, rulers and subjects. Only the final steps of this chain were observable to ordinary human beings. The higher causes lay in the realm of metaphysics, which could not be seen but still had to be debated.

Fruitful Dilemmas of Medieval Christian Philosophy

In moments of lucidity the schoolmen realized a contradiction between their philosophical efforts and the terms of their own faith. In his *Summa Contra Gentiles*, Thomas Aquinas (1225–1274) admitted, "For then alone do we know God truly, when we believe that he is far above all that man can possibly think of God." Joseph Campbell (1976a: 578–80) asks, if Aquinas really believed this, how could he devote his life to teaching the one truth about God? How could his church presume to know that its doctrines, dogmas and descriptions of God were the true ones? More strangely, how could the church presume to execute people for having concepts of God different from its own? But Aquinas's admission seems to have been a passing insight. He soon resumed his accustomed role as a mouthpiece for his Lord and church. In the *Summa Theologica* he went on to refute almost every "false" concept of God then circulating in Europe. Only near the end of his life did Aquinas suddenly stop and cease to teach or write theology. Perhaps Aquinas realized he could hardly presume to speak even for his earthly church, much less for God. Because in 1277, soon after Aquinas died, Pope John XXI issued a bull (or, as the Muslims would put it, a *fatwa*) condemning a total of 219 philosophical ideas then under discussion in the universities of Europe. The forbidden opinions included several held by Aquinas himself. Campbell provides a partial list of the banned concepts, including:

> That there are falsehoods and errors in the Christian religion as in all others.... That what the theologians say rests upon myths.... That man's good is in the natural sciences, from which knowledge flows the natural moral virtues described by Aristotle.... [That] the multiplicity of things [in the world] presupposes a multiplicity of ... causes [and not simply the hand of God controlling everything which occurs]. (1976a: 400–401)

The medieval universities of Europe were thus hotbeds of heretical thought, like the philosophical academies of classical Greece and Rome had been. But the classical schools had been silenced by church decree a thousand years before. The new medieval universities were supposed to teach orthodoxy, not stimulate controversy. The students were to receive truths from a realm beyond their own understanding. The universe had two levels, of mundane earth and godly heaven, and since the students' minds were confined to the lower level, they could not question or debate truths from above. So in a typical church service, the parishioners filed into the hall, the priests told them what to believe, and nobody argued back. But if universities were established in addition to churches, then professors and students were likely to engage in debates among themselves. Even if they turned all their practical questions into theological discussions of how God makes things happen, even if they confined moral philosophy to explaining why God wills the powerful to rule as they do, the schoolmen would still arrive at diverse explanations. It should have been obvious that these universities for teaching orthodoxy would conflict with the church hierarchy whenever professors and students tried to think for themselves.

So the university schoolmen were to teach and learn, but apparently not to think. They were told to uphold the philosophy of a two-story universe, in which their minds must be held subordinate to higher authority. But ordering the schoolmen to guard against their own intellectual freedom didn't work for long. In the 1330s the Franciscan professor William of Occam pointed out that the concepts of God as announced by the scholastic theologians were simply mental images existing inside their own heads. The chains of causation from heaven to earth were human constructions projected upon the universe.

Occam was a man of religion. But something moved him to demolish the whole second story of authority above his own mind. In Campbell's words (1976a: 582–83), Occam collapsed the earthly and heavenly worlds into one universe: where theology had been taken as an ultimate knowledge from the higher realms, Occam presented all religion as a matter of human psychology. Instead of dividing the universe horizontally in two, into fallen earth and spiritual heaven, Occam divided his universe into internal and external reality. Inside was the world of the human mind, with all its needs, beliefs, myths and social values, which had been almost the sum total of reality for cultures from the desert. Outside the human mind stretched the boundless natural universe, waiting to be explored on its own terms.

The Recollection of Classical Thought

By the 1200s, public memory of life before Christianity had been a virtual blank for centuries. The number of known copyists or teachers of classical knowledge were few. In his time, Joannes Scotus Erigena (810–877) seems to stand almost alone. Many orthodox Christians deemed it a real accomplishment that the memory of things past, including most knowledge of natural science, had been erased from living memory. The Renaissance was a time of remembrance. Recollection of pre-Christian thinking was the main innovation. When researchers and artists recalled their pre-Christian heritage in the 1400s, it was like a vista of the past appearing through a wall of clouds.

Until 1462, the ruins and artworks of classical Rome had been systematically mined for building stones and quicklime. Those overseeing the destruction apparently felt that nothing of value was being lost. Those buildings had been dedicated to devils and contained idols. But that year a "humanist" pope, Pius II, issued a church bull against the destruction of classical buildings (Smith 1976: 248). That bull perhaps symbolizes the end of the Middle Ages. For over one thousand years it had been heresy to proclaim the worth of any native European culture. Now even the pope expressed some pride in the pre-Christian past and eagerness to discover what it held. Of course, the church's memory of pagan heritage may have been selective, and imperial Rome was hardly representative of native European cultures. But the Renaissance also glorified purely Old European aspects of ancient culture. Classical sculptures of divinely beautiful nude men and women, or gods and goddesses, would formerly have been smashed as obscene idols. Now these artworks were treated with a reverence that had nothing to do with medieval civilization.

When books of Aristotle's philosophy were received from Arab hands in the 1200s, they appeared as the only major works on natural philosophy available. Many schools and universities therefore taught Aristotle as the leading authority on earthy matters for the next five hundred years, till the

1700s (Schmitt 1975: 488). At least the universities deemed some kind of learning about "earthly things" to be relevant.

"Humanism" began with the seemingly innocent process of translating books from one language to another. In the early 1400s, some professors started teaching Aristotle's philosophy in the original Greek. The students found that the received Latin translations had glossed over numerous distinctions of meaning in the Greek texts. In the resulting analysis of books in translation, some texts attributed to Aristotle were exposed as frauds, such as the *Secreta Sectarium*. Next, curious scholars began translating a series of formerly banned texts from ancient Greece. They found these works utterly different in tone or content from the Bible, or the works of medieval theologians. These rediscovered texts recorded the thoughts and theories of mortal humans, not the pronouncements of omnipotent authority from above. When Archimedes or Pythagoras had tried to define what was true, they had looked to the authority of nature, not to a voice from beyond. The ancients had also let their curiosity range over the whole world, rather than focusing on one set of authorized texts. The translators of these books were described as "humanists," because they dealt in "human" texts from the pagan past, rather than "divine" texts from the biblical past (Schmitt 1975: 508, 490–91).

As the number of translated classical books increased, Aristotle no longer seemed to stand alone as the voice of European antiquity. He was now one figure among many—Cynics, Stoics and mystics, all engaged in debate. The manuscript collectors and translators had unearthed a world of intellectual competition—the old Greek intellectual world, which St. Paul had deplored as a fountain of babble. The translated voices were largely from the old philosophical academies which had been officially closed in the 500s CE so that the clamor of merely human ideas would not distract from the words of God.

The translators next turned their gaze upon the Bible itself. Again flaws appeared in the official Latin text as compared with older Hebrew and Greek manuscripts. Human hands had tampered with the scriptures, and the clerical errors had been attributed to God (Schmitt 1975: 491). The humanists were now treading on ground deemed sacred by almost everyone. But if the Bible was sacred, how should its text be treated? Should the old Latin translations be accepted on faith? Or did the text merit closer analysis? Given time, the analysts went over the Bible with ever more powerful microscopes, inspecting the editorial fingerprints. By 1753, Jean Astuc would feel he could dissect the composite layers of Genesis like an archaeologist in an ancient city.

By the early 1400s, classical philosophy had been banned for so long that the ban itself was forgotten, and the church no longer stood on guard against dead classical pagans. Through unguarded doors the humanists brought "new" books from the pre-Christian past. Between 1495 and 1515 the Venetian publisher Aldo Manuzio printed 117 classical works, in a total of 120,000 copies (Crespo 1987: 174). Quite suddenly the church had lost control over book learning. If the intellectual life of Europe was like a drama in production, then the medieval stage had offered a limited set of biblical-style morality plays. Suddenly the stage started filling with new actors, most of them ghosts from the pagan past. These new actors presumed to speak, interrupting the play with entirely different lines, ignoring the stage directors' objections.

For most traditional Christians, the deeds of the so-called humanists were sacrilege, pure and simple. First the humanists had recovered and popularized the books of long-dead pagans, including the magical texts of Hermes Trismegistus, a "magi" and alchemist of ancient Egypt. Then they had printed edition after edition of these books, generating a market demand for heathenism. Finally they had turned on the Bible itself, trying to undercut faith in God's book by "proving" that it too was a human text and, by implication, no different in origin than pagan Greek or Egyptian books.

Many Christians therefore tried to "defend the sacred" from the humanists' profanity. This defense, however, presumed certain limitations on what could be considered sacred. According to the orthodox, anti-humanist or "pro-God" viewpoint, the Bible would not be divine if it had been written by human beings. Likewise, the laws of society would not be sacred if mere human beings had made them up. Things had to come from outside of nature and society before they could be considered holy. So prevalent was this orthodox presumption of polarity between the earthly and divine that many humanists also accepted the debate in those terms—they were rebels only in identifying with the earthly rather than the heavenly. Some humanists accepted that their eagerness to learn from pagan ancestors was indeed contrary to God's will and undertook their learning with a sense of rebellion. When texts from pagan times showed that the ancient Greeks and Romans had dabbled in democratic forms of government, some humanists found this an attractive anti-establishment notion. At the same time, they often accepted that since the Near Eastern societies of the Bible had been ruled by kings and emperors, this meant that autocratic forms of government were sanctioned by God. Thus down to the French revolution of the 1790s, and even the Spanish civil war of the 1930s, both autocrats and their classically-inspired democratic opponents tended to accept that the autocrats stood for religion, while the democrats upheld godlessness.

The Embattled Resumption of Ancient Learning

Most classical texts uncovered in the Renaissance do not seem controversial today. The "Corpus Hermeticum," an almost random compilation of pagan magical lore which sold out printing after printing, would seem like just one of thousands of occult works in modern bookstores. But in the late medieval world, such books were powerful threats to the social order. The humanists took the authorized texts taught in universities and churches, and asked if they were translated correctly. Next they questioned why certain books were designated as *the* authoritative texts rather than others. These were bookish questions raised by bookish men. The issues seemed relevant to scholars only, but in the atmosphere of late medieval times these questions were tools of major disruption. Humanist scissors were snipping at the great tapestry of belief woven for a thousand years to serve as the world view for all. The humanists were looking through the holes at the universe beyond, while the weavers cried at the destruction of their godly work.

When medieval universities began introducing pagan thinkers as "human authorities" in their curriculum, the schools picked up classical debates where the pagans had left off. The classical authorities tended to make faithful credulity and blind obedience seem stupid rather than virtuous. But the recovery of classical learning had implications beyond the questioning of authority. The works of Pythagoras, Hippocrates, Aristotle and Archimedes were not compilations of social mores handed down from God; they were collections of notes taken by people who were observing nature and trying to learn how it worked. These authors presumed that nature was an immense good, rather than a corrupt, fallen and hostile realm. For them, nature was the greatest teacher of all, far greater than any king or priest. When Renaissance students read these books, some simply tried to memorize the texts and treat them as received truth. But other students tried to understand how the authors had learned the things they wrote. Some of these students then became observers of nature in their own right. They started compiling their own notebooks, perhaps the most extraordinary being those of Leonardo. The note takers compared their observations with those of others. This new learning was called "humanism" also, because it relied on human observation rather than divine revelation. But this approach to learning departed from the original idea of humanism—learning from human rather that godly authorities. Instead, it was rudimentary natural science, which derived from the non-human authority of nature. With this transition, Europe returned to building on its own intellectual ground.

The Rebirth of Natural Science

For minds well schooled in biblical belief, the observation of nature was mainly a matter of reading biblical symbology into the universe. For this purpose it was sufficient, for example, to explain that eagles fly high in order to approach God. Pelicans reportedly picked the feathers on their breasts and fed their young on blood from the wound in order to symbolize the sacrifice of Christ (Attenborough 1987: 144). But new-style Renaissance thinkers hoped to learn *how* birds flew. They intended to perform natural science like the ancient philosophers. Unfortunately these budding scientists had little clue how to proceed. They questioned some things, failed to question others and generally had no idea how to test their ideas. In the sixteenth-century "vacuum debate," a series of thinkers such as Toletus and Domingo de Soto conducted "thought experiments" to test the Aristotelian belief that "nature abhors a vacuum." This was a controversial question, which challenged a long-accepted belief. The innovative thinkers asked, if a jar is filled with water and the lid sealed airtight, what would happen if the water froze? Some argued that as the water contracted into ice, a vacuum would indeed be created in the sealed jar. Others maintained that the jar must implode rather than allow a vacuum to exist. Both sides failed to observe that water actually *expands* on turning to ice (Schmitt 1967: 357–59).

At least these Renaissance debaters were able to point out that various accepted ideas were unproved beliefs. At least they vigorously debated how their environment worked, rather than how Europeans could more fully conform to ancient Near Eastern traditions. They no longer felt it sufficient to show that an idea came from a famous authority. Nature had emerged as a respectable standard of truth, by which the words of great men might be measured. Previously, if evidence from nature had seemed to contradict a great man, most people would have assumed that the great man spoke for God, and nature was wrong. All this was a fundamental shift in attitude. But even this degree of "critical thought" still made little difference to how people learned. The world was still full of sleeping assumptions. Hands-on work with earthly things was still delegated to peasants. Farmers and craftspeople learned from nature, but their knowledge concerned work the educated didn't do. "Humanist" thinkers conceived new explanations of cause and effect, but their ideas went untested. Nature was now taken as a teacher; but she was a professor with no paid position, who spoke no human language. The process of learning from nature, which Joseph Needham (1956) called "the discovery of the means of discovery," unfolded slowly in many small steps.

During the 1500s, as growing numbers of universities offered courses in "natural philosophy," various styles of debate went in and out of favor. Teachers like Jacobo Zabarella held that rhetorical excellence was just good packaging for an idea; the substance of a theory was best supported by evidence from *experiencia*. By this, Zabarella meant experience in general. He constructed his ideas and marshaled his arguments using whatever previous experience seemed relevant. Since he was an active man who traveled widely, he had many anecdotes to tell on almost any subject. But he did not perform *experimentia* (Schmitt 1969: 105). He worked in a debating hall, not a laboratory.

Galileo, on the other hand, definitely dabbled in *experimentia*. The trouble was, he didn't trust the results. He tried dropping objects of different densities, such as equal-sized pieces of wood and lead, to test his theory that the lightest body must fall the fastest. For some reason, however, these falling objects did not verify his calculations. He observed, "If one takes two different bodies, which have such properties that the first should fall twice as fast as the second, and one lets them fall from a tower, the first will not reach the ground appreciably faster." His conclusion was that these results were "contradictory and, so to speak, unnatural accidents." What could Galileo expect? His calculations were based on how perfect bodies would behave under perfect conditions. And he himself believed there are no perfect bodies in this corrupt world (Schmitt 1969: 115–17, 122). Instead of closely following the messy, erratic evidence provided by nature, Galileo tried to elevate the art of philosophical debate to a higher plane. In mathematical calculations he found an expression of pure logic, shorn of life's ambiguity. Like any aspiring prophet, Galileo hoped to know the mind of God. He would deduce it by doing mathematical thought experiments. In mathematics, he would "employ reasoning at all times rather than examples (for what we seek are the causes of effects, and these causes are not given to us by experience)" (Schmitt 1969: 111). If perfect bodies were not available for study in this world, Galileo would raise his eyes to the heavens. He would learn the laws of perfect motion from the heavenly bodies, observing the higher realms with his telescope. The church authorities despised him, because his claim of super-mundane knowledge rivaled theirs. Instead of seeking mere earthly knowledge like other scientists, Galileo aimed to outdo the church at its own game.

In the Renaissance and later Enlightenment, the sciences of greatest prestige were astronomy and physics. Men like Copernicus, Galileo and later Newton seemed to learn from nature at its highest level. Their wisdom was of clearly extra-mundane importance. Like the scholastic philosophers before them, these thinkers tried to understand how God controlled all things. In modern school textbooks, these men appear as founding fathers of the scientific age. But theirs was a culturally conditioned prestige. The astronomer-physicists stood out brightly in the firmament of knowledge partly because earthly learning was judged less important. As Leonardo da Vinci protested, "We know more about the movement of celestial bodies than about the soil underfoot" (in Suzuki 1997: 80). To the early modern mind, real science began with knowledge of the heavens. This was true, even though the astrophysics of the 1500s was of virtually no practical use to most Europeans. Only in Newton's day (the late 1600s) were the mathematical laws of physics combined with the know-how of craftspeople in any systematic way. After that, the astrophysicists slowly descended to earth, changing from high philosophers into practical engineers.

Of far less prestige were the so-called earth and life sciences. These were sciences related to crafts and farming, minerals, plants, beasts and human beings. This kind of knowledge did not seem to yield universal truths. Those studying organic life found they could not reduce life's infinite number of variables into mathematical formulas, though many tried. This untidy and lowly knowledge, however, had direct relevance for almost every European. During the Renaissance, over 90 percent of the people practiced some kind of farming. Everyone needed better nutrition and health care. The applications of botany, mineralogy, animal husbandry and medicine were more important for the evolution of modern Europe than astrophysics. In retrospect, perhaps the hierarchy of sciences should be reshuffled: the practitioners of plant breeding, explorers of the microscopic realm such as van Leeuwenhoek or the hands-on students of the human body such as Harvey might be spotlighted as the greatest pioneers of modern learning from nature (Schmitt 1973: 177).

The so-called scientific revolution was an unfolding of several different approaches to learning. Different scientific traditions emerged from different vocational and cultural circles. In one strand of scientific culture, the scholastic theologians imbibed classical philosophy, then reconceived God's laws as scientific principles governing the universe. Such scientists viewed their world through the spectacles of the Bible. As Carolyn Merchant (1989: 7) describes their beliefs, "Matter is dead and inert, remaining at rest or moving with uniform velocity in a straight line unless acted on by external forces." This "subject-object" context reflected the standpoint of socially dominant people, such as the educated sons of nobles. Their world was an arena subject to command from above, and science was the latest means of dominion over the earth. In the hands of such scientists, learning from nature evolved as "mechanistic science."

Meanwhile the know-how of farmers, craftspeople and healers evolved into "vitalistic science." In these practical circles, scientific questions concerned how living things could be helped to flourish. For most nature workers and vitalists, the world was alive—much as it was for the old pagans. As if to underline this pagan-vitalist link, the old antagonism of nature worshippers and Christians lived on in the controversies between orthodox Christians and vitalistic scientists. Later, when Darwin published his observations on biological evolution, many fundamentalist Christians reacted with passionate opposition. They knew a threat to their universe when they saw one: any vision of organisms in self-evolution contradicted the Christian belief that all creatures are objects controlled by one authority.

Like many scientists, Isaac Newton was split in his own mind between the mechanistic and vitalistic views of nature. On the one hand, he was the master interpreter of universal laws governing inanimate matter. On the other, he was an occultist fascinated with the animating spirits he saw in all things. As Frances Yates (1972: 201–2) writes of Newton, "The glimmerings of pattern he discovered in mathematics, physics and mechanics could never be, in his mind, adequate to explain the mystery and glory of life." In his later years, Newton leaned toward a more biblical faith and saw his work in physics as part of an ongoing revelation of God's laws. Yet it seems he never made a clear choice between his mechanistic and vitalistic views. Most likely, contemporary English society decided for him. As Martin Bernal points out, the mechanistic view had certain political and theological implications useful to the English Restoration monarchy. If matter and nature were seen as basically passive, requiring an external stimulus for all motion and order, then everything must originate and depend on a prime mover: "Otherwise, theologically, the universe would need no creator, or 'Grand Architect,' let alone a 'clock-minder,' while politically, England would need no King" (Bernal 1987: 272–73).

The Crisis of Near Eastern-Style Authority

In "Christendom," as medieval Europe was called, religious unity was more important than religious truth. Truth was determined by consensus in church council and approval by the chosen pope. Good and evil could be defined as conformity or deviation from the consensus. But the growing spirit of critical inquiry required that beliefs be measured by external standards of truth. If this was applied to religious orthodoxy, who would presume to sit in judgment on the church, correcting its mistakes like a teacher grading student papers? For better or worse, various reform-minded souls took this task upon themselves. Men like Martin Luther boldly claimed that the church's consensus was wrong—according to their own yardsticks of truth. The parishioners were forced to decide whom they believed. So the age of critical inquiry arrived within the church. Most people, however, still assumed there could be only one truth, one real community and one ultimate authority. As Pope John XXI feared in 1277, a period of intellectual strife was about to tear Christendom apart.

The Protestant revolt rose mainly in northwestern Europe, where the influence of Near Eastern culture was perhaps weakest. But in casting out the old unquestionable authorities, the leading rebels presumed they must impose new ones. And the "new" authorities were often older and more foreign than those they rebelled against. Luther and Calvin claimed to cut the ground from under the church hierarchy by making the Bible their ultimate authority. As Luther indignantly pointed out, many practices of the church, such as selling indulgences for forgiveness of sins, had no basis in scripture. The church, he claimed, had been meant to stick to its original traditions as established in the Holy Land and recorded in the Bible. Instead it had accommodated both clerical greed and local superstition. It had grown corrupt and gone native. The church allowed the sale of priestly offices and fostered cults of local saints, some of whom were also pagan deities, such as Bridget of Ireland and the Black Madonna. Apparently, all innovations by the European church were to be rejected. Once again, ancient Near Eastern tradition was proclaimed the standard to which all should conform.

The religious wars of medieval and Reformation times were basically crusades. The belligerent churches hoped to eliminate their religious competitors and secure a monopoly over cultural affairs. To do this they called on regional kings to lend their armies. As incentives to the rulers, the churches offered guarantees of salvation and divine sanction for the confiscation of heretic property. These were the same incentives offered to warriors in Middle Eastern holy wars. During the Reformation, Protestants and Catholics formed crusading alliances against one other. Protestant leaders appealed to their local princes, partly by urging them to confiscate Catholic church property. Protestant emissaries also pointed out that, in rejecting the international church, regional kings would emancipate themselves to become heads of their own national churches. No longer would they need to pass on tithes to Rome. As Protestant kings, they would be sovereign lords in both civil and religious affairs, like King David of Israel.

As the medieval church's unity came undone, each religious sect presented itself as the chosen party of God. Martin Luther proclaimed the issue in dualistic terms worthy of Zoroaster: "Once to every man and nation comes the moment to decide, in the war of truth with falsehood, for the good or evil side." The war of words began with mighty efforts at persuasion from both sides; later came recourse to violence. As the church-state alliances polarized, Europe split into two enemy camps, roughly north versus south. This was the pattern of demonization that had been seen before in the split of eastern and western Christendom, and would be seen in the twentieth century when Europe was once again divided into two ideological camps.

The Reformation religious wars burned on three fronts. On the main line of battle, the armies of north and south met in central Europe, turning Germany and Bohemia into killing fields. Behind the lines, both Catholic and Protestant authorities tried to exterminate heresy and treason on the home front. If their armies failed to defeat the Antichrist without, at least they could destroy the more defenseless enemy within. Thus the vast majority of deaths from witch hunts and Inquisitions happened during the period of Reformation conflict (roughly 1520–1648). A third front appeared in the world of books and ideas as the embattled churches sought to recover some measure of control over publishing, science and philosophy. This battle line was the least bloody. The emerging independent philosophers and scientists were seldom killed for heresy. They were, after all, educated and respectable men. It was usually sufficient to ban their works or to warn them to avoid controversy if they valued their careers. In some parts of Europe, notably Spain, such methods basically succeeded in rolling back the Renaissance.

Till the end of the Thirty Years War (1618–1648), Protestant and Catholic fanatics pursued these crusades within crusades in all seriousness. Each side strove to force a reunification of Christendom in its own image, and both sides failed. Armies of Christian soldiers pillaged and depopulated central Europe like legions from hell. Witch hunts raged through the war zones, till even avid prosecutors lost their appetite for killing women. The censorship of books proved quite effective in some regions such as Spain, but in the long run it also failed. All these crusades were eventually abandoned, and a vast disenchantment settled over the land, from which Christianity in Europe has never really recovered. The Protestants had declared war on organized religion for the sake of doctrinal purity and had ended up undercutting their own authority. As the number of breakaway sects multiplied, it grew obvious that no brand of religious orthodoxy could be imposed on a dissenting population. To that extent at least, the fanatics for the Bible enabled the new climate of free thought.

Before the Reformation the business of religious authorities was to define good and evil and enforce their standards on the local people. And this remains the business of most religious authorities in the Middle East today. Even after the Reformation many European priests and pastors continued trying to enforce orthodoxy with threats of eternal damnation. In the hands of such men, religion became mainly a series of rules to be followed in order to avoid punishment. But by the 1700s growing numbers of Europeans openly rejected this concept. The

English Lord Anthony Shaftesbury (1621–1683), for example, argued that morality was a means of achieving mental health, not of avoiding punishment after death. For Shaftesbury and his generation, wrongdoing was that which caused suffering and ill health, rather than that which deviated from religious orthodoxy. Increasingly, "the good" was something ordinary people defined for themselves in terms of their self-chosen goals. Even the law codes slowly changed in this direction, becoming collectively self-chosen "social contracts." The old lords and slightly reformed churches might still claim to represent a "higher authority," but the peasants and washerwomen now accused them of representing only themselves.

With the end of the religious wars, many educated Europeans saw Christianity as little more than a morass of hate-filled dogmatism. Some felt it their duty as enlightened people to defuse religious hatred and find something more humane to believe in. Perhaps, some reasoned, the old religions of classical and pagan times deserved another look. So in England during the 1660s and '70s, Ralph Cudworth advocated Platonic philosophy as a moderating antidote to the extremes of "Catholic superstition" and "Puritan enthusiasm." Bernal (1987: 191) explains that "Neo-Platonism and Hermeticism … were often philosophies espoused by moderates as attempts to transcend the raging political and religious battles of the time." In a similar spirit the "invisible colleges" of philosophers and scientists in ancient Greece were adopted as models for the Royal [scientific] Society of England. But such intellectual rejection of Christianity never gained a mass following. Instead, post-Reformation Christianity adapted to Europe. After the Reformation, priests and pastors in many areas had to compete for the affections of the local people. No longer could they presume to enforce their beliefs on the locals. Church leaders had to present themselves as champions of the people's own values, including the peasant farmers' regard for nature. Nature was increasingly presented as a blessing from the Lord, rather than an abode of the Devil. This softening of old biblical attitudes in Europe had been underway for centuries, as seen in St. Francis of Assisi's worshipful love of nature, or earlier in the words of Hildegarde of Bingen (1098–1179):

> The Earth is … mother of all that is natural, mother of all that is human. She is the mother of all, for contained in her are the seeds of all. The Earth … contains all moistness, all verdancy, all germinating power. It is in so many ways fruitful. All creation comes from it (in McLuhan 1994: 16).

So, without major announcements of theological change, many Christians came to regard both the Bible and the living book of nature as sources of grace and wisdom from God.

The Learning Economy

While Europe's medieval institutions of higher learning focused on the wisdom of the church, the peasants improved their farming skills. By trial and error their craft grew from lowly feudal labor into a folk science, an art and even a spiritual practice. Western Europe became a landscape in co-evolution, cultivated in ever more bio-intensive ways. Perhaps the kinds of changes under way can be represented by David Sabean's (1990) detailed survey of agriculture in the southern German village of Neckarhausen.

The medieval farming system in Neckarhausen involved heavy reliance on summer and winter grains, with some grapes for sour wine. The locals grew various grains in summer and spelt in the winter. The usual crop rotation took three years—one year for summer grain, one for winter grain and then a fallow year. The land was divided into small farms, with daughters and sons usually inheriting the land equally. But from the early 1700s the farmers evolved a much more complicated system, with new vegetables, fodder crops, fruit trees, and more livestock.

The fallow fields were the most important sites for innovation. Increasingly, these were planted in "fallow crops," such as the new potato, cabbage, clover or alfalfa. Such crops were "fallow" mainly in function because they restored elements to the soil that grain crops depleted. The result was a whole series of extra crops which fed on each other in a more bio-intensive way. These root and vegetable crops in the fallow fields were like an extra leg added to stabilize the farm economy. Now if the grain harvests were poor or the prices low, the root and other fallow crops could fill the gap. This was as crucial for early modern Europe as it was in Africa, where the addition of root crops such as cassava was a life-saver for some regions.

The fallow fields in Neckarhausen were traditionally pastures for village animals. This might seem to conflict with intensified cropping, but the early modern farmers made animal fodder one of their intensified crops. Instead of just letting grass grow and letting the village sheep or cows graze there, the villagers planted the best fodder plants they knew, and rationed the yield to the animals. The chosen fodder plants included clover and alfalfa, which had better nutritional value for the animals than local grass and greater benefit for the soil. These plants yielded more for the animals if they were protected from grazing until mature. So people started keeping their animals in fenced yards and carrying fodder to them. This also became a trend in modern Africa, where designated fields are often placed temporarily off limits for grazing, lest the animals nip the fodder plants at the bud. In Neckarhausen, this active production of fodder and protection of its growth meant that the village could provide more fodder from less pasture than before. The extra meat and milk enriched the local diet and provided another series of products for market. The extra manure enriched the soil (Sabean 1990: 437–42).

To round out the picture, the villagers increasingly filled the borders of their fields and other marginal places with fruit trees, usually apples. Then, instead of committing fields to vineyards for wine, the edges of fields became apple forests, and cider became the common alcoholic drink. By such means the villagers made their environment richer and more beautiful. If the game of adults is to make a real fairyland, Neckarhausen was on its way.

In the big picture, the forest cover of western Europe steadily shrank from about 95 percent of the land before the Middle Ages to around 20 percent in the 1800s. Whenever rising populations needed more food, they chopped down trees to extend the grain fields. Then in the early 1800s, the overall shrinkage of forest stopped. The demand for more food was finally being met through bio-intensification rather than mere extension of cropland. With smart but simple steps, the villagers learned to help the crops and improve the soil. They began using seed drills (an invention from village China) to plant their seed under the soil, instead of broadcasting it on the surface. Far more of the seed grew, and less went to feed the birds. Many farmers also started installing tile under-drainage in waterlogged fields. By the late 1800s the total land planted in grain across Germany, Britain and France started to shrink. With constant selection of seed from the best plants, the ratio of seed sown to seed harvested steadily improved. The rising yields allowed fewer acres to be devoted to grain (Price 1994: 80). Only farther east, especially in Russia, did field expansion remain the major means of increasing production.

In the hindsight of history, this series of local improvements has been lumped together and called "the first agricultural revolution." This "revolution" of gardeners roughly doubled crop yields from those recorded in medieval times, which was enough to move Europe beyond subsistence and virtually eliminate famine by the late 1800s (Ponting 1991: 245–46). All this happened first in northwestern Europe because the environment was conducive, and perhaps also because the culture and power of exploitative invaders was least dominant in that corner of the continent.

The "first agricultural revolution" happened through intensified labor, and the credit must go largely to village women. The nineteenth century was a period of renewed population pressure, when growing numbers of young men left the crowded villages for outside work. Many men filled the ranks of the rapidly growing armed forces, and many left for America. The departure of so many males brought a certain feminization of village work (Tranter 1994: 42). Adoption of foreign plants, selective breeding, active soil improvement, extra animal care, fallow crop weeding, and fodder gathering were all commonly seen as women's work. So women's economic importance resurfaced after the medieval age and the witch hunts. Once again in human history, women providers played a crucial role in expanding the means of sustenance.

Sure enough, the renewed economic role of women brought a "decline of the family" as defined by the church. Sabean (1990: 386) reports that the position of the German *hausherr* (lord of the house) came under threat as productive women sometimes fired their so-called managers. Most divorce cases recorded in Neckarhausen happened because "the woman provoked the split." The divorced men commonly tried to have the local authorities force their wives to take them back, partly from fear of having to support themselves without a wife. The church pastors were concerned by a rise in "illegitimate" pregnancies among women who wanted a child and a lover, but not a partly dependent, legally dominant husband.

Later, in twentieth-century Europe, the labor intensity of farming was reduced, and women's energies were increasingly directed to other vocations. But the founding principles of bio-intensive farming tended to remain. Wealthy, largely male emigrants to the Americas or the colonies had often imposed monocultural plantations on foreign villagers, but in Europe the disasters of the Irish potato famine and the French grape epidemic had underlined the ecological perils of monoculture. Most farmers in Europe had never given up their principles of crop diversity and ongoing soil-enrichment. The fruits of their efforts therefore increased on a generally sustainable footing. Even with growing consolidation of farms and a rising agro-chemical industry, the general practice of farming remained on a path of bio-intensification. In recent decades, Europe's farmers have grown ever more food on ever less land, which has allowed limited reforestation.

The Age of Makeshift

The European economic "takeoff" of the 1800s can be described as a vast experiment in diversification. First, many new crops and methods were used in farming. Next, people slowly learned to support themselves from an expanding range of other resources. Before the 1800s, innovations in farming and technology had usually happened in times of emergency, when population pressure on known resources was greatest. And the early 1800s was a time like that. People with too little or no land scrambled to piece together a living in any way they could. Sabean (1990: 156–57) says, "The working poor were on the margin of existence, such that even a family with all members contributing … had no cushion to fall back on." Diversification of crops, the making of more with less, part-time jobs off the farm, and migrations to America happened under Malthusian pressure. According to Roger Price (1994: 84), "the development of rural manufacture might be considered not as 'proto-industrialization' but … as part of a desperate effort to make ends meet." Across western Europe the subdivision of farm plots had rendered farms insufficient to support families; then the subdivision of farms tended to halt, and the entire surplus population had to find other work.

In hard times, people used what was formerly waste. Germany saw a rise of almost fantastic odd jobs, such as gathering manure from the roads for sale, collecting wool and horse hair from thorn bushes, catching rats, gathering resin to grease wagon wheels, selling herbal remedies door to door or offering magic spells as a kind of property insurance. Heinz Reif says, "The incomes of the poor [were] composed of too many component elements and imponderables to permit ready assessment" (in Sabean 1990: 157). This was the "economy of makeshift" out of which the modern economy grew.

In the 1800s, "overpopulated" Europe could have degenerated into another age of warlords, with the strongest seizing the means of life by force. And to a large extent this was the response of Europe's hereditary elites. The rulers staged wars of territorial expansion, either in Europe, or overseas in the race for colonies. The landed nobilities of Russia, Hungary and Romania increased taxes on the farmers and invested much of the money in modernized armies. The chief function of these armies was to hold down the growing threat of peasant revolt (Janos 1989: 347). Spain repeatedly exploded into civil and class war. In the impoverished countryside of southern Italy, the Mafia rose. The old dominators of Europe were alive and seeking to thrive in a time of want.

But this was not the response of most farmers or craftspeople. Most ordinary people were focused on meeting their needs through production, not coercion. The villagers of England before 1700 had traditionally produced cloth, iron tools, boats, paper, glass, housing, carriages and copper, all while tending their farms. Now the "surplus" people expanded these previously part-time crafts into full-time trades. The learning curve of the workers was steep: by the mid-1700s, homegrown rural industries produced gloves, stockings, fine furniture, lace, refined sugar, alum and brass goods (Lougheed 1994: 161).

Under similar pressure the Dutch had raised their networks of windmills and dikes to intensify farming in the coastal marshes. At the same time they expanded their fleet of ships, till at one point they owned half the world's shipping. As the Little Ice Age departed and the population rose again, the Dutch were more innovative than ever before. The feudal controls over workers were nearly gone, and people busily explored their options for survival (Van Houtte 1977: 316–17). Instead of suffering a population-slashing series of famines, plagues and wars, the Dutch managed to build up their biological wealth and productive machinery. Their thousands of waterwheels and windmills were put to use fulling cloth, tanning leather and making paper. In almost every community, numerous people displayed a capacity to invent and produce something new (Lougheed 1994: 161).

In Switzerland, textile industries evolved first as a winter craft to supplement farming. Then, as displaced farmers made weaving their full-time work, cottage industries tended to combine into local factories, powering their weaving machines with wheels on fast-flowing mountain streams (Foreman-Peck 1994: 220). Metal craft also grew as a part-time village business in the Swiss mountains. Over time, these cottage-industry skills evolved to the point of producing clocks and watches (Lougheed 1994: 172).

By such a gradual process of learning, the economies of Britain, Belgium, France and Switzerland all grew largely from the bottom-up at a pace of around 1.25 percent per year, from 1830 till 1910 (Foreman-Peck 1994: 254). The old hereditary nobles and landowners increasingly capitalized business as an investment, but the workers and managers who actually produced things usually came from farming, trade or merchant families. By contrast, nations such as Spain and Portugal, which relied heavily on colonies for farm products, silver and gold, fell behind in building their industrial skills. Their conservative elites practiced conquest and extraction, much like some of their Visigothic or Moorish ancestors. The nations which had to rely more on their own innovations, such as Germany and Switzerland, acquired wealth more slowly but finally became high-performance economies.

Exploring the Chain of Being

The rebirth of natural science began making a major difference in the 1800s by allowing a diversification of the materials society could live on. Previously, human beings had lived almost entirely off the Earth's fragile green film of photosynthetic plants. Now the growing number of amateur and professional scientists worked their way up and down the chain of being, "discovering" chemical compounds and bands of electromagnetic energy. Next they invented ways to use these resources to meet human needs. By the 1700s, early chemical engineers were concocting batches of sulphuric acid, hydrochloric acid, chlorine, "bleaching powder" (calcium hyperchlorate), and soda of salt as chemicals for treating paper and textiles. Several of Germany's major chemical firms started as family chemist shops, inventing dyes which didn't fade. Soon large numbers of people were supporting themselves by dealing in chemical substances; the possibilities and perils of a chemical economy had begun. Within a hundred years these new chemical, metallurgical and electrical industries would generate more income and more jobs than farming.

At first, industrial mining and metal smelting involved a massive new attack on the forests. In numerous cases from Wales to Germany, primitive iron mills opened in wooded countryside. Within several years the furnaces had consumed the trees, and the mills stood abandoned within a wasteland. The switch to coal as fuel at least diverted this destruction from the living forests. The metal industry instead turned on the earth's store of dead trees. This was simply another kind of biomass depletion, also done at a far faster rate than the Earth could renew. However, if the *living* forests were all consumed, there would be no possibility of biomass accumulation, period. The consumption of dead forests under the earth was a diversification, expanding the kinds of "food" the economy could devour. In view of the alternatives, it was a step from an endangered to an expanded resource base.

Whenever new resources in the chain of being were tapped, new dangers of unsustainable exploitation arose. If new resources were exhausted, the earth's food chain could be cut in new places. But a more diverse economy also had a greater flexibility to avoid overdepleting any one resource. The more links of the chain of being people could draw on, the less the danger that any one of them would be exploited beyond recovery. If every part of the world tree was used, there might be less likelihood of cutting a ring around the trunk.

Through the nineteenth century, applied physicists probed the spectrum of electromagnetic energy, tapping new resources. Their primary development was electric power. This was a new energy resource, but its generation depended on known sources such as coal and water power. Like other infant industries, the electric firms were like child prodigies, bearing enormous potential but often displaying frightful personality disorders. They were marvelously self-absorbed, exploring their own powers, while hardly noticing if they happened to deface the entire landscape. Simple arithmetic, which often passed for good sense among business engineers, indicated that the higher a dam was built, the more its power and profit would be multiplied. However, the environmental side-effects were multiplied as well. Instead of converting Europe's vast number of streamside mills into local generating plants, and rather than intensifying small-unit efficiency, the contractors usually went for maximum-sized contracts. Only slowly, after decade after decade of conflict with increasingly organized protest movements, would some sense of wider responsibility creep into the corporate offices of the electric giants. Then the uninsulated wiring strung all over Europe would have to be redone, and court cases started protecting watersheds by imposing height limits on dams. In incremental and often reluctant steps, the adolescent energy industries toyed with more efficient and sustainable means of power. Their proper goal, according to nineteenth-century German politician August Bebel, was the achievement of non-exhaustible energy. In the future, Bebel believed, "a few square miles in North Africa would suffice for the [solar energy] needs of a country like the German empire" (in Dominick 1992: 63).

The New Economy's "Excrements" and Their Uses

If the whole population of Europe had remained directly dependent on the land's organic resources, the continent would probably have been reduced to an ecological emergency zone, more or less like North India. As it was, the rapidly advancing industrial complex took some human pressure off the land, allowing whole sectors of society to live on other resources. Of course, every innovation had its side-effects, each requiring a solution. The new metal and chemical industries fed voraciously on the soil, rock and groundwater which underlay organic life. Instead of cutting forests for blast furnaces, much more fuel was obtained through open-pit mines which ripped whole landscapes out from under farms and forests. After these industries took their raw materials, they dumped the mineral or chemical leftovers ("excrements" as the German Greens later called them) usually straight into the public air and water supply. The nearby farmers generally needed no scientist to tell them the harmful effects on their fields.

Waste, of course, is partly in the eye of the beholder. The by-products of industry were materials which had always been in the world before. And as scientists had repeatedly shown, all matter and energy was a potential resource, depending on how it was used. "Excrements" therefore, were also potential resources. The more industries using non-organic materials, the more kinds of "wastes" they produced; the more uses found for non-organic substances, the more likely that one industry's waste would be discovered as another's raw material. In that case, the interests of public health and business profit could both suggest that industrial "wastes" be used for something besides chemical sewage.

One early attempt to weave such a "linkage" between industries was the production of chemical fertilizer. As the metal and chemical industries generated by-products such as sulphates, potash and phosphates, an agricultural use was found for some of these wastes. In that case, industrial excrements were being used the way Chinese cities had used human wastes. By the late 1800s the market for fertilizer was evident as farmers bought lime, marl, crushed bone or imported guano to build up their soil. Now farmers were being offered a range of new fertilizers made of leftovers from the chemical and metallurgical industries (Price 1994: 94). Many of the first chemical fertilizers could hardly be called "artificial," any more than rock powder could be considered artificial. Soil, of course, is originally mineral rock, pulverized by roots and microbes into a powder digestible to plants. And the early fertilizers were often nothing other than crushed rock of various minerals. In general, these additives enriched the soil, clearly raising crop yields beyond levels achieved in the first agricultural revolution.

After World War I, the enormous explosives industry tried to keep up demand for its products, partly by turning to production of chemical fertilizers. The industry developed an interest in selling as many chemicals to farmers as they could be induced to buy. These newer fertilizers were increasingly straight chemical additives. And instead of supplementing compost and manure, these chemicals started to replace organic fertilizer completely. Crops were offered straight infusions of potassium, phosphate or calcium—in dosages higher than most plants had ever known. The results seemed impressive at first, though feeding plants such food is like trying to produce supermen by feeding them synthesized vitamins and minerals rather than letting them eat food.

The chemical industry next began producing poisons for killing unwanted plants and animals. This pesticide and herbicide industry also grew partly out of wartime production, from the chemical warfare industry. The producers advertised these poisons as medicines, like antibiotics for farms, to kill certain pest infections and promote plant health. But though the lethal dose varied from species to species, these chemicals were poisonous to almost every living thing. The buildup of these toxins soon grew into a major environmental problem, and the quest was on for "safer pest-controls."

In all this relationship between farming and industry, a pattern emerged in which chemical products were constantly being "invented" and marketed to the public. After a time, the side-effects of these products grew evident and they might then be banned or modified, or put to alternative uses. Hopefully, this was a process of trial and error in which people would learn to use ever more things in ever better ways. If no fatal mistakes occurred, the experiments of industrial science might resemble the efforts of African women who long ago learned to treat and eat the poisonous manioc plant, thereby increasing the range of foods available for their families. But instead of a mother risking herself to try something new, the industrial scientists tended to use the whole public and biosphere as their guinea pigs in product tests of pesticides, herbicides, thalidomide, phosphate detergents, plutonium and bio-engineered organisms. So long as the onus was on the guinea pig to prove it had been harmed, the industrial process of trial and error was less like a series of controlled scientific experiments and more like a planetary game of Russian roulette.

Yet for all these risks, the scientific economy raced on. Hopefully the forces of discovery would tap an ever-widening spectrum of the Earth's resources in ways both sustaining and sustainable. By the end of the twentieth century, the fastest growing segment of the "developed world" economy was built on chips of silicon from sand. And if people could discover ways to live off sand, sun and communication, then maybe they could afford to let the plants and animals grow back again. Perhaps it would be not just technologically feasible but also economically fruitful to get off nature's back, as when the eighteen nations surrounding the Mediterranean Sea signed a clean-up agreement with $10 billion pledged to help make the region more healthy and beautiful for locals and tourists (*Christian Science Monitor*, July 18, 1980).

In the scientific economy, the whole world became a laboratory. But perhaps it had always been a laboratory. The trials and errors of ancient people had also changed the planet, however little they were aware of it. Maybe modern scientists just tend to suffer from a different myopia. They may fully intend to re-engineer or re-bio-engineer the world but sometimes forget that they and their families live within the hermetically sealed flask. Back at the beginnings of the scientific economy, many alchemists sensed they were working on nature and themselves at the same time. They felt that the upliftment of humanity must grow like a flower from the fulfillment of nature. A sixteenth-century alchemical text, the *Rosarium philosophorum* (Rose Garden of Philosophers) offered a context for future seekers of that fulfillment: "For this work you should employ venerable Nature, because from her and through her and in her our art is born and in naught else: and so our magisterium is the work of Nature and not of the Worker" (in Campbell 1976a: 265).

Chapter 14

The Dominator Backlash in Modern Europe

In Victorian times many Europeans believed that the new age of technical marvels had rendered scarcity obsolete. Therefore they asked, What cause could remain for war? If people's means of life were secure, why on earth would they want to keep killing each other? Of course the dominant leaders of Europe did not feel secure. They accelerated their drives for colonies, trying to stake out claims to all of the earth's remaining resources. Their rivalries led to numerous minor disputes and a massive arms race which exploded into World War I. The productive capacity of Europe was then channeled into all-out destruction of rival nations. To enable this seemingly Darwinian struggle for each nation's "place in the sun," young men of Europe were sacrificed by the millions, and rival military machines tried to cut off food to "enemy populations." In the wake of that holocaust, the winning alliance stripped the losers of their colonies and gutted their armed forces. The Victorian optimism about a future free from rivalry or scarcity now seemed like a pipe dream.

In the face of such trauma, Europe's mass media echoed with calls for return to tradition. But what tradition did they want? The radio amplified the voices of pastors and politicians into every community. Many spoke of Europe's past greatness, blamed traditional enemies for present problems and spoke of the "faith of our fathers." But which fathers did they emphasize? Through the tangled web of Europe's history, various kinds of traditionalists marked their chosen lines of heritage. The selection of guiding traditions could be like tracing rightful kings through the tangled bloodlines of contending noble families. All voices of tradition recall specific periods of history. If psychological archetypes are at work in people's minds, they probably come from the experiences of real ancestors. So the modern voices for tradition presumed an identification with certain ancestors, as opposed to others.

In a gradually more egalitarian Europe, some orators gave vent to the frustrations of the old dominators. Their words reflected the passions and ideas of the ancient military elites. Adolph Hitler in particular expressed such sentiments, often with a shocking authenticity, as if channeling the voices of ancient conquerors. Through such shame and pride-driven men, Europe's history-long clash of warlords and cultures could be conjured again. Through the mass-media echo chamber, that whole traumatic history could be called up and replayed as a psychodrama in the streets.

The modern fascist movements were coalitions of people who identified with the greatness of the old military nobility. Their backlash had gathered force for nearly two hundred years, as the old elites built an ideological case against their detractors. This building of a new dominator ideology is perhaps best explored by Martin Bernal (1987) in his volumes of *Black Athena.*

The Resurgence of Aryan Identity

Possibly till the French revolution of 1789, the nobles of Europe had seen themselves more as families than as an ethnic class. Then the great commoner revolt prodded them to reformulate their claims to privilege. The rising business community also feared the revolutionary mob. Successful business people generally aspired to join the upper classes, not to pull the nobles down. So, as celebrated in thousands of pulp novels, the old nobles and the newly wealthy business people tended to marry. They wove their ties of blood and upheld civilization as they knew it.

In 1795, a few years after the revolution, Professor J.F. Blumenbach of Gottingen University published a book called *De Generis Varietate Nativa.* This was the first "scientific study" of human racial types. It provided zoological-style classifications of races and attempted to isolate the "pure" origins of each race. In this book, the term "Caucasian" was first used to describe the White race. Through the galaxy of human relations, Blumenbach tried to trace a lineage for his own kind of people, somewhat as the patrilineage of Israel is traced in the Bible. Blumenbach, however, was working on the larger scale of races across the world. His genealogy seemed to transcend the ethnocentric view of folklore and become a science. He traced the ancestry of the White race to a pure heartland in the Caucasus Mountains of Inner Asia. There, it was commonly believed, Noah's ark had come to rest near the peak of Mt. Ararat. Also according to legend, from there had come the virile warriors who became the ruling classes of White Europe (Bernal 1987: 219).

Linguistics provided better evidence that Europe's "finest people" had come from Inner Asia. In 1786, Sir William Jones announced that the similarities between Sanskrit, Latin, Persian and German could be no accident: there had to be a common source behind them all. Jones described the series of related languages, spoken in a semi-circle of borderlands around Inner Asia, as the "Indo-European family" of languages. Some German linguists theorized a common homeland from which these Indo-European speakers came, calling it the "*Urheimat.*" The original bringers of Indo-European speech they called the "Aryans."

Within Europe, Germany was obviously closer to the *Urheimat* than the lands further west, and the German language bore a closer relation to the Aryan tongues of India and Persia. At first, Sanskrit was hailed as the most ancient of Aryan languages. Later, that honour was moved closer to Europe, as Ferdinand de Saussure claimed the "Proto-Indo-European language" was most likely Lithuanian. In 1823, the German linguist H.J. Klaproth proposed the name "Indogermanish" for the source behind most European languages (Bernal 1987:

372). By the mid-1800s it was widely accepted as scientific fact that the Aryan race had come from the steppes of Inner Asia, and these people were responsible for all that was higher and stronger in the civilizations of Europe, the Near East, Persia and India.

The "Aryan Model" of History

When Ernst Curtius wrote volume one of his *History of Greece* in 1857, the theory of an Aryan source of civilization underlay his thinking. Curtius emphasized a racial distinction between the ancient Indo-European Hellenes and the more primitive aboriginal Pelasgians. "The Pelasgian times," he said, "lie in the background—a vast period of monotony: impulse and motion are first communicated by Hellen and his sons, and with their arrival history commences" (in Bernal 1987: 333). Curtius already ascribed to what Martin Bernal calls the "Aryan Model" of European origins, as opposed to the "Ancient Model" of the classical historians of Greece and Rome, who had commonly looked to Africa and the Near East for much of their cultural heritage. The Greeks particularly emphasized their debt to Egypt. Bernal shows how an emerging racism in nineteenth-century Europe overrode the views of classical historians. In an "ethnic cleansing" of the history books, the new racists claimed purely Indo-European roots for Western civilization. Connop Thirwall, for example, utterly dismissed the ancient accounts of early Egyptian colonies in Greece. In volume one of his *History of Greece* (1835) he said, "Settlers of purely Egyptian blood, crossing the Aegean and founding maritime cities [in Greece], appears inconsistent with everything we know about national characters" (in Bernal 1987: 326). Bernal points out that, as Thirwall wrote these lines, the newspapers of Europe were full of reports from the war in Greece, where a powerful Egyptian navy and army under Mohamed Ali fought to maintain a centuries-old Ottoman administration there.

Bernal is concerned to undo such ethnic cleansing of history. He debunks racist scholarship and punctures the Aryan Model like a balloon. But in stressing the reality of African and West Asian influence on classical Europe, Bernal seems to imply the unreality of Indo-European influence. In that case, the Dorian invasion of ancient Greece might seem a figment of imagination, invented to satisfy the self-imagery of Aryan racists. Bernal, however, admits that classical Greece was influenced from every side—by Egyptians, Hebrews, Turks, Dorians, Slavs and Romans (Bernal 1987: 331). Later, people crafted versions of history in which the roles of different ethnic groups were emphasized according to the storyteller's preference.

"Pure" and "Mongrelized" Greece

In the Enlightenment of the 1700s and afterwards, it had become fashionable to downplay Europe's biblical heritage and to draw inspiration instead from classical Greece and Rome. So from the United States to Russia, Greco-Roman buildings sprouted like mushrooms. Ancient Greece was more popular than

Rome, and Athens, it seemed, had been the very fountainhead of native European genius. Therefore the movement to liberate Greece from Ottoman rule (in the 1820s) became a new crusade. According to Prussian Education Minister Wilhelm von Humboldt, the early Greeks were a chosen race destined to be the light of the world. They were the original Europeans, unsullied by the later influence of other races and cultures. As von Humboldt put it to his students in the mid-1800s:

> Our study of Greek history is ... a matter quite different from our other historical studies. For us the Greeks step out of the circle of history.... We fail entirely to recognize our relationship to them if we dare to apply the standards to them which we apply to the rest of world history.... From the Greeks we take something more than earthly—almost godlike. (in Bernal 1987: 287)

Von Humboldt tried to model Prussian education on the old Greek gymnasium schools. His German classrooms featured military-style discipline, with a well-rounded curriculum of natural sciences, arts, and body-toughening sports. Such Prussian gymnasiums were probably quite true to the spirit of education in the Hellenic empire after Athens and Sparta had been thrown together under the Macedonians.

This popular fascination with classical culture could have lead directly to the rediscovery of Old Europe. The old values of communal democracy, esteem for women, and knowledge of nature were alive in classical times and documented in ancient literature. It might have been obvious that these values represented a different culture from that of the military elites. And for a few students of the classics, a "matriarchal" past did seem obvious. The modern popularization of classical culture has fostered increased support for democracy, nature protection and women's rights. But the old elites and nobles of Europe tended to see their own sides of the classical heritage.

In the light of linguistic discoveries, the Greek dialects most directly related to "Indogermanish" were Dorian dialects. And when members of the Prussian upper class looked for their own cultural roots in Greece, they sensed their closest kinship with the "Dorian race." This ethnic group was singled out for attention in Karl Mueller's *Die Dourer* (1830), translated into English as *The History and Antiquities of the Doric Race*. In Hesiod's legend of Greek history, the Dorians were described as a "race of iron" who had descended from the north as raiders. Now early-modern historians credited these Dorians with virtually all the technological, artistic and intellectual achievements of ancient Greece. This race was presented as a vigorous, heroic and almost godlike people standing at the headwaters of Western history.

Those who claimed that ancient Greece had risen up in a blaze of glory also had to explain its later decline to a poor province ruled by the Ottoman empire. How had the sons of Hercules and Alexander fallen so far? The racist answer was that they had intermarried with lesser types, and the mongrelized children lost their parent's inborn virtues. According to Compte de Gobineau, they became "blackened" and "Semitized." Gobineau carefully distinguished the various ethnic pollutants he saw corrupting the Greeks. He listed "aborigines" (which is what he called pre-Dorian Europeans), "Black Hamitics," Phoenicians, Hebrews, Arab Semites, Philistines, Libyans and Cretans. In other words, Greece had fallen from its height of glory under Alexander the Great because its people had become the cosmopolitans of Alexander's dreams. All this was depressing reading for a European gentleman of relatively pure lineage. But Gobineau quickly turned to the good news. The Indo-Aryan people of northern Europe, such as Gobineau himself, had never been so mongrelized. They remained pure Indo-European Caucasians who were generally aloof from interbreeding with lesser types (in Bernal 1987: 360–62). The Aryans of northern Europe were therefore the true inheritors of all superior qualities which the Dorians had lost. Gobineau found this boost to his self-image intoxicating. In the future, he felt sure, the cream must rise to the top.

The Aryan Christian Church

As Martin Bernal (1987: 289) explains it, the romantic movement of the 1800s was a quest to recover both the original message of Christianity and Europe's pagan past. So in English schools, boys were taught to be Christian gentlemen, partly by studying the pagan classics of Greece and Rome. Such a contradiction in sentiments troubled Matthew Arnold. He felt it disturbing that the Victorian public remained so biblical in culture. Since the Protestant Reformation, he complained, the British had seemed to be more "Hebraic" than the Roman Catholics. Arnold felt it unnatural: England should look more to its native European heritage. "Hellenism," he argued in his 1869 book *Culture and Anarchy*, "is an Indo-European growth. England is an Indo-European nation. Thus it would seem to belong naturally to the movement of Hellenism" (in Bernal 1987: 347).

Most other European culture-boosters were far less willing to ditch the Bible and Christianity. On the continent, Compte de Gobineau continued the Romantic struggle to entertain both Christianity and Indo-Aryan "Hellenism." He was, as Bernal (1987: 343) puts it, "torn between his conservative support for the church and his excitement at the new theory of racism." In the Bible, which Gobineau claimed to believe in, there was only one creation of humanity. But Gobineau also believed that the world's races were different species with different origins. His own theories of ethnic degeneration in Greece must have suggested that the different "species" of people were sexually compatible, but he did not wish to dwell on that point.

Gobineau's impulse to recover his heritage was governed by who he identified with. According to his historical theory, the "aboriginals" of Europe were among the polluters of Indo-Aryan blood. This was just how the Dorian Spartans had viewed relations with "aboriginal" helots. Gobineau identified with the ancient invaders of Europe and shared their contempt for the Old European peasantry. He no more identified with Old Europe than the Spartans had. He had no more respect for Europe's non-noble peasants than most settlers in the New World had for the conquered native Americans.

Still, Gobineau wished to claim Christianity as part of his own heritage and to mix Aryan exclusivity with church membership. And in this desire he was far from alone. "Visigothic Christianity" (or the tribal Christianity of the invaders of Rome) was back. The descendants of feudal lords would now attempt to cleanse their religion of all non-Aryan influence. So Matthew Arnold proposed in a private letter:

> Bunsen used to say that our great business was to get rid of all that was purely Semitic in Christianity and to make it Indo-Germanic, and Schleiermacher that in the Christianity of us Western nations there was really much more of Plato and Sokrates than of Joshua and David. (in Bernal 1987: 348)

It was hoped that a lily-white Christianity could be made to appear from behind the tarnished veneer of church history. Only one fact seemed to truly block the quest: Jesus was a Jew. But where there was will, there was a way. Paul Legarde proposed that Jesus was really an "Aryan Jew" who had been crucified by "Semitic Jews" (in Bernal 1987: 349). Other churchmen claimed that Jesus was not a Jew at all, but a Christian Aryan—and the Jews killed him because he exposed their inferiority.

Soon it became an ongoing practice to isolate and purge the elements of "Semitic" culture from Europe's national religions. Ernst Renan wrote in 1855, that Semitic civilizations contain no

> mythology, nor epic, nor science, nor philosophy, nor fiction, nor plastic arts, nor civil life.... The Semitic race is to be recognized almost entirely by negative characteristics.... There is a complete absence of complexity, subtlety or feeling, except for unity. It has no variety in its monotheism. (in Bernal 1987: 346)

Such "Christians" were rapidly setting the stage for a literal purge of non-Aryan influence. For fifteen hundred years, Europe had been culturally dominated by an originally Middle Eastern religion. Now an enormous romantic and nationalistic backlash was gathering force. Who should be blamed for oppressing Europe's native soul? Which European traditions should be thrust forward as the truest and best? The quest for what German romantics called "*zeitgeist*" (the authentic spirit of the land and its people) seemed to involve a violent reaction against certain "outside elements." More accurately, it involved an identification with traditions originating from Central Asian invaders, rather than traditions from the Middle East. Such a *zeitgeist* celebrated the spirit of Europe's old dominators more than that of its indigenous villagers.

Later, in a more democratic Europe, a real *zeitgeist* of the land could give rise to bio-regionalism. But first, the old Indo-European nobles, or would-be nobles, of Germany, Italy, Spain and Hungary rallied in a last bid for dominance over the land. They wrapped themselves in their national flags and claimed the blessings of both the church and the old Indo-European gods. With mounting boldness they urged an Aryan Christian crusade for the ethnic-cleansing of Europe.

The Master Race

After claiming the support of tradition, country and religion, those who would dominate Europe next claimed the support of science. For many traditional people, of course, Darwin's theory of evolution seemed a virtually Satanic insult to the dignity of humankind. If Darwin's theories were true, there would be no pure bloodlines—only blood from the beasts. If people were descended from animals, didn't that mean they were and are animals? And if this was believed, how could people be expected to treat each other any better than they treated their animals?

Later, however, Herbert Spencer reinterpreted Darwin's theory in a way flattering to humanity. He turned the tables on the naysayers, arguing that if evolution progressed through the survival of the fittest, then humankind was the winner of that struggle. Humans had won dominion over all the creatures of the Earth, not by passively receiving it from God, but through hard struggle and sheer superiority (Ponting 1991: 148–49). When Darwin's theory was presented in this way, it suddenly became popular. Perhaps evolution was simply God's way of separating the wheat from the chafe, and the saved from the losers. For people of a certain frame of mind, the next question was obvious. If humankind had won in free competition with all other species, then which group of humans would win the next round of competition?

Various versions of "Social Darwinism" competed as programs for the further evolution of humanity. Of course, the very idea of social evolution was still dubious to those who took creation as fixed, and society as accountable to an eternal social code. And such a fixed vision was still considered orthodox for Christians. But this too could evolve. With a mental dexterity comparable to that required for the creation of an Aryan church, the Christians of industrial-age Europe began incorporating social progress through competition into their basic beliefs. If life on Earth was an endless race for the survival of the fittest, then it would be fitting to God's plan if Christian civilization was victorious over all others. Perhaps such competition was simply how God's hand worked in history. In that case, military victory and economic success were signs of God's favor. Perhaps the ancient Indo-European warriors and their dominator ancestors had been right all along: it was not that the most holy would inherit the Earth, but that the strongest were the most holy.

By such reasoning, Hitler could claim to speak for European tradition, Christianity and scientific realism, all in one breath. The lesson he drew from all these sources was one—that might does make right. One would like to believe that Hitler was the last national leader to seriously believe this. Later Western presidents and prime ministers would draw the line at saying profit makes right.

The prophets of fascism, most prominently Franco, Mussolini and Hitler, considered themselves repressed idealists in revolt. In the wake of World War I, as most Europeans turned their backs against militarism, and some countries slashed military spending with a vengeance, these military men felt themselves stifled beyond endurance. For them, the new democratic establishment was a denial of all that was heroic in Europe's past. As far as they could see, it had always been the striving for superiority that had made men great. And nothing could be more destructive to greatness than the abolition of superiority as a goal. But postwar society seemed ready to reject the whole struggle for survival of the fittest. Democrats and socialists spoke of "equality," as if elimination of superiority was the aim of life. Should the inheritors of a noble warrior tradition surrender to this? Or should they make themselves central to European civilization once more, by any means necessary? It was still plausible that the old dominators could make a comeback. In most of Europe the traditions of democracy were still paper-thin.

The Nazis claimed to represent the pure cultural essence of Germany. They said they were the voice of all ethnically pure Germans, including the "Nordic peasantry." The claim of common Aryan ancestry had some truth. No doubt many old Indo-Aryan invaders had intermarried with farming villagers, so that in recent centuries no clear ethnic line remained between descendants of the Old Europeans and the steppe invaders. Possibly the peasant population had absorbed the invaders, save for the old elite of princes and landlords. But Nazi history made no mention of Aryans being culturally absorbed. Its spokesmen presumed that the whole population of ethnic Germans was Indo-Aryan. They acknowledged the original Aryan homeland as Inner Asia but made no mention of any Old Europeans in Germany before the Aryans came. With that simplification, the Nazis proceeded to attribute Indo-Aryan warrior values to the villagers. Richard Darre, the Nazi Minister of Agriculture, said that traditional German farmers used to take all "cowards, those unfit for fighting, and people with contemptible bodies" and drown them in the swamps (Dominick 1992: 95). So he presented the zeal for breeding warriors and weeding out "inferior types" as a simple expression of traditional values rising from the people of the soil.

In their drive to recover Indo-Aryan heritage, the Nazis dredged up a shockingly ancient strata of collective memory. Anyone who has seen films of Hitler at his rallies must at least suspect that something came over or through him: many traditional societies would attribute such powerful outbursts to possession by ancestral spirits. Watching Hitler's tirades, some observers immediately thought of Attila the Hun, standing before their eyes in the twentieth century. Out of his mouth poured violent passions, judgments and commands which could have come from a Kurgan warlord.

Riane Eisler (1987: 182) says "Nazi Germany was a return to Kurgan times, not only in its myths, but in its realities." In the confiscation of property from murdered Jews and the wholesale transfer of wealth from conquered nations, naked plunder returned as a central part of the economy. German society was transfigured into a war machine, as completely as Mongolia under Genghis Khan. Close to a hundred years after serfdom had been abolished in Russia, Hitler reimposed slavery across Europe. The slave-camp workers were non-Aryans who existed to serve the conquering race. Like the Kurgans, Hitler proposed rewarding his war heroes with extra wives. Aryan warriors were given a sexual motive to fight, and women were officially designated as war booty. So the loyal thanes would breed a race of conquerors with maximum efficiency (Eisler 1987: 183).

Nazi foreign policy was a series of escalating, all-or-nothing risks. The high command repeatedly gambled on utter domination of other countries, or else the destruction of Germany. The high-risk behaviour resembled that of ancient nomadic tribesmen on the steppes, who often needed to capture war booty to survive the winter. But the German people were in no such position of stark need. It might be assumed that the Nazis' shocking risks were simply a matter of ambition and greed, but that is not what Hitler and Goebbels said. They seriously believed that their path was an inescapable necessity dictated by the laws of life. In the battle of evolution, finally only one race would inherit the earth. It must be the German race or some other. Nazi leaders showed an utterly consistent submission to this belief. As the armies of the Allies closed in on Berlin, Hitler turned to cursing the weakness of his people. Clearly they had proved themselves inadequate to the challenge of life. If they were too weak to prevail over their enemies, then they deserved to die and be replaced by a stronger race. So Hitler, having dictated that no Aryan must be allowed to surrender, took his woman with him to the grave.

The Nazis had interpreted Darwin's theory of evolution as a Kurgan warlords would—as a law of kill or be killed, rule or be ruled. Of course, minds less pervaded by the dominator mentality would see whole other dimensions to evolution. So Lynn Margulis and Dorian Sagan showed how cooperation between one-celled organisms had enabled multicellular life to evolve. From there, evolution had continued mainly through mutual support and symbiosis among multicelled creatures. According to these experts (1986: 248), "The brutal destroyers always end up destroying themselves—automatically leaving those who get along better with others to inherit the world."

Chapter 15

The Possible Ecological Age

DURING THE CLOSING YEARS OF WORLD WAR II IN GERMANY, ADOLPH Metternich wrote a manuscript later published as *The Deserts Threaten*. In this work, he described the course of environmental destruction in ancient times, how modern industry and plantation agriculture had accelerated the process, and how a terminal degradation of nature would be the ultimate threat to human survival (in Dominick 1992: 148–49). Building on Metternich's warnings, we can summarize some of the social costs of desertification as follows:

- The less biological wealth that remains, the more people tend to compete for what is left.
- The more desperate the competition for organic matter, the more society is fragmented along ethnic, racial and gender lines.
- The more lifeless the environment, the less experience of common ground people share.
- The more natural resources are exhausted, the more people are cast adrift as economic refugees.
- Where large numbers of people cannot find adequate sustenance, either the rule of law breaks down, or the law of the strongest prevails.
- Where the health of the environment declines, human health breaks down as well.

During the 1950s the traditional German support for "nature protection" began changing into something more urgent. Growing numbers of conservationists felt that the issue was no longer just the preservation of natural beauty; it was the survival of humanity. By 1970, when the newspaper *Die Welt* printed the headline, "20 MILLION DRINK WATER FROM GERMANY'S LARGEST SEWER" (the Rhine), it was becoming fairly clear to the broad majority of Europeans that the death of nature would leave no place to hide. As Raymond Dominick (1992: 41) put it, "The demise of Homo sapiens—now that was something that could provoke most people to protest!"

Protest they did. And at times, especially in the 1970s, various prophets of environmental doom seemed to dominate the Green movements. These radicals issued scenarios predicting complete disaster and called for enforced limits to growth. Later, some members of the more radical "Deep Ecology" movement seemed to assume that nature and the modern economy are inherently incompatible, and that the scientific economy must be dumped. The news media often put their spotlights on such extreme views. But the real news was the spread of a larger and more popular, practical and optimistic kind of environmentalism among farmers, mothers, scientists, engineers, teachers and local leaders.

For extremists, this mainstream environmentalism was a hopelessly flawed compromise which would only succeed in slightly postponing doomsday. But moderate environmentalism had its own great hope. Basically, it was hope that practical solutions could be found and that the solutions could involve better kinds of growth rather than no growth. "Radicals" tended to stress that a finite system has limits and we must stay within them. Pragmatists tried to solve various problems as they appeared, and to do so in ever better ways. Of course the radicals were right in a way. No doubt there are bottom-line requirements for nature and humanity to survive. But are there upper limits to the growth and evolution of life? What, after all, are the limits to broadening the spectrum of usable energy and materials? What are the limits to efficiency in doing more with less? If we make our work benefit nature, how biologically rich can this planet become?

For farmers experimenting with no-till planting, the short-range goal is to reduce the speed of erosion. The real long-term goal is to grow back the soil richer than before. Tradition and science both demonstrate that ecological capital can increase over time if it is managed and reinvested well. For most of our history and over most of the Earth, the soil has grown deeper and more fertile over time. Only recently, and over fairly limited areas, did we help jack up the rates of erosion, till they sometimes exceeded soil formation by 300 percent (Eisenburg 1998: 30; Suzuki 1997: 100). It is entirely possible for farmers, ranchers, city councils and landscape architects to restore the long-term trend.

Until several decades ago, Western civilization was still fundamentally hostile to nature. On taking possession of forested land, North American settlers had generally "tamed the wilderness" by clearing it of trees. On the Great Plains they nearly eliminated the local grass. North American men commonly used to carry firearms and shoot wild animals on sight—as if beasts were weeds. So the American frontier became a kind of battle line in a war against nature. Only after the war was virtually "won" did a great change of heart set in. "What change?" we might ask. Well, we now commonly complain about the sorry state of the environment. We bad-mouth other people who degrade the countryside. Of course, our complaints are cheap and seldom lead to any action. But all this seemingly empty talk generates an atmosphere of intolerance for the destruction of nature and a desire to heal it where we can.

Another thing we commonly do is plant gardens. This too is a private act, so miniscule in scale as to have no possible effect on the planet—unless tens or hundreds of millions of others also want to make their place in the sun a little natural paradise. Gardening has its fashions, and the recent trends are mainly toward naturalism. Instead of highly domesticated square flower beds, clipped hedges and mowed lawns, many people want a wilder look, featuring rocks, trees or creeping vines, like a plot of wilderness plunked down at their door. Those who own country cottages commonly let the property go back to natural climax vegetation, which is why much of New England has reverted to a well-populated forest.

For many modern people, a garden is a personal statement. It shows the owner's relation with the Earth, like a wardrobe shows the wearer's relation to society. The exploding growth of the gardening business suggests a popular dream, the dream of nature and culture growing together. Evan Eisenburg (1998: 177) comments that, "for some people, that is what paradise is: a small piece of the earth's surface that can be made over to match our dreams." Perhaps that is the most practical thing most of us feel able to do for the Earth—to beautify whatever plot of land we can call our own. If a dedicated gardener was to issue a vision statement, it might read like this: "I want the place I live to be green and beautiful. When I leave this place, the ground will be more fertile than when I came. Year by year I will make it richer and more beautiful here, because it pleases me." If such sentiments sound familiar, perhaps it is because they echo the attitudes of traditional villagers around the world.

Our public gardens are called parks, be they yard-sized city flower beds or regions of protected forest. The old conservation movements tried to fence off chunks of wilderness. They presumed, from experience, that humans were natural enemies of nature and would destroy it completely unless halted by a higher authority. By the 1990s about 3.2 percent of the world's land was enclosed in national parks. No doubt the planet does need a network of wilderness regions to maintain its health, and probably most people feel that much more land should go into parks. But when traditional conservationists focus on saving the wilderness, they tend to overlook two bigger possibilities for renewing the planet.

The first larger opportunity lies in simply *lowering our defenses against nature and reducing our efforts to suppress its regeneration.* For example, over the past century or so, the overgrazed grasslands of North and South America have been sprouting a new cover of tough, thorny, cow-resistant plants such as mesquite trees with cactus. As these hardy thickets of mesquite have spread, many landowners have undertaken a difficult and expensive war to suppress them. The weapons used include fire, poison, chainsaws and giant tractors that drag chains through the ground to rip out trees by the roots. The per-acre costs of this battle often exceed the market value of the land. And even after that expense, almost every bit of mesquite root left in the ground tends to sprout a new tree.

So we commonly fight the regeneration of nature, often spending many millions of dollars on chemical or mechanical means of extermination. But whatever living things we fight, we could befriend. If we have already befriended wild dogs, emmer wheat, potatoes and penicillium mold, we can also find mutually beneficial relations with mesquite trees, cactus, buffalo grass, antelopes and nitrogen-fixing bacteria. Befriending more species of plants and animals has always been a primary basis for progress.

A mature mesquite tree sprouts about sixteen kilograms of beans each year, which contain around 19 percent sugar and 13 percent protein (Kingsolver et al. 1977: 109). The Southwest Amerindians grind the beans for flour to make a sweet bread. Beside mesquite trees, the prickly pear cactus commonly appears, bearing its own abundant crop of succulent red fruits. All this is good food, but only to those who are culturally accustomed to eating it. Otherwise, cultural blinders may lead us to presume that these and other life forms are useless weeds. With such blinders, the European explorers of the New World gazed upon the forests and plains of the Americas and basically said, "I don't see anything I like." At this point in history, perhaps we had best befriend whatever species are thriving, rather than strive to eliminate even more of the species we don't respect.

The other huge possibility conservationists often ignore is that of *making our settled areas more nature-friendly*. Old-fashioned conservationists seemed to settle for a vision of guarding the wilderness against civilization, rather than changing civilization into something good for nature. In many traditional cultures around the world, the villages were as rich in plant and animal life as the surrounding countryside. Perhaps a future high-tech, high-nature civilization will look like that again. Then if raccoons, rabbits and deer infest the urban woods, so much the better.

I want to close this book with two hopeful stories, one returning to the theme of thorn forests in the American southwest, and the other about a village women's group in Kenya. The first story describes nature's powers of self-renewal, and the second shows traditional farmers as a force of nature. Both stories focus on small plots of ground and on some very ordinary kinds of healing.

Story 1: Nature's "Dot Matrix" in the American Desert

In the Chihuahua desert of western Texas, plants tend to grow in scattered little clumps. Over vast stretches of land one or two creosote bushes will appear within each area the size of a football field (Michener 1985: 563–65). Between the clumps of bush and cactus, the ground is commonly as naked as a dirt road, especially where too many cattle were fenced in (Holmes 1992: 13).

A cactus drops no leaves to build the soil. It enriches its spot of ground only when it dies. Then a new cactus will likely rise on that spot. Over many years a thorny clump of prickly pear or barrel cactus may appear. The clump depends on a slowly generated island of enriched earth. Each cactus clump tends to draw a small community of creatures to itself. Pack rats and pocket gophers like to burrow within the thorn-encrusted maze. Harris ground squirrels get their shelter and about two-thirds of their food from prickly pears (Holmes 1992: 13), and the manure of these animals accumulates around the cactus cluster.

Birds are also drawn to cactus. In Arizona the common cactus-dwelling birds include Western Kingbirds, Cactus Wrens, Roadrunners, Mourning Doves, Curve-billed Thrashers, English Sparrows, House Finches, Crested Flycatchers, Purple Martins, Screech Owls, Elf Owls, Red-tailed Hawks, Gila

Woodpeckers and Mearns Gilded Flickers. These birds excrete various seeds from the surrounding chapparel into the shade of the cactus clump, and they also help pollinate the cactus flowers.

Ants, lizards and snakes also find food and shelter in the cactus. They borrow into the earth, loosening it so the roots and the rain can penetrate down. When it rains, the soil of a cactus clump absorbs water better than the bare desert floor. And where soil retains both moisture and organic refuse, a chain reaction begins. The organic matter sucks up water like a sponge, then rots, generating carbonic acid. The carbonic acid acts on minerals in the soil, slowly dissolving them into molecular form, which renders the micro-nutrients digestible to plants (Faulkner 1963: 9–10).

If a mesquite or palo verde tree grows up through a cactus clump, it will push its taproots down and draw the groundwater up. A mature mesquite's roots may reach 50 meters down (Page 1984: 92). These trees, like cactus, are water hogs. Their lateral roots "rob" moisture and nutrition from the surrounding area, but beneath their branches they create a more livable microclimate. Even the usually solitary creosote bush creates a tiny island where annual flowers may bloom even in a drought year (Barbour et al. 1977: 240). Mesquite and palo verde trees are legumes, with all that suggests for soil building. They shower the clumps of vegetation beneath them with leaves and beans. The leaves provide nitrogen-laden fertilizer. The beans feed more animals such as black-tailed jackrabbits, cactus mice and kangaroo rats (Mares 1977: 141–44).

In a mature clump of desert plants, an understory of shrubs, such as sage, may appear between the tree branches and the cactus on the ground. At this stage, the clump is a densely populated little village, struggling to grow in a poor country. In some areas of the Southwest, such clumps of desert plants have slowly merged into fairly continuous thorny thickets. On the northern outskirts of the Chihuahua desert, wide areas are covered in such bush. Richard Phelan (1976: 210, 215) describes it as "a prickly hot jungle full of fruits and seeds and cactus apples, swarming with wildlife." It is a tangle of mesquite, catclaw acacia, huisache, prickly pear, yucca, rat-tail cactus, retama, and black chaparral, often growing six meters high and so thick you can hardly see through it. Such jungles of tough desert plants moved into U.S. territory during the nineteenth century and spread northward like a flood of illegal migrants. These plants used to appear mainly around the streambeds. Then, moving from south to north, they started filling in the overgrazed countryside between streams.

If much of the world is growing hotter and dryer, perhaps the rise of drought-resistant plants suggests that the plant world is adjusting before our eyes. Over much of the Old and New Worlds, harsher conditions and human predation have virtually eliminated the plant life of the past. Now, perhaps a new wave of tougher plants will take those deserted regions as their open frontiers. Prickly pear and mesquite are also spreading on the fringes of the Sahara, in the Near East and in regions surrounding the Red Sea.

Story 2: Mama Ndolo's Women

My first inspiration for this book came from meeting a group of women in Kenya. "Mama" Benedetta Ndolo leads a village women's group in the Iveti hills of Machakos district. From the top of the hill in her village you can see for kilometers to the northwest, over the dusty countryside stretching towards Somalia. Ndolo and her friends taught me what is probably the main lesson this book can offer: Whatever land you have, take care of it.

For a whole afternoon, Mama Ndolo took me all around her village, showing off her group's various accomplishments. We toured the hill slopes terraced by village work parties. We examined cement rainwater jars, paid for one at a time by funds from the women's group garden. Then we looked at the many small nurseries of fruit-tree seedlings. Towards evening we went to her house, and there I saw what impressed me most.

Mama Ndolo's latrine was out her back door, through her grove of banana trees. I went there just as darkness was falling. The valley was deep in shadow, with orange sunbeams still streaming over the hilltop, lighting the wisps of cloud overhead. Mama Ndolo's outhouse had no roof, only reed walls covered with morning glory and passion fruit vines in full flower. Inside, I watched as the stars winked on and the moon appeared above the trees. A roof on that latrine would have been a disaster. Instead of a private flower-garden planetarium it would have been a dark little cell with flies buzzing inside. But the best thing about the latrine was the sound of the wind in the trees.

Three years before, Mama Ndolo's friends had started planting tree nurseries of mango, eucalyptus and other seedlings. Years before, at the U.N. Conference on Desertification in Nairobi, several African governments had proposed planting two great belts of forest, one across North Africa and the other south of the expanding Sahara. After the conference, most governments did little about it, perhaps because they were under pressure to cut spending and pay on their loans. In Kenya the government advocated tree planting, but it was the village women who were most concerned to save the land beneath their feet. Nobody paid these women, or counted the cost of their reforestation efforts. The trees were their pay. And now the new forests of Mama Ndolo's village reached nearly four meters tall.

When you sat in the latrine, you could hear the breeze sifting through a whole hillside of young trees. It was a sound like whispering, or the purring of cats, as if the trees had moods and sighing their happiness. They seemed full of confidence, as if sure that while Mama Ndolo's women are here, this place will never be a desert.

Bibliography

Adam, Andre. 1972. "Berber Migrants in Casablanca." In Ernest Gellner and Charles Micaud (eds.), *Berbers and Arabs: From Tribe to Nation in North Africa*. Lexington: Lexington Books.

Agarwal, Anil. 1989. "Nature and Society in Modernizing India." In Carla M. Borden (ed.), *Contemporary Indian Tradition: Voices on Culture, Nature, and the Challenge of Change*. Washington, D.C.: Smithsonian Institution Press.

Ahmed, Akbar S. 1988. *Discovering Islam: Making Sense of Muslim History and Society*. New York: Routledge and Keegan Paul.

Akhavi, Shahrough. 1987. "Iran: Implementation of an Islamic State." In John L. Esposito (ed.), *Islam in Asia: Religion, Politics and Society*. New York: Oxford University Press.

Allchin, Bridget and Raymond. 1982. *The Rise of Civilization in India and Pakistan*. London: Cambridge University Press.

Allchin, Bridget, Andrew Goudie, and Karunarka Hedge. 1978. *The Prehistory and Paleogeography of the Great Indian Desert*. New York: Academic Press.

Allegro, John M. 1971. *The Chosen People*. London: Hodder & Stoughton.

Allen, Thomas B. 1996. "Xinjiang." *National Geographic*. (March).

Alport, E.A. 1972. "The Mzab." In Ernest Gellner and Charles Micaud (eds.), *Berbers and Arabs: From Tribe to Nation in North Africa*. Lexington, Mass.: Lexington Books.

Altorki, Soraya. 1986. *Women in Saudi Arabia: Ideology and Behavior Among the Elite*. New York: Columbia University Press.

Armstrong, Karen. 1993. *A History of God: The 4,000-Year Quest of Judaism, Christianity and Islam*. New York: Alfred A. Knopf.

Attenborough, David. 1987. *The First Eden: The Mediterranean World and Man*. Boston: Little Brown & Co.

Badran, Margot, and Miriam Cooke (eds.). 1990. *Opening the Gates: A Century of Arab Feminist Writing*. London: Virago Press.

Barb, A.A. 1971. "Mystery, Myth, and Magic." In J.R. Harris (ed.), *The Legacy of Egypt*. Second edition. Oxford: Clarendon Press.

Barber, Elizabeth Wayland. 1994. *Women's Work: The First 20,000 Years: Women, Cloth and Society in Early Times*. New York: W.W. Norton.

Barbour, M.G., et al. 1997. "The Structure and Distribution of Larrea Communities." In T.J. Mabry, J.H. Hunziker and D.R. Di Feo Jr. (eds.), *Creosote Bush: Biology and Chemistry of* Larrea *in New World Deserts*. Stroudsburg, Pa.: Dowden, Hutchinson, Ross.

Barich, Barbara E. 1987. "Adaptation in Archaeology: An Example from the Libyan Sahara." In Angela E. Close (ed.), *Prehistory of Arid North Africa, Essays in Honor of Fred Wendorf*. Dallas: Southern Methodist University Press.

Barraclough, Geoffrey, and Norman Stone (eds.). 1989. *The Times Atlas of World History*. Third edition. Toronto: Stoddart.

Bat Ye'or. 1996. *The Decline of Eastern Christianity Under Islam: From Jihad to Dhimmitude*. Trans. from French by Miriam Kochan and David Littman. London: Associated University Presses.

Bernal, Martin. 1987. *Black Athena, The Afroasiatic Roots of Classical Civilization, Volume 1: Fabrication of Ancient Greece, 1785–1985*. London: Free Association Books.

Bloom, Harold. 1990. *The Book of "J."* Trans. by David Rosenburg. New York: Grove Weidenfield.

Borowiec, Andrew. 1998. *Modern Tunisia: A Democratic Apprenticeship*. London: Praeger.

Boyce, Mary. 1979. *Zoroastrians: Their Religious Beliefs and Practices*. London: Routledge & Keegan Paul.

Briard, Jacques. 1979. *The Bronze Age in Barbarian Europe: From the Megaliths to the Celts*. Trans. by Mary Turton. London: Routledge & Keegan Paul.

Brown, Lester R., project director. 1998. *State of the World, 1998: A Worldwatch Institute Report*. New York: W.W. Norton & Co.

Bryson, Reid A., and Thomas J. Murray. 1977. *Climates of Hunger: Mankind and the World's Changing Weather*. Madison: University of Wisconsin Press.

Bumiller, Elisabeth. 1990. *May You Be the Mother of a Hundred Sons: A Journey Among the Women of India*. New York: Fawcett Columbine.

Butt, Gerald. 1997. *The Arabs: Myth and Reality*. London: I.B. Tauris.

Cable, Mildred, and Francesca French. 1984. *The Gobi Desert*. London: Virago Press.

Cai Junsheng. 1995. "Myth and Reality: The Projection of Gender Relations in Prehistoric China." In *The Chalice and the Blade in Chinese Culture*. Beijing: Chinese Partnership Research Group, China Social Sciences Publishing House.

Campbell, Joseph. 1976a. *Creative Mythology: The Masks of God*. New York: Penguin.

_____. 1976b. *Occidental Mythology: The Masks of God*. New York: Penguin.

_____. 1976c. *Oriental Mythology: The Masks of God*. New York: Penguin.

Central Arid Zone Research Institute (CAZRI), Jodhpur. 1977. *Case Study on Desertification, Luni Development Block, India*. U.N. Conference on Desertification.

Chandler, Wayne. 1989. "Of Gods and Men: Egypt's Old Kingdom." In Ivan Van Sertima (ed.), *Egypt Revisited*. Second edition. London: Transaction Publishers.

Chauhan, Kamala. 1988. *Cultural History of Northern India (Prior to Medieval Invasion)*. Delhi: Pratibha Prakashan.

Claiborne, Robert. 1970. *Climate, Man, and History*. New York: W.W. Norton & Co.

Cleary, Thomas (trans. and ed.). 1989. *Immortal Sisters: Secrets of Taoist Women*. Boston: Shambhala.

Cloudsley-Thompson, J.L. 1984. *The Sahara*. Key Environments Series. Oxford: Pergamon.

Collins, Robert. 1972. *The Medes and Persians: Conquerors and Diplomats*. New York: McGraw-Hill.

Cornford, F.M. 1923. *Greek Religious Thought from Homer to the Age of Alexander*. New York: Dutton.

Crespo, Virgilio Pinto. 1987. "Thought Control in Spain." In Stephen Haliczer (ed. and trans.), *Inquisition and Society in Early Modern Europe*. London: Croom Helm.

Cross, Nigel, and Rhiannon Barker (eds.). 1991. *At the Desert's Edge: Oral Histories from the Sahel*. London: Panos.

Crossan, John Dominic. 1991. *The Historical Jesus: The Life of a Mediterranean Jewish Peasant*. San Francisco: Harper.

Das, Veena. 1989. "Difference and Division as Designs for Life." In Carla M. Borden (ed.), *Contemporary Indian Tradition: Voices on Culture, Nature, and the Challenge of Change*. Washington, DC: Smithsonian Institution.

Davidson, Basil. 1991. *Africa in History, Themes and Outlines*. Second edition. New York: MacMillan.

_____. 1989. "The Ancient World and Africa: Whose Roots?" In Ivan Van Sertima (ed.), *Egypt Revisited*. Second edition. London: Transaction.

de Bary, William Theodore (ed.). 1958. *Sources of Indian Tradition*, vol. 1. New York: Columbia University Press.

de Bary, William Theodore, Wing-tsit Chan, and Burton Watson (eds.). 1960. *Sources of Chinese Tradition*, vol. 1. New York: Columbia University Press.

Dimmitt, Cornelia. 1986. "Sita: Fertility Goddess and Sakti." In John Sratton Hawley and Donna Marie Wulff (eds.), *The Divine Consort: Radha and the Goddesses of India*. Boston: Beacon.

Diop, Cheikh Anta. 1989. "Origin of the Ancient Egyptians." In Ivan Van Sertima (ed.), *Egypt Revisited*. Second edition. London: Transaction.

Dolukhanov, Paul M. 1979. *Ecology and Economy in Neolithic Eastern Europe*. London: Gerald Duckworth.

Dominick, Raymond H., III. 1992. *The Environmental Movement in Germany, 1871–1971: Prophets and Pioneers*. Bloomington: Indiana University Press.

Du Fangqin. 1995. "The Rise and Fall of the Zhou Rites: A Rational Foundation for the Gender Relationship Model." In *The Chalice and the Blade in Chinese Culture*. Beijing: Chinese Partnership Research Group, China Social Sciences Publishing House.

Du Jinpeng. 1995. "The Social Relationships of Men and Women in the Xia-Shang Era." In *The Chalice and the Blade in Chinese Culture*. Beijing: Chinese Partnership Research Group, China Social Sciences Publishing House.

Dudley, Donald. 1970. *The Romans*. London: Hutchinson.

Dunn, Ross E. 1972. "Berber Imperialism: The Ait Atta Expansion in Southeast Morocco." In Ernest Gellner and Charles Micaud (eds.), *Berbers and Arabs: From Tribe to Nation in North Africa*. Lexington, Mass: Lexington Books.

Ebrey, Patricia Buckley (ed.). 1981. *Chinese Civilization and Society: A Sourcebook*. New York: The Free Press.

Eck, Diana L. 1986. "Ganga: The Goddess in Hindu Sacred Geography." In John Stratton Hawley and Donna Marie Wulff (eds.), *The Divine Consort: Radha and the Goddesses of India*. Boston: Beacon.

Eisenburg, Evan. 1998. *The Ecology of Eden*. Toronto: Random House of Canada.

Eisler, Riane. 1987. *The Chalice & the Blade: Our History, Our Future*. San Francisco: Harper & Row.

Eisler, Riane, Davis Loye, and Kari Norgaard. 1995. *Women, Men & the Global Quality of Life*. Report of the Gender Equity and Quality of Life Project of the Center for Partnership Studies, Pacific Grove, California.

Encyclopaedia Britannica. 1988. Fifteenth edition, vol. 1, Micropaedia, "Alexandria, School of."

Encyclopedia of Islam. 1955. Second edition. "Hidjab" [Veil]. Leiden.

Faruki, Kemal A. 1987. "Pakistan: Islamic Government and Society." In John L. Esposito (ed.), *Islam in Asia: Religion, Politics and Society*. New York: Oxford University Press.

Faulkner, Edward H. 1963. *Plowman's Folly*. Norman: University of Oklahoma Press.

Finch, Charles S. 1989. "Science and Symbol in Egyptian Medicine: Commentaries on the Edwin Smith Papyrus." In Ivan Van Sertima (ed.), *Egypt Revisited*. Second edition. London: Transaction.

Flon, H. 1979. "Can Climate Repeat Itself? Possible Climatic Warming and the Case for Paleoclimatic Warm Phases." In W. Bach, J. Pankranth, and W.W. Kellogg (eds.), *Man's Impact on Climate: Proceedings of an International Conference Held in Berlin, June 14–16, 1978*. New York: Elsevier Scientific Publishing.

Fluehr-Lobban, Carolyn. 1993. "Personal Status Law in the Sudan." In Donna Lee Bowen and Evelyn A Early (eds.), *Everyday Life in the Muslim Middle East*. Bloomington: Indiana University Press.

Foreman-Peck, James. 1994. "Foreign Trade and Economic Growth." In Derek, K. Aldercroft and Simon P. Ville (eds.), *The European Economy 1750–1914; A Thematic Approach*. New York: Manchester University Press.

Frawley, David. 1992. "On the Banks of the Sarasvati: The Ancient History of India Revisited." *The Quest* (Autumn).

French, Marilyn. 1985. *Beyond Power: On Women, Men and Morals*. New York: Ballantine Books.

Freud, Sigmund. 1938: "The Psychology of Everyday Life." In A.A. Brill (trans.), *The Basic Writings of Sigmund Freud*. New York: Modern Library.

Friedl, Erika. 1989. *Women of Deh Koh: Lives in an Iranian Village*, Washington D.C.: Smithsonian Institution.

Fukuoka Mansanubu. 1989. *The Road Back to Nature: Regaining the Paradise Lost*. Trans. by Frederic P. Metreaud. New York: Japan Publications.

Gabor, D., U. Colombo, A. King, and R. Galli (eds). 1981. *Beyond the Age of Waste: A Report to the Club of Rome*. Oxford: Pergamon Press.

Gadgil, Madhav. 1989. "Husbanding India's Natural Resources: The Tradition and the Prospects." In Carla M. Borden (ed.), *Contemporary Indian Tradition: Voices on Culture, Nature, and the Challenge of Change*. Washington D.C.: Smithsonian Institution Press.

Gao Shiyu. 1995. "A Fixed State of Affairs and the Mispositioned Status: Gender Relations During the Sui, Tang, Five Dynasties, and Song Dynasty." In *The Chalice and the Blade in Chinese Culture*. Beijing: Chinese Partnership Research Group, China Social Sciences Publishing House.

Garcea, Elena A. 1993. *Cultural Dynamics in the Saharo-Sudanese Prehistory*. Rome: Gruppo Editoriale Internazionale.

Gautier, Achilles. 1987. "Prehistoric Men and Cattle in North Africa." In Angela E. Close (ed.), *Prehistory of Arid North Africa: Essays in Honor of Fred Wendorf*. Dallas: Southern Methodist University Press.

Geller, Ernest. 1972. "Introduction." In Ernest Gellner and Charles Micaud (eds.), *Berbers and Arabs: From Tribe to Nation in North Africa*. Lexington, Mass.: Lexington Books.

Ghaliounhui, P. 1973. *The House of Life: Magic and Medical Science in Ancient Egypt*. Amsterdam: B.M. Israel.

Ghani, Ashraf. 1987. "Afghanistan: Islam and Counterrevolutionary Movements." In John L. Esposito (ed.), *Islam in Asia: Religion, Politics and Society*. New York: Oxford University Press.

Gimbutas, Marija. 1991. *The Civilization of the Goddess: The World of Old Europe*. San Francisco: Harper & Row.

_____. 1989. *The Language of the Goddess*. San Francisco: Harper & Row.

_____. 1977. "The First Wave of Eurasian Steppe Pastoralists into Copper Age Europe." *Journal of Indo-European Studies* 5 (Winter).

_____. 1971. *The Slavs*. London: Thames and Hudson.

Golden, Peter B. 1990. "The Peoples of the South Russian Steppes." In Denis Sinor (ed.),*The Cambridge History of Early Inner Asia*. New York: Cambridge University Press.

Grainger, Alan. 1990. *The Threatening Desert: Controlling Desertification*. London: Earthscan.

Grant, Michael. 1992. *Greeks and Romans: A Social History*. London: Weidenfeld and Nicholson.

Graves, Robert. 1960. *The Greek Myths*. Vol. 1, revised edition. Hammondsworth: Penguin.

Griffith, Samuel B. (trans. and ed.). 1863. *Sun Tzu, The Art of War*. Oxford: Oxford University Press.

Haliczer, Stephen (trans. and ed.). 1987. *Inquisition and Society in Early Modern Europe*. London: Croom Helm.

Hamady, Sania. 1960. *Temperament and Character of the Arabs*. New York: Twayne.

Hammond, H.G.L. 1959. *A History of Greece, to 322 BC*. Oxford: Clarendon.

Harlan, Lindsay. 1992. *Religion and Rajput Women: The Ethic of Protection in Contemporary Narratives*. Berkeley: University of California Press.

Harris, J.R. 1971. "Medicine." In J.R. Harris (ed.), *The Legacy of Egypt*. Second edition. Oxford: Clarendon Press.

Hart, David M. 1972. "The Tribe in Modern Morocco: Two Case Studies." In Ernest Gellner and Charles Micaud (eds.), *Berbers and Arabs: From Tribe to Nation in North Africa*. Lexington, Mass.: Lexington Books.

Hassan, Fekri A., and G. Timothy Gross. 1987. "Early Halocene at Siwa Oasis, Northern Egypt." In Angela E. Close (ed.), *Prehistory of Arid North Africa: Essays in Honor of Fred Wendorf*. Dallas: Southern Methodist University Press.

Hawkes, Jacquetta. 1973. *The First Great Civilizations: Life in Mesopotamia, the Indus Valley, and Egypt*. New York: Alfed A. Knopf.

Hawley, John Stratton, and Donna Marie Wulff (eds.). 1986. *The Divine Consort: Radha and the Goddesses of India*. Boston: Beacon Press.

Hazleton, Lesley. 1980. *Where Mountains Roar: A Personal Report from the Sinai and Negev Desert*. New York: Holt, Rinehart and Winston.

Hein, Norvin. 1986. "Comments: Radha and the Erotic Community." In John Sratton Hawley and Donna Marie Wulff (eds.), *The Divine Consort: Radha and the Goddesses of India*. Boston: Beacon Press.

Henderson, Hazel. 1996. *Building a Win-Win World: Life Beyond Global Economic Warfare*. San Francisco: Berrett-Koehler.

_____. 1981. *The Politics of the Solar Age: Alternatives to Economics*. Garden City, N.Y.: Anchor Press/Doubleday.

Hillel, Daniel. 1994. *Rivers of Eden: The Struggle for Water and the Quest for Peace in the Middle East*. Oxford: Oxford University Press.

Hitti, P.K. 1965. *The Arabs: A Short History*. London: MacMillan.

Holmes, Anita. 1992. *Cactus: The All American Plant*. New York: Four Winds Press.

Hopkirk, Peter. 1984. *Foreign Devils on the Silk Road: The Search for the Lost Cities and Treasures of Chinese Central Asia*. Oxford: Oxford University Press.

Hurst, Thomas Daniel. 1994. *Exploring Ancient Native America: An Anthropological Guide*. New York: MacMillan.

Huzayyin, Soliman. 1972. "Changes in Climate and Vegetation, and Human Adjustment in the Saharo-Arabian Belt, with Special Reference to Africa." In William L. Thomas Jr. (ed.), *Man's Role in Changing the Face of the Earth*.

Chicago: University of Chicago.
Jagchid, Sechin. 1970. "Trade, Peace and War Between the Nomadic Altaics and the Agricultural Chinese." *Bulletin of the Institute of China Border Studies* 1 (Taiwan).
Jagchid, Sechin, and Paul Hyer. 1979. *Mongolia's Culture and Society*. Boulder, Colo.: Westview Press.
James, T.G.H. 1984. *Pharaoh's People: Scenes from Life in Imperial Egypt*. London: The Bodley Head.
Janos, A.C. 1989. "The Politics of Backwardness in Continental Europe, 1780–1945." *World Politics* 41.
Jansson, AnnMari, Monica Hammer, Carl Folke, and Robert Costanza (eds.). 1994. *Investing in Natural Capital: The Ecological Economics Approach to Sustainability*. International Society for Ecological Economics. Covelo, Calif.: Island Press.
Jordan, Robert Paul. 1981. "Somalia's Hour of Need." *National Geographic* 159, no. 6 (June).
Josephy, Alvin M. (ed.) 1971. *The Horizon History of Africa*. New York: American Heritage.
Karanga, Maulana. c. 1989. "Towards a Sociology of Maatian Ethics: Literature and Context." In Ivan Van Sertima (ed.), *Egypt Revisited*. London: Transaction.
Kenyatta, Jomo. 1966. *My People of Kikuyu*. London: Oxford University Press.
Khazanov, Anatoly M. 1994: *Nomads and the Outside World*. Second edition, trans. by Julia Crookenden. Madison: The University of Wisconsin Press.
Kingsolver, J.M., et al. 1977. "Prosopis Fruits as a Resource...." In B.B. Simpson (ed.), *Mesquite: Its Biology in Two Desert Biosystems*. Stroudsburg, Pa.: Dowden, Hutchinson.
Kinsley, David. 1988. *Hindu Goddesses: Visions of the Divine Feminine in the Hindu Religious Tradition*. Berkeley: University of California Press.
Kohl, Larry. 1989. "Above China." *National Geographic* 175, no. 3 (March).
Korten, David C. 1995. *When Corporations Rule the World*. West Hartford, Conn.: Kumerian Press.
Kramer, Samuel Noah. 1990. *History Begins at Sumer*. Third edition. Philadelphia: University of Pennsylvania.
Kwok, D.W.Y. 1999. Book review of Manufacturing Confucianism: Chinese Traditions and Universal Civilization by Lionel M. Jenses (Durham, N.C.: Duke University Press, 1997.) *Journal of World History* 10, no. 1 (Spring).
Lacey, Robert. 1981. *The Kingdom: Arabia and the House of Sa'ud*. New York: Harcourt Brace Jovanovich.
Lamy, L. 1981. *Egyptian Mysteries*. New York: Crossroad Books.
Lane-Poole, Stanley. 1925. *A History of Egypt, Volume VI: The Middle Ages*. Fourth edition. Dublin: Trinity College.
Langewische, William. 1996. *Sahara Unveiled: A Journey Across the Desert*. New York: Pantheon.
Lanzhou Institute of Glaciology, Cryopedology and Desert Research. 1982. "The Establishment of Forest Shelter Belts in Oases: Principles and Technology." In James Walls (ed.), *Combating Desertification in China*. Nairobi: UNEP.
Lao Tzu. 1963. *Tao Te Ching*. Hammondsworth: Penguin.
Laroui, Abdallah. 1977. *A History of the Maghreb*. Trans. by Ralph Manheim. Princeton, N.J.: Princeton University Press.
Laurence, Margaret. 1963. *The Prophet's Camel Bell*. Toronto: The Canadian Publishers.
Lefkowitz, Mary R., and Guy MacLean Rogers. 1996. *Black Athena Revisited*. Chapel

Hill: University of North Carolina Press.
Lessner, Erwin. 1955. *Cradle of Conquerors: Siberia.* New York: Doubleday.
Liu Ruzhen. 1995. "Women's Status and Gender Relations in the Liao, Jin, and Yuan Dynasties." In *The Chalice & the Blade in Chinese Culture.* Beijing: Chinese Partnership Research Group, China Social Sciences Publishing House.
Lougheed, Alan. 1994. "Industry and Technical Change." In Derek K. Aldercroft and Simon P. Ville (eds.), *The European Economy, 1750–1914: A Thematic Approach.* Manchester: Manchester University Press.
Luckenbill, D.D. 1926–27. *Ancient Records of Assyria and Babylonia.* Chicago: University of Chicago Press.
Mabry, T.J., J.H. Hunziker, D.R. Di Feo Jr. and M.G. Barbour (eds.). 1977. *Creosote Bush: Biology and Chemistry of* Larrea *in New World Deserts.* Stroudsburg, Pa.: Dowden, Hutchinson, Ross.
Macfarlane, Alan. 1985. "The Root of All Evil." In David Parkin (ed.), *The Anthropology of Evil.* London: Basil Blackwell.
Mack, Burton L. 1995. *Who Wrote the New Testament? The Making of the Christian Myth.* San Francisco: Harper.
_____. 1993. *The Lost Gospel: The Book of Q and Christian Origins.* San Francisco: Harper.
MacKerras, Colin. 1990. "The Uighurs." In Denis Sinor (ed.), *The Cambridge History of Early Inner Asia.* Cambridge: Cambridge University Press.
Makkai, Laszlo. 1990. "The Hungarians' Prehistory, Their Conquest of Hungary, and Their Raids to the West." In Peter F. Sugar (ed.), *A History of Hungary.* Bloomington: Indiana University Press.
Manor, James. 1989. "India: State and Society Diverge." *Current History* (December).
Mansfield, Peter. 1991. *A History of the Middle East.* New York: Penguin Viking.
Mares, M.A. 1977. "Prosopis as a Niche Component." In B.B. Simpson (ed.), *Mesquite: Its Biology in Two Desert Biosystems.* Stroudsburg, Pa.: Dowden, Hutchinson, Ross.
Margulis, Lynn, and Dorian Sagan. 1986. *Microcosmos: Four Billion Years of Evolution from our Microbal Ancestors.* New York: Summit Books.
McLuhan, T.C. 1994. *The Way of the Earth: Encounters with Nature in Ancient and Contemporary Thought.* New York: Simon & Schuster.
Melyukova, A.I. 1990. "The Scythians and Sarmatians." Trans. by Julia Crookender. In Denis Sinor (ed.), *The Cambridge History of Early Inner Asia.* Cambridge: Cambridge University Press.
Merchant, Carolyn. 1989. *Ecological Revolutions: Nature, Gender, and Science in New England.* Chapel Hill: University of North Carolina Press.
Mernissi, Fatima. c. 1991. *The Veil and the Male Elite: A Feminist Interpretation of Women's Rights in Islam.* Trans. by Mary Jo Lakeland. Reading, Mass.: Addison-Wesley. Quotations reprinted by permission of Addison-Wesley-Longman Inc.
_____. 1987. *Beyond the Veil: Male-Female Dynamics in Modern Muslim Society.* Revised edition. Bloomington: Midland Books.
Metzner, Ralph. 1997. "Clashing Cultures and Hybrid Mythologies." In Joan Marler (ed.), *From the Realm of the Ancestors: An Anthology in Honor of Marija Gimbutas.* Manchester, Conn.: Knowledge, Ideas & Trends.
Michener, James A. 1985. *Texas.* New York: Random House.
Milburn, M. 1984. "Archeology and Prehistory." In J.L. Cloudsley-Thompson (ed.), *Sahara Desert, Key Environments Series.* Oxford: Pergamon Press.
Min Jiayin. 1995. "Conclusion." In *The Chalice and the Blade in Chinese Culture: Gender*

Relations and Social Models. Beijing: Chinese Partnership Research Group, China Social Sciences Publishing House.
Minai, Naila. 1981. *Tradition and Transition in the Middle East*. New York: Seaview Books.
Modupe, Prince. 1958. *I Was a Savage*. London: Museum Press.
Moorhouse, Geoffrey. 1990. *Apples in the Snow: A Journey to Samarkand*. London: Hodder and Stoughton.
_____. 1974. *The Fearful Void*. London: Hodder and Stoughton.
Morgan, David. 1987. *The Mongols*. New York: Basil Blackwell.
Ms. 1992. "Niger: Blame Rains Down on Women." *Ms*. 3, no. 2.
Munson, Henry Jr. 1993. *Religion and Power in Morocco*. New Haven: Yale University Press.
Naipaul, V.S. 1991, *India—A Million Mutinies Now*. New York: Viking.
_____. 1984. *Finding the Centre: Two Narratives*. London: Deutsch.
_____. 1981. *Among the Believers: An Islamic Journey*. New York: Knopf.
Needham, Joseph, with Wang Ling. 1956. *Science and Civilization in China*, vol. 2. London: Cambridge University Press.
Newell, R.E. 1971. "The Amazon Forest and Atmospheric General Circulation." In W.H. Matthews, W.W. Kellog, and G.D. Robertson (eds.), *Man's Impact on the Climate*. Cambridge, Mass.: MIT Press.
Norris, N.T. 1986. *The Arab Conquest of the Western Sahara*. Harlow: Longman.
_____. 1984. "Indigenous Peoples of the Sahara." In J.L. Cloudsley-Thompson (ed.), *Sahara Desert—Key Environments Series*. Oxford: Pergamon Press.
Norris, Richard A., and Williston Walker. 1985. *A History of the Christian Church*. Fourth edition. New York: Scribner.
O'Connell, Robert L. 1995. *Ride of the Second Horseman: The Birth and Death of War*. New York: Oxford University Press.
O'Flaherty, Wendy Doniger. 1989. "Impermanence and Eternality in Indian Art and Myth." In Carla M. Borden (ed.), *Contemporary Indian Tradition: Voices on Culture, Nature, and the Challenge of Change*. Washington, D.C.: Smithsonian Institution Press.
Olivia, Pavel. 1981. *The Birth of Greek Civilization*. London: Orbis.
O'Neil, Mary. 1987. "Magical Healing, Love Magic and the Inquisition in Late Sixteenth-century Modena." In Stephen Haliczer (eds. and trans.), *Inquisition and Society in Early Modern Europe*. London: Croom Helm.
Page, Jake. 1984. *Arid Lands*. Planet Earth Series. Alexandria, Va,: Time/Life Books.
Pagels, Elaine. 1995. *The Origin of Satan*. New York: Random House.
_____. 1989. *The Gnostic Gospels*. New York: Vintage Books.
_____. 1981. *Adam, Eve, and the Serpent*. New York: Vintage Books.
Palter, Robert. 1996. "Black Athena, Afrocentrism, and the History of Science." In Mary R Lefkowitz and Guy MacLean Rogers (eds.), *Black Athena Revisited*. Chapel Hill: University of Carolina Press.
Panikkar, K.M. 1963. *A Survey of Indian History*. Fourth edition. London: Asia Publishing House.
Parkin, David (ed.). 1985. *The Anthropology of Evil*. London: Basil Blackwell.
Parrack, Dwain W. 1969. "An Approach to the Bioenergetics of Rural West Bengal." In Andrew P. Vayda (ed.), *Environment and Cultural Behavior*. Garden City, N.J.: Natural History Press.
Patai, Raphael. 1990. *The Hebrew Goddess*. Third edition. Detroit: Wayne State University Press.
_____. 1983. *The Arab Mind*. Revised edition. New York: Charles Scribners Sons.

_____. 1962. *Golden River to Golden Road: Society, Culture and Change in the Middle East.* Philadelphia: University of Pennsylvania Press.

Perry, Mary Elizabeth. 1987. "Beatas and the Inquisition in Early Modern Seville." In Stephen Haliczer and Croom Helm (eds and trans), *Inquisition and Society in Early Modern Europe.* London: Croom Helm.

Phelan, Richard. c. 1976. *Texas Wild.* New York: E.P. Dutton.

Phillips, Wendell. 1966. *Unknown Oman.* London: Longmans, Green & Co.

Ponting, Clive. 1991. *A Green History of the World: The Environment and the Collapse of Great Civilizations.* Hammondsworth: Penguin Books.

Praux, Claire. 1971. "Graeco-Roman Egypt." In J.R. Harris (ed.), *The Legacy of Egypt.* Oxford: Clarendon Press.

Price, Roger. 1994. "The Transformation of Agriculture." In Derek K. Aldercroft and Simon P. Ville (eds.), *The European Economy, 1750-1914: A Thematic Approach.* Manchester: Manchester University Press.

Pulleyblank, E.G. 1983. "The Chinese and Their Neighbors in Prehistory and Early Historic Times." In David K. Keightley (ed.), *The Origins of Chinese Civilization.* Berkeley: University of California Press.

Qadir, C.A. 1988. *Philosophy and Science in the Islamic World.* London: Croom Helm.

Randhawa, M.S. 1980. *A History of Agriculture in India.* New Delhi: Indian Council for Agricultural Research.

Ranke-Heinemann, Uta. 1990. *Eunuchs for the Kingdom of Heaven.* New York: Doubleday.

Renouf, P.Le Page. 1880. "The Hibbert Lectures, 1879: Lectures on the Origin and Growth as Illustrated by the Religion of Ancient Egypt." London: Williams and Northgate.

Reudy, John. 1992. *Modern Algeria: The Origins and Development of a Nation.* Bloomington: Indiana University Press.

Rifkin, Jeremy with Ted Howard. 1989. *Entropy: Into the Greenhouse World.* Rev. Ed. New York: Bantam.

Roberts, C.H. 1971. "The Greek Papyri." In J.R. Harris (ed.), *The Legacy of Egypt.* Second edition. Oxford: Clarendon Press.

Roberts, David. 1995. "Egypt's Old Kingdom." *National Geographic* (January).

Rohrlich, Ruby. 1980. "State Formation in Sumer and Subjugation of Women." *Feminist Studies* 6 (Spring).

Roset, Jean-Pierre. 1987. "Paleoclimatic and Cultural Conditions of Neolithic Development in the Early Holocene of Northern Niger." In Angela E. Close (ed.), *Prehistory of Arid North Africa: Essays in Honor of Fred Wendorf.* Dallas: Southern Methodist University Press.

Roux, Georges. 1980. *Ancient Iraq.* Second edition. Hammondsworth: Penguin.

Ruthven, Malise. 1984. *Islam in the World.* New York: Oxford University Press.

Sabean, David Warren. 1990. *Property, Production and Family in Neckarhausen, 1700–1870.* Cambridge: Cambridge University Press.

_____. 1984. *Power in the Blood: Popular Culture and Village Discourse in Early Modern Germany.* New York: Cambridge University Press.

Said, Edward W. 1979. *Orientalism.* New York: Vintage Books.

Sandars, N.K. (trans.). 1962. *The Epic of Gilgamesh.* New York: Penguin.

Sandray, Peggy Reeves. 1988. *Female Power and Male Dominance—On the Origins of Sexual Inequality.* Cambridge: Press Syndicate of the University of Cambridge.

Schaar, Stuart. 1971. "The Barbary Coast." In Alvin M. Josephy (ed.), *The Horizon History of Africa.* New York: American Heritage.

Schmitt, Charles B. 1975, "Philosophy and Science in Sixteenth-century Universities: Some Preliminary Comments." In J.E. Murdoch and E.D. Sylla (eds.), *The Cultural Context of Medieval Learning*. Dordrecht and Boston: D. Reidel.

_____. 1973. "Towards a Reassessment of Renaissance Aristotelianism." *History of Science* 11.

_____. 1969. , "Experience and Experiment: A Comparison of Zabarella's View with Galileo's in *De Moto*." *Studies in the Renaissance* 16.

_____. 1967. "Experimental Evidence for and Against a Void: The Sixteenth Century Arguments." *Isis* 58.

Schneider, Jane. 1991. "Spirits and the Spirit of Capitalism." In Eric R. Wolf (ed.), *Religious Regimes and State Formation: Perspectives from European Ethnography*. Albany, N.Y.: State University of New York Press.

Schneider, Stephen H., and Randi Londer. 1984. *The Co-evolution of Climate and Life*. San Francisco: Sierra Club Books.

Schomer, Karine. 1986, "Where Have All the Radhas Gone?: New Images of Woman in Modern Hindi Poetry." In John Sratton Hawley and Donna Marie Wulff (eds.), *The Divine Consort: Radha and the Goddesses of India*. Boston: Beacon Press.

Shaaban, Bouthaina. 1988. *Both Right and Left Handed: Arab Women Talk About Their Lives*. London: Women's Press.

Shafiq, Duriya. 1952. "Islam and the Constitutional Rights of Women." *Al-Misri*, May 14.

Shahabuddin, Syed, and Theodore P. Wright. 1987. "India's Muslim Minority: Politics and Society." In John L. Esposito (ed.), *Islam in Asia: Religion, Politics and Society*. New York: Oxford University Press.

Shinnie, P. L. 1971. "The Legacy to Africa." In J.R. Harris (ed.), *The Legacy of Egypt*. Second edition. Oxford: Clarendon Press.

Shiva, Vandana. 1988. *Staying Alive: Women, Ecology, and Development*. London: Zed Books.

Shore, A.F. 1971. "Christian and Coptic Egypt." In J.R. Harris (ed.), *The Legacy of Egypt*. Second edition. Oxford: Clarendon Press.

Simpson, B.B. (ed.). 1977. *Mesquite: Its Biology in Two Desert Biosystems*. Stroudsburg, Pa.: Dowden, Hutchinson, Ross.

Singh, Mahendra Pratap. 1981. *Life in Ancient India, 800–1200*. Varnasi: Vishwavidyalaya Prakashan.

Sinkiang (Xinjiang) Autonomous Region, Office of Environmental Protection. 1977. *Tame the Wind, Harness the Sand and Transform the Gobi*. Associated case study. United Nations Conference on Desertification, Nairobi, 1977.

Sinor, Denis. 1990a. "Introduction, The Concept of Central Asia." In Denis Sinor (ed.), *The Cambridge History of Early Central Asia*. Cambridge: Cambridge University Press.

_____. 1990b. "The Establishment and Dissolution of the Turk Empire." In Denis Sinor (ed.), *The Cambridge History of Early Central Asia*. Cambridge: Cambridge University Press.

_____. 1990c. "The Hun Period." In Denis Sinor (ed.), *The Cambridge History of Early Central Asia*. Cambridge: Cambridge University Press.

Sitwell, N.H.H. 1984. *The World the Romans Knew*. London: Hamish Hamilton.

Smil, Vaclav. 1993. *China's Environmental Crisis: An Inquiry into the Limits of National Development*. Armonk, N.Y.: M.E. Sharpe.

Smith, David. 1988. *Prisoners of God: The Modern Day Conflict of Arab and Jew*. London: Quartet.
Smith, John Holland. 1976. *The Death of Classical Paganism*. London: G. Chapman.
Sorin, Vladimir. 1992. "Last of an Ancient People." *Cultural Survival Quarterly* (Winter).
Sowell, Thomas. 1996. *Migrations and Cultures: A World View*. New York: Basic Books.
Spain, James W. 1963. *The People of the Khyber: The Pathans and Pakistan*. New York: Praeger.
Spence, Jonathan D. 1991. *The Search for Modern China*. New York: Norton.
Spivey, Nigel, and Simon Stoddart. 1990. *Etruscan Italy*. London: Batsford.
Starhawk. 1987. *Truth or Dare: Encounters with Power, Authority and Mystery*. San Francisco: Harper & Row.
Stewart, Desmond. 1979. *Early Islam*. New York: Time-Life Books.
Stone, Martin. 1997. *The Agony of Algeria*. London: Hurst & Co.
Sun Xiao and Pan Shaoping. 1995. "Order and Chaos: The Social Position of Men and Women in the Qin, Han, and Six Dynasties Period." In *The Chalice and the Blade in Chinese Culture*. Beijing: Chinese Partnership Research Group, China Social Sciences Publishing House.
Suzuki, David. 1997. *The Sacred Balance: Rediscovering Our Place in Nature*. Vancouver: Douglas & McIntyre.
Swift, Jeremy, and the editors of Time/Life. 1978. *The Sahara*. Amsterdam: Time-Life Books.
Szadeczky-Kardoss, Samuel. 1990. "The Avars." In Denis Sinor (ed.), *The Cambridge History of Early Inner Asia*. Cambridge: Cambridge University Press.
T'an Ssu-t'ung. 1960. "On the Study of Humanity." In William Theodore de Bary, Wing-tsit Chan and Burton Watson (eds.), *Sources of Chinese Tradition*. Vol. 2. New York: Columbia University Press.
Thompson, Jane Werner. 1981. *Deserts of the World, Future Threat or Promise?* New York: Philomel Books.
Toomer, G.J. 1971. "Mathematics and Astronomy." In J.R. Harris (ed.), *The Legacy of Egypt*. Second edition. Oxford: Clarendon Press.
Tranter, Neil L. 1994. "Population, Migration and Labor Supply." In Derek H. Aldercroft and Simon P. Ville (eds.), *The European Economy, 1750-1914: A Thematic Approach*. Manchester: Manchester University Press.
Trimingham, J. Spencer. 1962. *A History of Islam in West Africa*. London: Oxford University Press.
Turnbull, Colin M. 1972. *The Mountain People*. New York: Touchstone Books.
Van Houtte, J.A. 1977. *An Economic History of the Low Countries, 800–1800*. London: Widenfeld and Nicholson.
Van Sertima, Ivan. 1989. "Black Dynasties and Rulers—Introduction." In Ivan Van Sertima (ed.), *Egypt Revisited*. Second edition. London: Transaction.
Vessels, Jane. 1992. "The Simpson Outback." *National Geographic* (April).
Vinogradov, Amal R. 1972. "The Socio-Political Organization of a Berber Taral Tribe—Pre-Protectorate Morocco." In Ernest Gellner and Charles Micaud (eds.), *Berbers and Arabs: From Tribe to Nation in North Africa*. Lexington, Mass.: Lexington Books.
Waley, Arthur. 1939. *Three Ways of Thought in Ancient China*. New York: MacMillan.
Walker, Barbara G. 1996. *The Women's Encyclopedia of Myths and Secrets*. Edison, N.J.: Castle Books.

Walls, James. 1980. *Land, Man and Sand: Desertification and its Solution*. New York: MacMillan.

Walton, K. 1969. *The Arid Zones*. London: Hutchinson.

Ward, Allen, Fritz, M. Heichelheim, and Cedric A. Yao. 1999. *A History of the Roman People*. Third edition. Upper Saddle River, N.J.: Prentice Hall.

Warren, Andrew, and Judith K. Maizels. 1977. "Ecological Change and Desertification." In *Desertification: Its Causes and Consequences*. Secretariat of the United Nations Conference on Desertification, Nairobi, 1977. Oxford: Pergamon Press.

Waterbury, John. 1972. "The Coup Manque." In Ernest Gellner and Charles Micaud (eds.), *Arabs and Berbers: From Tribe to Nation in North Africa*. Lexington, Mass.: Lexington Books.

Watterson, Barbara. 1991. *Women in Ancient Egypt*. New York: St. Martin's Press.

Wilber, Ken. 1979. *No Boundary: Eastern and Western Approaches to Personal Growth*. Los Angeles: Center Publications.

Willis, John Ralph. 1971. "The Spread of Islam." In Alvin A. Josephy Jr. (ed.), *The Horizon History of Africa*. New York: American Heritage.

Wolfram, Herwig. 1985. *Treasures on the Danube*. Vienna: H. Böhlaus.

Wong, Jan. 1996. *Red China Blues*. Toronto: Doubleday/Anchor Books.

Wright, Karen. 1998. "Empires in the Dust." *Discover: The World of Science* (March).

Wright, Robin B. 1985. *Sacred Rage: The Wrath of Militant Islam*. New York: Simon and Schuster.

Yates, Frances A. 1972. *The Rosecrucian Enlightenment*. London: Routledge and Keegan Paul.

Yu Ying-shih. 1990. "The Hsiung-nu." In Denis Sinor (ed.), *The Cambridge History of Early Inner Asia*. Cambridge: Cambridge University Press.

Zhang Zhijing. 1995. "Changes in Women's Status as Reflected in Ancient Chinese Law." In *The Chalice & the Blade in Chinese Culture*. Beijing: Chinese Partnership Research Group, China Social Sciences Publishing House.

Zhao Zhewei. 1995. "Goddesses Worshipped by the Chinese." In *The Chalice & the Blade in Chinese Culture*. Beijing: Chinese Partnership Research Group, China Social Sciences Publishing House.

Zia Ud-din Barni. 1958. "Fatwa-yi-Jahandari." In Wm. Theodore de Bary (ed.), *Sources of Indian Tradition*. Vol. 1. New York: Columbia University Press.

Index